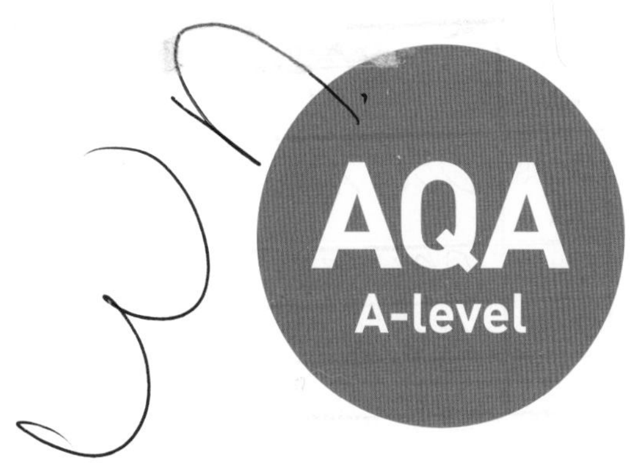

Computer Science

Includes AS and A-level

Bob Reeves

Approval message from AQA

This textbook has been approved by AQA for use with our qualification. This means that we have checked that it broadly covers the specification and we are satisfied with the overall quality. Full details of our approval process can be found on our website.

We approve textbooks because we know how important it is for teachers and students to have the right resources to support their teaching and learning. However, the publisher is ultimately responsible for the editorial control and quality of this book.

Please note that when teaching the ***AQA A-level Computer Science*** course, you must refer to AQA's specification as your definitive source of information. While this book has been written to match the specification, it cannot provide complete coverage of every aspect of the course.

A wide range of other useful resources can be found on the relevant subject pages of our website: www.aqa.org.uk

The Publishers would like to thank the following for permission to reproduce copyright material:

P.11 © chombosan - Fotolia.com; **P.24** © VvoeVale - iStock via Thinkstock.com; **P.69** Courtesy of Wikipedia, The Opte Project (http://creativecommons.org/licenses/by/2.5/); **P.111** © Hodder & Stoughton; **P.136** *middle* © Sergey Kamshylin - Fotolia.com, *bottom* © mark huls - Fotolia.com; **P.137** © Jenny Thompson - Fotolia.com; **P.142** screenshot from TRANSYT from TRL Software (trlsoftware.co.uk); **P.214** © ra3rn - Fotolia.com; **P.217** © davemhuntphoto - Fotolia.com; **P.218** © Bob Reeves; **P.231** *top* © TheVectorminator - Fotolia.com, *bottom* © R+R - Fotolia.com; **P.267** *top* © Maksym Yemelyanov - Fotolia.com, *bottom* © finallast -Fotolia.com; **P.271** © KarSol - Fotolia.com; **P.289** © Igor Mojzes - Fotolia.com; **P.295** Courtesy of Wikimedia Commons, author Ordercrazy, Creative Commons CC 1.0 (http://creativecommons.org/publicdomain/zero/1.0/deed.en); **P.313** © Maxim Pavlov - Fotolia.com

Every effort has been made to trace all copyright holders, but if any have been inadvertently overlooked the Publishers will be pleased to make the necessary arrangements at the first opportunity.

Although every effort has been made to ensure that website addresses are correct at time of going to press, Hodder Education cannot be held responsible for the content of any website mentioned. It is sometimes possible to find a relocated web page by typing in the address of the home page for a website in the URL window of your browser.

Hachette UK's policy is to use papers that are natural, renewable and recyclable products and made from wood grown in sustainable forests. The logging and manufacturing processes are expected to conform to the environmental regulations of the country of origin.

Orders: please contact Bookpoint Ltd, 130 Milton Park, Abingdon, Oxon OX14 4SB.
Telephone: (44) 01235 827720. Fax: (44) 01235 400454. Lines are open 9.00–17.00, Monday to Saturday, with a 24-hour message answering service. Visit our website at www.hoddereducation.co.uk

First published in 2015 by
Hodder Education
An Hachette UK Company,
Carmelite House
50 Victoria Embankment
London EC4Y 0DZ
www.hoddereducation.co.uk

Impression number 5 4 3

Year 2019 2018 2017 2016

Cover photo © LaCozza – Fotolia

A catalogue record for this title is available from the British Library

ISBN 978 1 447 183951 1

Contents

Introduction

What is computer science?

The world of computer science continues to develop at an amazing rate. If you had spoken to an A-level student embarking on a computer science course just ten years ago they might not have believed that in the year 2015 we would all be permanently connected to the Internet on smart phones, watching movies in high definition on 55-inch curved-screen TVs, streaming our favourite music to our phones from a database of millions of tracks stored in 'the cloud' or carrying round a tablet that has more processing power than the flight computer on the now decommissioned space shuttle.

No-one really knows where the next ten years will take us. The challenge for you as a computer scientist is to be able to respond to this ever-changing world and to develop the knowledge and skills that will help you to understand technology that hasn't yet been invented!

Studying A-level computer science gives you a solid foundation in the underlying principles of computing, for example: understanding how algorithms and computer code are written; how data are stored; how data are transmitted around networks; and how hardware and software work. It also provides you with a deeper level of understanding that goes beyond the actual technology. For example, you will learn about how to use computation to solve problems and about the close links between computer science, mathematics and physics.

You might be surprised to learn that many of the key principles of computing were developed before the modern computer, with some concepts going back to the ancient Greeks. At the same time, you will be learning about the latest methods for solving computable problems in today's world and developing your own solutions in the form of programs or apps.

Studying computer science at A level is challenging, but it is also highly rewarding. There are very few jobs that do not involve the use of computers and having a good understanding of the science behind them will effectively prepare you for further study or employment.

Course coverage and how to use this book

This book has been written to provide complete coverage of the AQA Computer Science specifications for AS and A level that are taught from September 2015. The content of the book is matched and sequenced according to the specification, and organised into sections in accordance with the main specification headings used by AQA.

Students studying A level need to be familiar with all of the content of the AS specification and in addition need to cover those sections highlighted throughout the text and are flagged up as A level only. There is support for every section of the specification including the written papers and coursework element.

The main objective of the book is to provide a solid foundation in the theoretical aspects of the course. Further support and practical examples of coded solutions are provided on line via Dynamic Learning.

Chapters contain:

Specification coverage
Taken directly from the specification, it shows which elements of AS and A level are covered within each chapter.

KEYWORDS
All of the keywords are identified with concise definitions. These form a glossary, which is useful for revision and to check understanding.

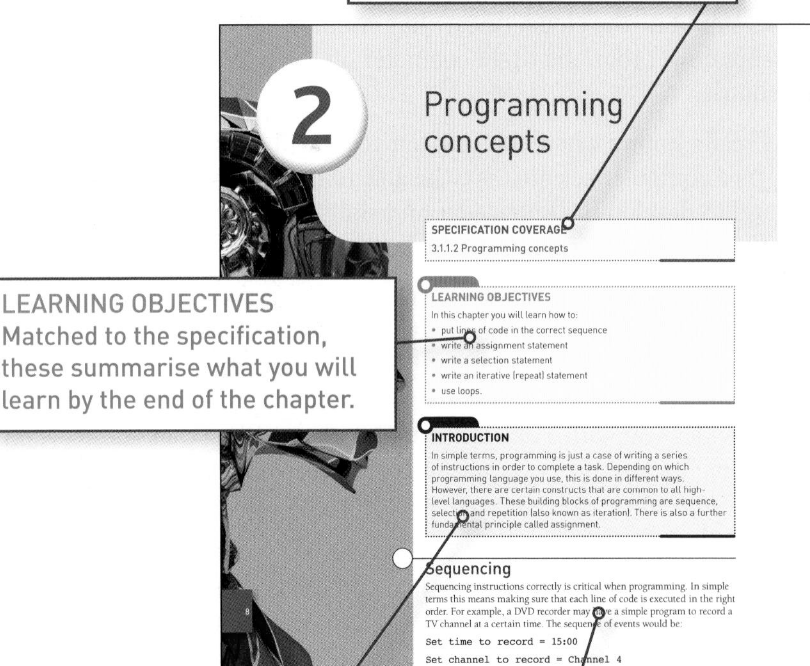

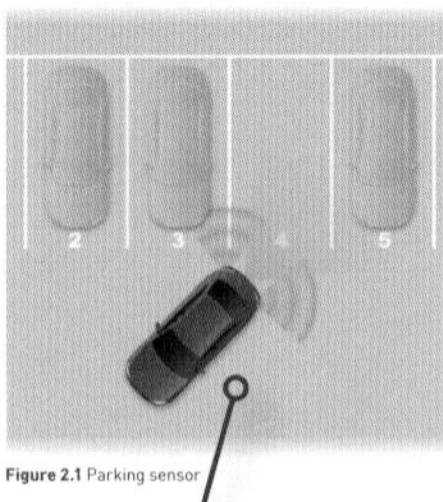

Figure 2.1 Parking sensor

LEARNING OBJECTIVES
Matched to the specification, these summarise what you will learn by the end of the chapter.

INTRODUCTION
This is a concise introduction to set the scene.

The main text
This contains detailed definitions, explanations and examples.

Diagrams and images
The book uses diagrams and images wherever possible to aid understanding of the key points.

Acknowledgements

Dave Fogg for producing the VB code examples used.

Matthew Walker for producing the Python code example used.

Paul Varey for his initial proofread.

Dedicated to Tommy and Eli.

Code examples
Where relevant there are examples of pseudo-code or actual code to demonstrate particular concepts. Code examples in this book are mainly written using the Visual Basic framework. Visual Basic 2010 Express has been used as this is available as a free download. The code can also be migrated into other versions of VB. Note that code that is longer than one line in the book is shown with an underscore (_). It should be input as one line in VB.

KEY POINTS
All of the main points for each chapter are summarised. These are particularly useful as a revision aid.

TASKS
These are activities designed to test your understanding of the contents of the chapter. These may be written exercises or computer tasks.

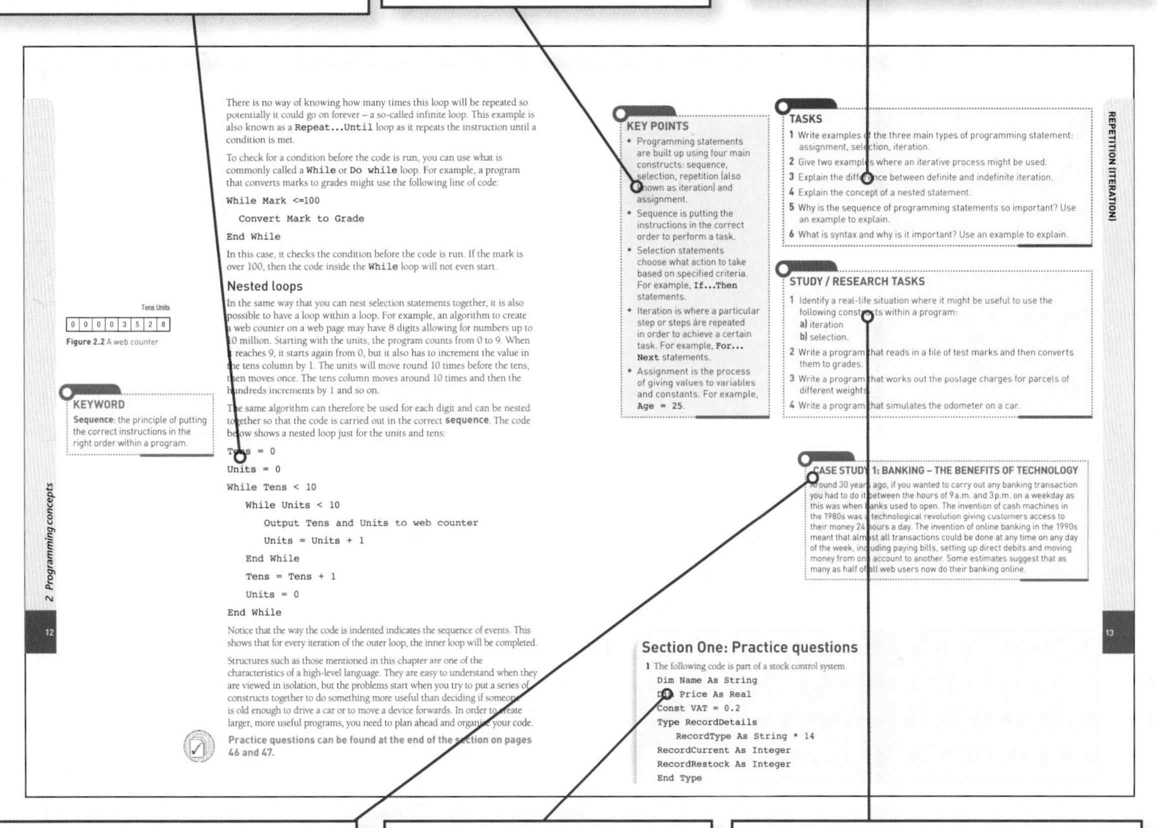

There is no way of knowing how many times this loop will be repeated so potentially it could go on forever – a so-called infinite loop. This example is also known as a **Repeat...Until** loop as it repeats the instruction until a condition is met.

To check for a condition before the code is run, you can use what is commonly called a **While** or **Do while** loop. For example, a program that converts marks to grades might use the following line of code:

```
While Mark <=100
  Convert Mark to Grade
End While
```

In this case, it checks the condition before the code is run. If the mark is over 100, then the code inside the **While** loop will not even start.

Nested loops

In the same way that you can nest selection statements together, it is also possible to have a loop within a loop. For example, an algorithm to create a web counter on a web page may have 8 digits allowing for numbers up to 10 million. Starting with the units, the program counts from 0 to 9. When it reaches 9, it starts again from 0, but it also has to increment the value in the tens column by 1. The units will move round 10 times before the tens, then moves once. The tens column moves around 10 times and then the hundreds increments by 1 and so on.

The same algorithm can therefore be used for each digit and can be nested together so that the code is carried out in the correct **sequence**. The code below shows a nested loop just for the units and tens:

```
Tens = 0
Units = 0
While Tens < 10
   While Units < 10
      Output Tens and Units to web counter
      Units = Units + 1
   End While
   Tens = Tens + 1
   Units = 0
End While
```

Notice that the way the code is indented indicates the sequence of events. This shows that for every iteration of the outer loop, the inner loop will be completed.

Structures such as those mentioned in this chapter are one of the characteristics of a high-level language. They are easy to understand when they are viewed in isolation, but the problems start when you try to put a series of constructs together to do something more useful than deciding if someone is old enough to drive a car or to move a device forwards. In order to create larger, more useful programs, you need to plan ahead and organise your code.

Practice questions can be found at the end of the section on pages 46 and 47.

Tens Units

0	0	0	0	3	5	2	8

Figure 2.2 A web counter

KEYWORD
Sequence: the principle of putting the correct instructions in the right order within a program.

2 Programming concepts

12

KEY POINTS
- Programming statements are built up using four main constructs: sequence, selection, repetition (also known as iteration) and assignment.
- Sequence is putting the instructions in the correct order to perform a task.
- Selection statements choose what action to take based on specified criteria. For example, **If...Then** statements.
- Iteration is where a particular step or steps are repeated in order to achieve a certain task. For example, **For... Next** statements.
- Assignment is the process of giving values to variables and constants. For example, **Age = 25**.

TASKS
1 Write examples of the three main types of programming statement: assignment, selection, iteration.
2 Give two examples where an iterative process might be used.
3 Explain the difference between definite and indefinite iteration.
4 Explain the concept of a nested statement.
5 Why is the sequence of programming statements so important? Use an example to explain.
6 What is syntax and why is it important? Use an example to explain.

STUDY / RESEARCH TASKS
1 Identify a real-life situation where it might be useful to use the following constructs within a program:
a) iteration
b) selection.
2 Write a program that reads in a file of test marks and then converts them to grades.
3 Write a program that works out the postage charges for parcels of different weights.
4 Write a program that simulates the odometer on a car.

CASE STUDY 1: BANKING – THE BENEFITS OF TECHNOLOGY
Around 30 years ago, if you wanted to carry out any banking transaction you had to do it between the hours of 9 a.m. and 3 p.m. on a weekday as this was when banks used to open. The invention of cash machines in the 1980s was a technological revolution giving customers access to their money 24 hours a day. The invention of online banking in the 1990s meant that almost all transactions could be done at any time on any day of the week, including paying bills, setting up direct debits and moving money from one account to another. Some estimates suggest that as many as half of all web users now do their banking online.

Section One: Practice questions

1 The following code is part of a stock control system.

```
Dim Name As String
Dim Price As Real
Const VAT = 0.2
Type RecordDetails
   RecordType As String * 14
RecordCurrent As Integer
RecordRestock As Integer
End Type
```

REPETITION (ITERATION)

13

CASE STUDY
These provide real-life examples of the applications of computing.

Practice questions
These are revision questions designed to check understanding of the topics covered across the whole section.

STUDY/RESEARCH TASKS
These questions go beyond the specification and provide a further challenge designed to encourage you to 'read around the subject' or develop your skills and knowledge further.

Section One: Fundamentals of programming

1 Programming basics

SPECIFICATION COVERAGE

3.1.1.1 Data types

3.1.1.2 Programming concepts

3.1.1.6 Constants and variables in a programming language

3.5.1.2 Integers

LEARNING OBJECTIVES

In this chapter you will learn:

- the basic principles of writing instructions in the form of programming code
- what constants and variables are and how to use them
- what the main data types are
- how to store data using meaningful names.

KEYWORDS

Memory: the location where instructions and data are stored on the computer.

Algorithm: a sequence of steps that can be followed to complete a task and that always terminates.

Syntax: the rules of how words are used within a given language.

INTRODUCTION

In its simplest form a computer program can be seen as a list of instructions that a computer has to work through in a logical sequence in order to carry out a specific task. The instructions that make up a program are all stored in **memory** on the computer along with the data that is needed to make the program work.

Programs (also known as applications or apps) are created by writing lines of code to carry out algorithms. An **algorithm** is the steps required to perform a particular task and the programming code contains the actual instructions written in a programming language. This language is in itself an application that has been written by someone else to enable you to write your own programs.

In the same way that there are lots of different languages you can learn to speak, there are also lots of programming languages, and in the same way that some languages have many different dialects, there are also different versions of some of the more popular programming languages.

Another similarity with natural languages is that each programming language has its own vocabulary and rules that define how the words must be put together. These rules are known as the **syntax** of the language. The difference between learning a foreign (natural) language and a computer language is that there are far fewer words to learn in a computer language but the rules are much more rigid.

Naming and storing data

In addition to instructions, the computer program also needs data to work with. For example, to add two numbers together requires an **add** instruction and then the two numbers that need to be added. You need to give these two data items names so that the computer will know which data to use.

The data are stored in memory along with the instructions. You could view memory rather like a series of pigeon-holes, each having a unique address, known as a **memory address**.

> **KEYWORD**
>
> **Memory address**: a specific location in memory where instructions or data are stored.

It is a really good idea to use names that indicate the purpose of the data – in the case of the example above the two numbers might be called `Number1` and `Number2`. Using meaningful names will help you when they are trying to trace bugs and it also allows other programmers to follow the code more easily. It is good practice to adopt a common naming convention. In this case the first character in upper case and the rest in lower case.

This process of giving data values is called 'assigning', and it looks something like these two:

```
Number1 ← 23
Name ← "Derek"
```

The ← means 'becomes' or 'equals'. `Number1` is an example of a variable. In the example above it has been given a value of `23`, though this value will change while the program is being run. `Name` is another example of a variable and has been given the value `"Derek"`.

> **KEYWORD**
>
> **Assignment**: the process of giving a value to a variable or constant.

Different programming languages have slightly different ways of assigning values. For example, you may need to use the equals sign to make the **assignment** in the code you are writing. So a simple algorithm to add two numbers together might look like this:

```
Number1 = 2
Number2 = 3
Answer = Number1 + Number2
```

Figure 1.1 shows a simplified visualisation of how this program is handled. There will be millions of memory addresses, of which just three are shown in this diagram.

Memory address	1000	1001	1002
Variable	Number1	Number2	Answer
Data	2	3	5

Figure 1.1

Constants and variables

> **KEYWORDS**
>
> **Constant**: an item of data whose value does not change.
>
> **Variable**: an item of data whose value could change while the program is being run.

Data are stored either as **constants** or as **variables**. Constants (as you'd expect from the name) have values that are fixed for the duration of a program. For example, if you were writing a program that converted miles into kilometres you could set the conversion rate as a constant because it will never change. In this case we could call the constant `ConvertMilestoKm` and assign it a value of 1.6 as there are approximately 1.6 km to the mile. Then whenever we want to convert a distance in miles to its metric equivalent we would multiply it by the constant `ConvertMilestoKm`.

Notice that the name given to the constant is self-explanatory. We could have just called it `Constant1`. However, by giving it a meaningful name it makes the code easier to work with as the program gets bigger. It also makes it easier for anyone else that looks at the code, to work out what the program is doing. This is important for three main reasons:

KEYWORD

Debug: the process of finding and correcting errors in programs.

- It makes it easier to find and correct errors/bugs in code. This is called **debugging**.
- There are many occasions where there are several programmers working on the same program at the same time, so having a sensible naming convention makes it easier for everyone to understand.
- It will be easier to update the code later on when further versions of the program are created.

The value of variables can change as a program is being run. For example, the same conversion program will require the user to type in the number of miles they want to convert. This number will probably be different each time the user enters data. Therefore, you need to have a variable that you could call `NumberOfMiles`.

There are lots of other examples – the number of answers a pupil has got right in a test would (hopefully) increase as they work their way through a test so the data would have to be stored as a variable. The password a user uses to access a network can be changed at any time, so it would also be classed as a variable.

Variable and constant declaration

Declaring a constant or variable means that when you are writing code you describe (or declare) the variables and constants that you are going to use before you actually use them in your program.

Some programming languages force you to declare the variables and constants you intend to use in your program before you start writing any code. The benefits of doing this are that it forces you to plan first and the computer will quickly identify variables it does not recognise.

KEYWORD

Declaration: the process of defining variables and constants in terms of their name and data type.

There are two parts to a **declaration**. You need to supply a suitable name for the constant/variable and you need to specify the data type that will be used. The declarations might look something like this:

```
Dimension Age As Integer

Dimension Name As String

Dimension WearsGlasses As Boolean
```

`Dimension` or `Dim` is one of the command words used in Visual Basic to indicate that a variable is being declared. Once you have declared a variable it starts with a default value. In the above examples `Age` will start as zero, `Name` as nothing (also known as the empty string) and `WearsGlasses` will start with the value False. Other languages may use different default values so it is good practice to assign an initial value to the variable just to make sure it is correct.

Data types

It is important to consider how you want your program to handle data. For example, to create the miles to kilometres conversion program, you have to tell the program that miles and kilometres both need to be stored as numbers.

KEYWORD

Data type: determines what sort of data are being stored and how it will be handled by the program.

KEYWORD

Integer: any whole positive or negative number including zero.

There are lots of **data types** you might need to use and you need to think carefully about the best type to use. For example, if storing numbers, how accurate do you need the number to be? Will a whole number be accurate enough or will you need decimals? In addition to numbers, you will probably want to store other data such as a person's name, their date of birth or their gender.

All programming languages offer a range of data types but the actual name of the data type may vary from language to language. Here are some of the most common data types:

- Integer: This is the mathematical name for any positive or negative whole number. This might be used to store the number of cars sold in a particular month or the number of pupils in a class. The range of numbers that can be stored depends on how much memory is allocated. For example, an **integer** in Visual Basic can store numbers between –2 147 483 648 through to +2 147 483 647.
  Declaring a number as an integer means that the program will then handle the data accordingly. For example, 2 + 3 will equal 5. In some languages, if you did not set it to integer 2 + 3 would equal 23 (two three).
- Real/Float: This is a number that has a fractional or decimal part, for example 3.5 or $3\frac{1}{2}$. In our miles to kilometres conversion program, you would need to store both miles and kilometres using this data type as the user might want to convert a number that is not a whole number. Other examples might include a person's height in metres or their weight in kilograms.
- Text/String: This data type is used to store characters, which could be text or numbers. For example, you could use this to store a person's name or address. Some programming languages refer to this data type as alphanumeric because you can actually store any character you want in a string whilst text implies it can only store letters. Text or string variables are normally shown in quotation marks. For example you might assign the name Frank to a variable like this: `Name ← "Frank"`. House numbers and phone numbers are often stored as text / string as although they are numbers, you would never need to carry out any calculations on them and in the case of telephone numbers the leading zero is important and would be omitted if stored as a number.
- Boolean: The simplest data type is a simple yes/no or true/false. This is called a Boolean data type. It is named after George Boole who discovered the principles behind logic statements. Boolean data types can be used to store any kind of data where there are two possible values.
- Character: This data type allows you to store an individual character, which might be a letter, number or symbol. All computers have a defined character set, which is the range of characters that it understands. This would commonly be all the upper and lower case letters, plus other keyboard characters and any special characters.
- Date/Time: This will store data in a format that is easily identifiable as a date or time, e.g. 30.04.2014 or 12:30. The program will then handle the data accordingly. For example, if you added 5 to the date, it would tell you the date in five days' time. 30.05.2014 + 5 would become 04.06.2015. If you did not declare it as a date you may get the wrong answer, for example 30.05.2019.
- Pointer/Reference: This data type is used to store a value that will point to or reference a location in the memory of the computer. If you think of memory

KEYWORD

Pointer: a data item that identifies a particular element in a data structure – normally the front or rear.

KEYWORDS

Array: a set of related data items stored under a single identifier. Can work on one or more dimensions.

Element: an single value within a set or list – also called a member.

Record: one line of a text file.

as a series of pigeon-holes or addresses where instructions and data are stored, the **pointer**/reference is used in a program to go to a specific address. For example, you could set up a pointer called `Pointer1` and put address 1001 in it. The program would then go to memory address 1001 and take the data from it. In the example below it would be the data assigned to `Number2`. Other lines of code will then be needed to tell the program what to do with the data it finds there.

Figure 1.2 shows how a pointer is used to reference an item of data.

`Pointer1` = 1001 → 1001

Memory address	1000	1001	1002	1003
	Number1	Number2	Add	Answer

Figure 1.2

- Array: An **array** is a collection of data items of the same type. For example, if you wanted to store a collection of names in a school register, you could call this `Register` and each item of data would be stored as text. Each individual name in the array is called an **element**. Every element is numbered so that `Register(2)` would be the second person in the array, `Register(4)` the fourth person and so on. Note that 0 is often used as the first element of an array, rather than 1. If this was the case then `Register(2)` would actually be the third person in the array, `Register(4)` the fifth and so on.

  Figure 1.3 shows a simple array with six elements. `Register(2) = Hussain`, `Register(4) = Schmidt` (assuming array indexing starts at 1).

Brown
Hussain
Koening
Schmidt
Torvill
West

Figure 1.3

- Records: This is used to store a collection of related data items, where the items all have different data types. For example, you might set up a **record** called `Book`, which is used to store the title, author name and ISBN of a book. `Title` and `Author` are text whereas the `PublicationDate` is set as a Date data type.

  You could write it like this:

```
Book = Record
  Title, Author As Text * 50
  ISBN As Text * 13
  PublicationDate As Date
```

  When the program is run, every time data are entered for the book, the user will type in up to 50 characters of text for the title and author and then the ISBN. A variable could now be set up using this record data type and this variable would contain all of this data.

Built-in and user-defined data types

Built-in data types are those that are provided with the programming language that you are using. The list of built-in types varies from language to language, but all will include versions of the types listed above.

Most programming languages allow users to make up their own data types, usually by combining existing data types together. These are simply called user-defined data types. For example, if you were making a program to store user names and IDs, you may create a user-defined data type called **Logon** made up of a set number of characters and numbers.

```
Type Logon
   UserName As String * 10
   UserID As Integer * 5
End Type
```

This code will set a new data type called **Logon**, which will be made up of a 10-character user name followed by a 5-digit user ID. In total, the data type will have 15 characters/digits to store the data.

The reasons for creating user-defined types are mainly to do with efficiency. As you start to write your own programs you will find that they can get very long and complex and that debugging can be very time-consuming. Most programmers try to make their code as organised and efficient as possible as this will save them time as the program develops. For example, it is easier to reuse a block of code rather than have to write it all over again.

Most programmers aim to create code that is 'elegant'. This means that it does exactly what it is supposed to do as efficiently as possible. Often this means writing as few lines of code as possible with no repeated coding.

Using a user-defined data type is just one example of where it is possible to be more efficient. With our example of storing **Logon** information using a user-defined variable, because all the data are stored in one variable rather than two, when the program needs this information, we only need to access one variable rather than two.

Practice questions can be found at the end of the section on pages 46 and 47.

KEY POINTS

- Programming languages are used to write applications (apps).
- An algorithm is a sequence of instructions that can be followed to complete a task. Algorithms always terminate.
- Programming code is made up of algorithms that are implemented within a programming language.
- Instructions are stored in memory along with the data required by the program.
- Data are stored and named according to certain conventions.
- Variables and constants are used to store data and must be declared in some languages.
- There are several data types built in to every programming language and the programmer can also define their own.

TASKS

1 Give two reasons why it is a good idea to use meaningful variable names.

2 Use examples to explain the difference between a constant and a variable.

3 Why is it important to declare all variables and constants at the beginning of a program?

4 Explain the difference between a value and a variable.

5 Suggest suitable data types and variable names for:

  a) the current rate of VAT

  b) today's date

  c) the total takings from a shop

  d) a person's date of birth

  e) which wrist a person wears a watch on.

STUDY / RESEARCH TASKS

1 A list of data is also known as a one-dimensional array. Find out what two- and three-dimensional arrays are and give examples of where you might use each.

2 Identify the built-in data types for the main programming language that is used in your school or college.

3 Research data types that are specifically used to store sound and video data. How do they differ from other data types?

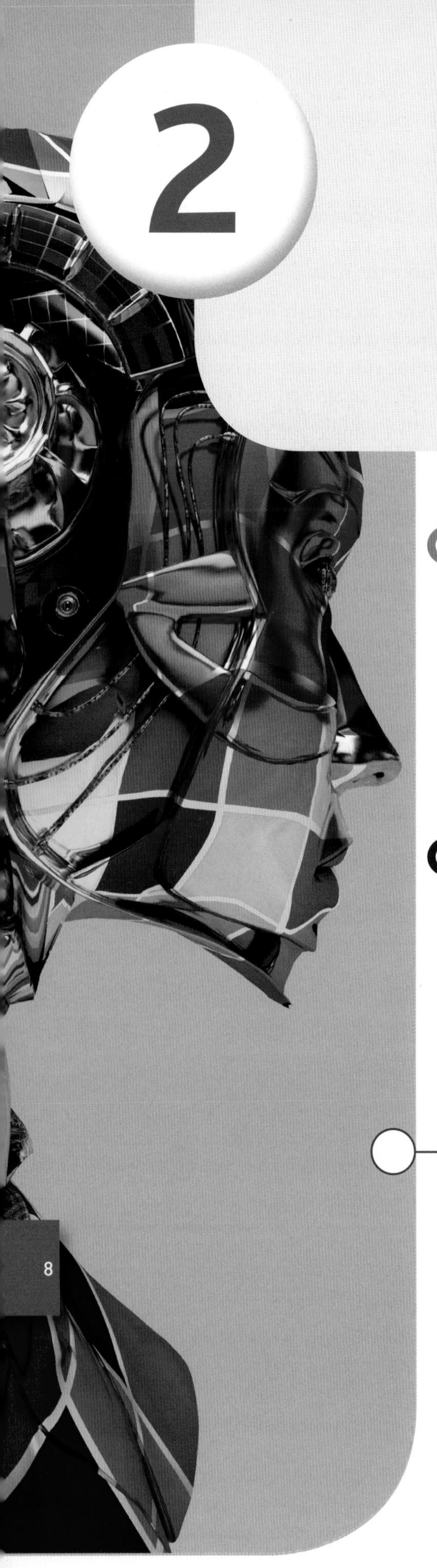

2 Programming concepts

SPECIFICATION COVERAGE

3.1.1.2 Programming concepts

LEARNING OBJECTIVES

In this chapter you will learn how to:

- put lines of code in the correct sequence
- write an assignment statement
- write a selection statement
- write an iterative (repeat) statement
- use loops.

INTRODUCTION

In simple terms, programming is just a case of writing a series of instructions in order to complete a task. Depending on which programming language you use, this is done in different ways. However, there are certain constructs that are common to all high-level languages. These building blocks of programming are sequence, selection and repetition (also known as iteration). There is also a further fundamental principle called assignment.

Sequencing

Sequencing instructions correctly is critical when programming. In simple terms this means making sure that each line of code is executed in the right order. For example, a DVD recorder may have a simple program to record a TV channel at a certain time. The sequence of events would be:

```
Set time to record = 15:00
Set channel to record = Channel 4
Check time
If time = 15:00 Then Record
```

If any of these instructions were wrong, missed out or executed in the wrong order, then the program would not work correctly.

KEYWORD

Syntax: the rules of how words are used within a given language.

The actual process of writing statements varies from one programming language to another. This is because all languages use different **syntax**. Common usage of the word syntax refers to the way that sentences are structured to create well-formed sentences. For example, the sentence 'Birds south fly in the winter' is syntactically incorrect because the verb needs to come after the noun. Programming languages work in the same way and have certain rules that programmers need to stick to otherwise the code will not work.

Assignment

We met the concept of an assignment statement in Chapter 1. Assignment gives a value to a variable or constant. For example you might be using a variable called `Age` so the code:

```
Age ← 34
```

will set the variable `Age` to have the value `34`.

The value stored in the variable could change as the program is run. For example, a computer game might use a variable called `Score`. At the beginning of the game the value is set to 0. Each time the player scores a point, the assignment process takes place again to reset the value of `Score` to 1 and so on.

Assigning values will take place over and over again while a program is being run. Initially, the programmer will assign a value to the variable. Then as the program runs, the algorithms in the program code will calculate and then return (re-assign) the latest value. Assignments are the fundamental building blocks of any computer program because they define the data the program is going to be using.

Selection

KEYWORD

Selection: the principle of choosing what action to take based on certain criteria.

The **selection** process allows a computer to compare values and then decide what course of action to take. For example you might want your program to decide if someone is old enough to drive a car. The selection process for this might look something like this:

```
If Age < 17 Then
    Output = "Not old enough to drive"
Else
    Output = "Old enough to drive"
End If
```

In this case, the computer is making a decision based on the value of the variable `Age`. If the value of `Age` is less than 17 it will output the text string `"Not old enough to drive"`. For any other age it will output the text string `"Old enough to drive"`. The `If` statement is a very common construct. In this case it is used to tell the program what to do if the statement is true using the `If...Then` construct. If the statement is false, it uses the `Else` part of the code. This is a very simple selection statement with only two outcomes.

KEYWORD

Nesting: placing one set of instructions within another set of instructions.

Nested selection

You can carry out more complex selections by using a **nested** statement. For example, a program could be written to work out how much to charge to send parcels of different weights. This could be achieved using the following sequence of selection statements:

```
If Weight >= 2000 Then
  Price = £10
Else If Weight >= 1500 Then
  Price = £7.50
Else If Weight >= 1000 Then
  Price = £5
Else
  Price = £2.50
End If
```

When the weight is input, it works through the lines of code in the `If` statement and returns the correct value. For example, if the parcel weighs 1700 g it will cost £7.50 as it is between 1500 g and 1999 g. If it weighed 2000 g or more, the `If` statement would return £10.

In some languages complex selections can be implemented using constructs such as this `Case` statement. The following example shows a section of code that allows the user to type in a country code to identify where a parcel is being sent to:

```
Select Case ParcelDestination
  Case 1
      WriteLine ("Mainland UK")
  Case 2
      WriteLine ("Europe")
  Case 3
      WriteLine ("USA")
  Case Else
      WriteLine ("Rest of the World")
End Select
```

This routine takes the value of the variable `ParcelDestination` and compares it against the different criteria. So if `ParcelDestination` is 1 then `Mainland UK` will be printed to the screen.

KEYWORD

Iteration: the principle of repeating processes.

Repetition (Iteration)

It is useful to be able to repeat a process in a program. This is usually called **iteration**. For example you might want to count the number of words in a block of text or you may want to keep a device moving forward until it reaches a wall. Both these routines involve repeating something until a condition is met – either you run out of words to count or the device comes to a wall. An iterative process has two parts – a pair of commands that show the start and finish of the process to be repeated and some sort of condition.

> **KEYWORDS**
>
> **Definite iteration**: a process that repeats a set number of times.
>
> **Indefinite iteration**: a process that repeats until a certain condition is met.
>
> **Loop**: a repeated process.

There are two basic forms of iteration – definite and indefinite. **Definite iteration** means that the instructions are repeated in a loop a certain number of times. **Indefinite iteration** means that the instructions are repeated in a loop until some other event stops it. Iteration statements are often referred to as **loops**, as they go round and round. Let's look at an example of each.

Definite iteration

If you want a process to be carried out a set number of times you will need to use definite iteration. For example, the following code could be used to operate a robotic device. It will move a device forward 40 units:

```
For Counter = 1 To 40
   Move forward 1 unit
Next
```

After it has moved forward 40 units it will stop. It will try and move irrespective of whether it meets an obstacle or not. This is known as a `For...Next` loop as it will carry out the instruction for a set number of times.

Indefinite iteration

In this case the loop is repeated until a specified condition is met – so it uses a selection process to decide whether or not to carry on (or even whether to start) a process.

This routine moves a device forward until the sensor detects an obstacle:

```
Repeat
   Move forward 1 unit
Until Sensors locate an obstacle
```

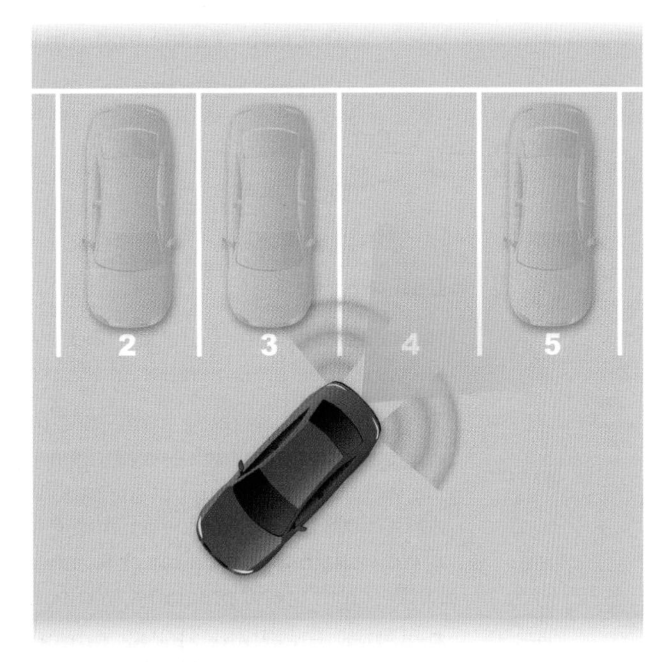

Figure 2.1 Parking sensor

There is no way of knowing how many times this loop will be repeated so potentially it could go on forever – a so-called infinite loop. This example is also known as a **Repeat...Until** loop as it repeats the instruction until a condition is met.

To check for a condition before the code is run, you can use what is commonly called a **While** or **Do while** loop. For example, a program that converts marks to grades might use the following line of code:

```
While Mark <=100
  Convert Mark to Grade
End While
```

In this case, it checks the condition before the code is run. If the mark is over 100, then the code inside the **While** loop will not even start.

Nested loops

In the same way that you can nest selection statements together, it is also possible to have a loop within a loop. For example, an algorithm to create a web counter on a web page may have 8 digits allowing for numbers up to 10 million. Starting with the units, the program counts from 0 to 9. When it reaches 9, it starts again from 0, but it also has to increment the value in the tens column by 1. The units will move round 10 times before the tens, then moves once. The tens column moves around 10 times and then the hundreds increments by 1 and so on.

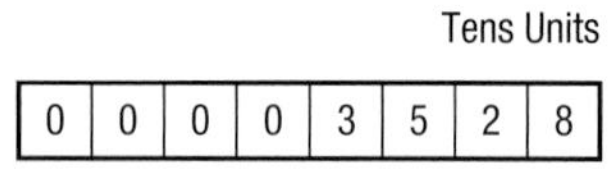

Figure 2.2 A web counter

KEYWORD

Sequence: the principle of putting the correct instructions in the right order within a program.

The same algorithm can therefore be used for each digit and can be nested together so that the code is carried out in the correct **sequence**. The code below shows a nested loop just for the units and tens:

```
Tens = 0
Units = 0
While Tens < 10
   While Units < 10
      Output Tens and Units to web counter
      Units = Units + 1
   End While
   Tens = Tens + 1
   Units = 0
End While
```

Notice that the way the code is indented indicates the sequence of events. This shows that for every iteration of the outer loop, the inner loop will be completed.

Structures such as those mentioned in this chapter are one of the characteristics of a high-level language. They are easy to understand when they are viewed in isolation, but the problems start when you try to put a series of constructs together to do something more useful than deciding if someone is old enough to drive a car or to move a device forwards. In order to create larger, more useful programs, you need to plan ahead and organise your code.

Practice questions can be found at the end of the section on pages 46 and 47.

KEY POINTS

- Programming statements are built up using four main constructs: sequence, selection, repetition (also known as iteration) and assignment.
- Sequence is putting the instructions in the correct order to perform a task.
- Selection statements choose what action to take based on specified criteria. For example, `If...Then` statements.
- Iteration is where a particular step or steps are repeated in order to achieve a certain task. For example, `For... Next` statements.
- Assignment is the process of giving values to variables and constants. For example, `Age = 25`.

TASKS

1 Write examples of the three main types of programming statement: assignment, selection, iteration.

2 Give two examples where an iterative process might be used.

3 Explain the difference between definite and indefinite iteration.

4 Explain the concept of a nested statement.

5 Why is the sequence of programming statements so important? Use an example to explain.

6 What is syntax and why is it important? Use an example to explain.

STUDY / RESEARCH TASKS

1 Identify a real-life situation where it might be useful to use the following constructs within a program:

a) iteration

b) selection.

2 Write a program that reads in a file of test marks and then converts them to grades.

3 Write a program that works out the postage charges for parcels of different weights.

4 Write a program that simulates the odometer on a car.

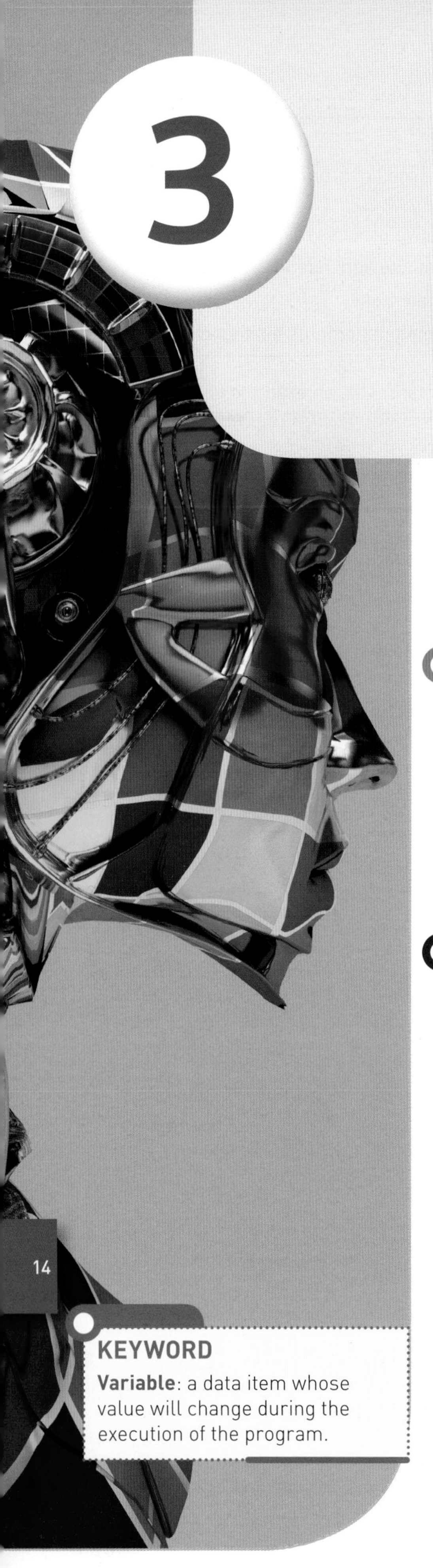

3 Basic operations in programming languages

SPECIFICATION COVERAGE

3.1.1.3 Arithmetic operations in a programming language

3.1.1.4 Relational operations in a programming language

3.1.1.5 Boolean operations in a programming language

3.1.1.7 String-handling operations in a programming language

3.1.1.8 Random number generation in a programming language

LEARNING OBJECTIVES

In this chapter you will learn:

- the correct syntax for writing basic programming code
- how to construct arithmetic operations, Boolean operations and relational operations
- how to handle basic string operations
- how Visual Basic, Python and C# implement these operations.

INTRODUCTION

There are a number of basic operations that you can perform on text and numeric data when programming. These fall into four main categories: arithmetic operations, relational operations, Boolean operations and string handling. In this chapter all of these basic operations are explained with simple examples to illustrate each. The examples are based on Visual Basic and there is also a look-up table at the end the chapter to show these how these basic operations could be implemented in Python and C#.

When programming, the syntax for each operation will vary depending on which language you are using. In practice, when creating full programs you will be using many of these operations in combination to perform particular tasks. In Sections Two and Three, you will see many of these basic operations being used in context.

When programming, values are likely to come from a **variable** or constant, or be generated as part of the program. For example, an assignment statement that adds two numbers together may use three variables called `Answer, FirstNumber and SecondNumber`:

```
Answer = FirstNumber + SecondNumber
5 = 3 + 2
```

KEYWORD

Variable: a data item whose value will change during the execution of the program.

Note that in some languages these operations can be carried out on numeric values or text strings. For example:

```
txtAnswer = txtFirstVariable + txtSecondVariable
```

```
DavidSmith = David + Smith
```

This has implications for the programmer as you might expect a simple addition of 2 + 2 to equal 4. However, 2 + 2 could also result in the answer 22 if the programmer has not defined the values as integers. This is one reason why it is good practice to declare variables at the beginning of every program, so that the program knows how to handle the data.

Arithmetic operations

KEYWORD

Arithmetic operation: common expressions such as +, –, /, ×.

Most of these are the standard mathematical operations that you use every day such as add, subtract, multiply and divide.

- Addition: The sum of two or more values. Example: 5 = 3 + 2 or `Answer = FirstNumber + SecondNumber`.
- Subtraction: One value minus another. Example: 2 = 5 – 3 or `Answer = FirstNumber - SecondNumber`.
- Multiplication: The product of two values. Example: 6 = 3 * 2 or `Answer = FirstNumber * SecondNumber`.
- Division of real numbers: A real number is one with a fractional part so may result in an answer with a fractional part. Example: 3.1 = 6.2/2 or `Answer = FirstNumber / SecondNumber` (where all variables have been declared as Real or Float).
- Division of integers: An integer is a whole number and therefore may generate a number with a remainder. Example: 3rl = 7/2 or `Answer = FirstNumber / SecondNumber` (where all variables have been declared as Integer). The DIV operation can also be used in the format `Answer = FirstNumber DIV SecondNumber` in which case the quotient and remainder are calculated simultaneously.
- Modulo operation: The modulo or MOD operator is used to divide one number by another to find the remainder. Example: 1 = 7 MOD 2 or `Answer = FirstNumber MOD SecondNumber` as 7/2 = 3rl.
- Exponentiation: Repeated multiplication of a base number in the form B^n where B is the base number and n is the number of times to repeat the multiplication. For example 2^4 is 2 × 2 × 2 × 2. Example: 16 = 2 ^ 4 or `Answer = FirstNumber ^ SecondNumber`.
- Rounding: Replacing the real value with a simpler representation that is close to the original value. For example, 2.315432 becomes 2.3. There are various methods for **rounding** within each programming language such as rounding up and down, or rounding to a specific number of decimal places. Example: 2 = Round(2.3) or `Answer = Round(FirstNumber)`.
- Truncating: Shortening a value by cutting it off after a certain number of digits. It is the equivalent of rounding down. There are various methods for **truncating** within each programming language. Example: 2 = Truncate (2.345) or `Answer = Truncate(FirstNumber)` where `FirstNumber` is a decimal value.
- Random number generation: Creating a number to be used in a program that is random. There are several methods of doing this. Often the number is set to be generated between two fixed values. There are

KEYWORDS

Rounding: reducing the number of digits used to represent a number while maintaining a value that is approximately equivalent.

Truncating: the process of cutting off a number after a certain number of characters or decimal places.

various methods within each programming language. For example: `0.123 = Rnd ()` or `Answer = Rnd()`.

Random numbers are a very useful tool for programmers. Typical applications include:

- Creating a range of test data to be used on a new program
- Producing data to use in computer simulations
- Creating random events and movements in computer games
- Selecting a random sample from a dataset.

However, as most **random number generation** techniques used in programming languages start from a seed value and then use an algorithm to create the random number, it means that the number cannot be truly random as the algorithm used will produce an element of structure to the results. Consequently, random numbers generated by programming languages are often referred to as **pseudo-random numbers**. This is perfectly adequate for the purposes listed above but in other circumstances, such as encryption, this level of randomness would not be sufficient.

KEYWORDS

Random number generation: a function that produces a completely random number.

Pseudo-random number generator: common in programming languages, a function that produces a random number that is not 100% random.

Relational operations

Relational operations work by making comparisons between two data items. They consist of operands and operators where the operands are the values and the operator is the comparison being made. For example, in the operation A > B, A and B are the operands and > is the operator. Most programming languages recognise the standard method for representing these operators as shown in Table 3.1.

KEYWORD

Relational operations: expressions that compare two values such as equal to or greater than.

Table 3.1 Table of relational operators

Relational operator	Sign
Equal to	= or ==
Not equal to	<> or !=
Less than	<
Greater than	>
Less than or equal to	≤
Greater than or equal to	≥

Relational operations are often performed in order to create selection statements. For example: `If A > 1 Then...` means if A is 2 or more then the next action is carried out. In common with all operations, the comparisons could also be made between textual data as well as numerical data.

Boolean operations

Boolean operations are those which result in either a TRUE or a FALSE answer. Boolean algebra is used in logic circuits and is an underlying principle to how modern computers work. It is also fundamental to the process of searching data whether that is in a database, or on the web. Once the Boolean operation has been evaluated, a further action is then taken. For example, on a database search, a subset of data would be created containing records that met the search criteria. The examples below are based on a scenario where an online dataset is being searched to find a new car.

KEYWORD

Boolean operations: expressions that result in a TRUE or FALSE value.

KEYWORDS

AND: Boolean operation that outputs true if both inputs are true.

OR: Boolean operation that outputs true if either of its inputs are true.

NOT: Boolean operation that inverts the result so true becomes false and false becomes true.

XOR: Boolean operation that is true if either input is true but not if both inputs are true.

The four basic operations are:

- **AND**: This is known as a conjunction as it adds together the data. For example, using the search phrase "Four Door AND Less than 3 years old" would return a value of TRUE only if both conditions were met, so the car would need to have four doors AND be less than 3 years old.
- **OR**: This is known as a disjunction, which means that a TRUE result is produced if any of the conditions are met. For example, in the search phrase "Four Door OR Three Door", only one of the conditions needs to be met to get a TRUE result, so all three- and four-door cars would be listed.
- **NOT**: This is known as a negation as it reverses the input. For example, "NOT Ford" would result in data that did NOT contain the word Ford.
- **XOR**: This is known as an exclusive OR and means that a TRUE result is produced when one or the other condition is met but not both. For example, "Sunroof XOR Air conditioning" would result in data where the car either had a sunroof or air conditioning, but not both. XOR operations are used extensively when creating logic gates and there is more on this in Chapter 30.

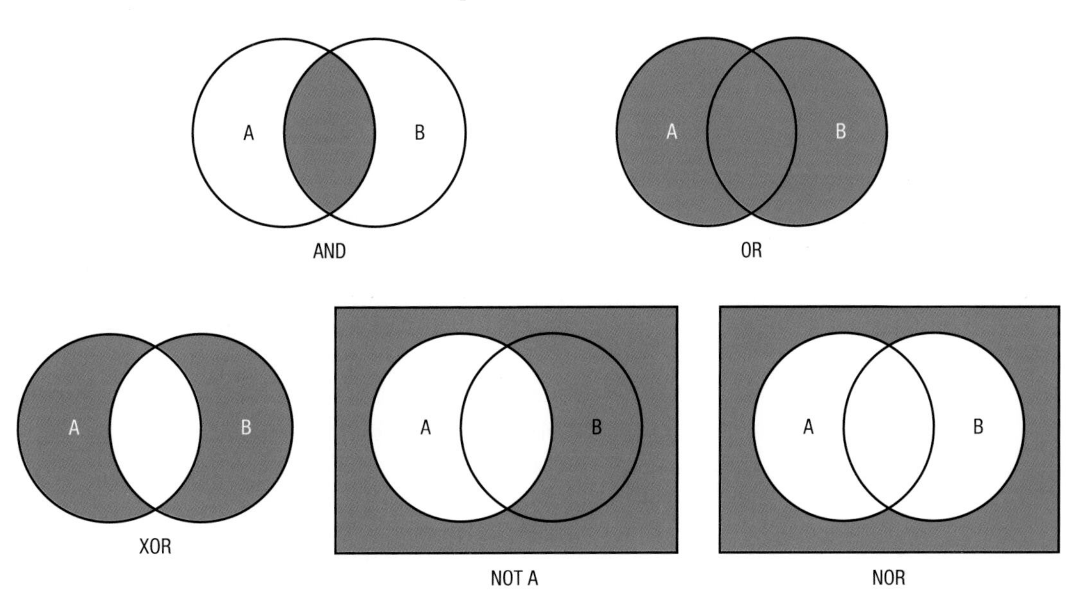

Figure 3.1 Venn diagrams to represent the four basic Boolean operations.

You may have noticed that it is possible to embed relational operators within Boolean operations. For example, "Four Door AND Less than 3 years old" uses the `less than` operator. It is also possible to join lots of Boolean operations together to produce the desired outcome. For example, a very specific search might be: "Four Door AND Less than 3 years old AND Ford OR Vauxhall NOT Fiat".

String-handling functions

KEYWORD

String-handling functions: actions that can be carried out on sequences of characters.

At the beginning of the chapter we saw that it is possible to carry out operations on numbers and text data. This section looks specifically at the way in which text can be handled. To be more precise, this section looks at how strings can be handled. A string is a sequence of characters and can actually be made up of text, numbers and symbols. There are many situations where you will need to work with strings to produce the desired outcome, for example, searching for strings of characters or combining strings together.

- Length: The length of the string is how many characters are used to store a particular item of data. The string length is often variable although there is usually an upper limit placed on its size. There are various operations that you can carry out using the string length. For example, you may want to set the maximum length or calculate the length of a particular string of data. For example: `Dim LastName As String` is used to define a string data type and `Len (Variable1)` calculates the length of data value stored in `Variable1`.
- Position: Within a text string it is possible to identify the position of every character. This is useful when you want to extract particular characters from the string or identify substrings within the data. There are various operations that you can carry out. For example, to find the start position of a particular string of characters within another string:
```
Txt = "JohnSmith22HighStreetLeicester"
AddressPosition = InStr(txt,"22HighStreet")
```
This would return a value of 10 in this example as that is the position where the address data starts within the string being searched. This assumes that the start position is 1. Some languages take the start position as 0, in which case this would return a value of 9.
- Substring: A substring is a string contained within another string as shown in the example above. Various techniques can be used to extract data from anywhere in a string to create a substring providing the start and end position are known or the start position and length are known. For example:
```
txt= "JohnSmith22HighStreetLeicester"
txtAddress = str.Substring (10,21)
```
This would create the substring "22HighStreetLeicester" and store it in a variable called `txtAddress`. It does this by starting at position 10 of the string and then extracting the next 21 characters.
- Concatenation: This is the process of adding strings together to create another string. For example:
```
txtFirstName = "John"
txtLastName = "Smith"
txtFullName = txtFirstName + txtLastName
```
This would create the value "JohnSmith" stored in a variable called `txtFullName`.
- Character codes: Every character that you can use on a computer including all the keyboard characters has a corresponding **character code**, which is used to identify it. This might be an ASCII code or Unicode (see Chapter 26). This can be used in various ways, for example, if you need to convert a text value to a numeric value in order to carry out a calculation on it. You might do this when encrypting data. For example:
`asc(Variable1)` returns the ASCII code value of the value stored in `Variable1` where `Variable1` is a text character.
`chr(Variable1)` returns the text character where `Variable1` is an ASCII code.
`chrW(Variable1)` returns the Unicode code value of `Variable1` where `Variable1` is a text character.

In addition to converting strings to character codes, there are a number of other conversions that a programmer might need to do in order to manipulate the data further. Most programming languages include specific functions to carry out these conversions.

KEYWORD

Character code: a binary representation of a particular letter, number or special character.

- String to Integer / Integer to String: An integer is a whole number. Some programming languages convert between the two automatically if the two variables are declared correctly. For example:

```
Dim i as Integer
Dim s as String
i = 1
s = i
```

  This would result in `s` (a string) becoming 1 (an integer). The same code could be used to reverse the process.
- String to Float / Float to String: A float is also called a real number and is any number including those with a fractional part. In Visual Basic, a function exists to carry out this conversion:
  `Convert.ToDouble(x)` will convert the text string `x` into a Double data type, used by Visual Basic to store real numbers.
  `Convert.ToString(n)` will convert the real number `n` into a string.
- String to Date-time / Date-Time to String: Date-time is usually stored with built-in formatting such as dd.mm.yyyy and hh:mm:ss. To manipulate individual parts of the data, it can be converted into a string. Most programming languages have built-in functions. For example, in Visual Basic:
  `DateTime.ToString(date)` converts the date and time into a string
  `String.ToDateTime(String)` converts a string into date-time format.

Examples of common operations in Python and C#

Table 3.2 gives some examples of how these common operations can be executed in both Python and C#. Note that there will be other ways of implementing these operations based on the specific requirements of the program being written.

You will notice that there are some commonalities between programming languages and some operations that are handled completely differently.

4 Subroutines, local and global variables

SPECIFICATION COVERAGE

3.1.1.9 Exception handling

3.1.1.10 Subroutines (procedures/functions)

3.1.1.11 Parameters of subroutines

3.1.1.12 Returning a value/values from a subroutine

3.1.1.13 Local variables in subroutines

3.1.1.14 Global variables in a programming language

LEARNING OBJECTIVES

In this chapter you will learn:

- what a subroutine is and how they are used to create programs
- how to create subroutines
- what a function is and how to create them
- what parameters and arguments are and how they are used within a function
- what local and global variables are.

A-level students will learn:

- what an exception is and how a program should deal with it.

KEYWORDS

Subroutine: a named block of code designed to carry out a specific task.

Procedure: another term for a subroutine.

Subprogram: another term for a subroutine.

Routine: another term for a subroutine.

Local variable: a variable that is available only in specified subroutines and functions.

Global variable: a variable that is available anywhere in the program.

INTRODUCTION

In programming a **subroutine** is a named block of programming code that performs a specific task. All programs therefore are made up of a series of subroutines. They provide structure to programs in the same way that chapters provide structure to a book. Subroutines are also called **procedures**, **subprograms** or **routines**.

Subroutines use variables that can either be local or global. **Local variables** are those that can only be used within that subroutine whereas **global variables** are accessible throughout the program.

- String to Integer / Integer to String: An integer is a whole number. Some programming languages convert between the two automatically if the two variables are declared correctly. For example:

```
Dim i as Integer
Dim s as String
i = 1
s = i
```

  This would result in `s` (a string) becoming 1 (an integer). The same code could be used to reverse the process.
- String to Float / Float to String: A float is also called a real number and is any number including those with a fractional part. In Visual Basic, a function exists to carry out this conversion:

  `Convert.ToDouble(x)` will convert the text string `x` into a Double data type, used by Visual Basic to store real numbers.

  `Convert.ToString(n)` will convert the real number `n` into a string.
- String to Date-time / Date-Time to String: Date-time is usually stored with built-in formatting such as dd.mm.yyyy and hh:mm:ss. To manipulate individual parts of the data, it can be converted into a string. Most programming languages have built-in functions. For example, in Visual Basic:

  `DateTime.ToString(date)` converts the date and time into a string

  `String.ToDateTime(String)` converts a string into date-time format.

Examples of common operations in Python and C#

Table 3.2 gives some examples of how these common operations can be executed in both Python and C#. Note that there will be other ways of implementing these operations based on the specific requirements of the program being written.

You will notice that there are some commonalities between programming languages and some operations that are handled completely differently.

Table 3.2 Common operations in Python and C#

Operation or function	Python example	C# example
Add	`c = a + b`	`c = a + b`
Subtract	`c = a – b`	`c = a – b`
Multiply	`c = a * b`	`c = a * b`
Divide real number	`c = a / b`	`c = a / b`
Divide integer	`c = a//b`	`Int c = a / b`
Modulo	`c = a % b`	`c = a % b`
Exponentiation	`c = a ** b` or `exp(n)`	`math.Pow (a,b)`
Round	`round (x[,n])`	`math.Round()`
Truncate	`round (x[, n])` Truncates according to the size input	`math.Truncate (a, n)` Truncates the value of *a* to *n* places
Random number generation	`random()`	`random()`

Substring	`var1 = "JohnSmith"`
	`print var1[0 : 4]`
	Prints "John"
		`string input = "JohnSmith"`
		`string sub =`
		`input.Substring (0, 4)`
		Returns the value "John"

Concatenation	`c = a + b`	`c = a + b`
Convert character to character code	`Chr()` gives the character code value of the character `Ord()` gives the integer value of the character	`Encoding,ASCII,GetBytes ()` converts a character to an ASCII code `convert.ToChar()` converts an ASCII code to a character
Convert string to integer	`int()` converts a string to an integer `str()` converts an integer to a string	`.ToInt32` converts a string to a 32-bit integer `.ToString` converts an integer to a string where the variable before the dot is an integer
Convert string to date-time	`time.strftime(format[,t])` converts a string into a time with a specified format	`ConvertToDateTime(dateString)` converts date-time to string `DateTime.Parse()` converts the string contained in the brackets into a date-time format
Convert string to float	`float()` converts string to float `str()` converts float to string	`.ToFloat` converts a string to a float where the value before the dot is a string `.ToString` converts a float to a string where the variable before the dot is an integer

Practice questions can be found at the end of the section on pages 46 and 47.

TASKS

1 Write an example of a calculation using each of the arithmetic operators.
2 What is the difference between a division of a real/float and the division of an integer?
3 Most calculations will get their values from variables. Why are variables used in programming rather than just typing the raw values?
4 Use examples to explain the difference between truncation and rounding.
5 Why might random numbers be used?
6 What is the difference between an OR statement and an XOR statement? Give an example.
7 How can you create a substring from a string?
8 What formats can strings be converted into?
9 Why are random numbers generated in programming languages not entirely random?

KEY POINTS

- The syntax of a language describes the rules that you must follow.
- Arithmetic operations include common processes such as add, subtract, multiply and divide.
- Other arithmetic operations include rounding, truncating and exponentiation.
- Most languages include a random number generator.
- Relational operations compare two or more values to produce a result.
- Boolean operations return a true or false value and include AND, OR, NOT and XOR.
- Different types of operations can be combined to create more complex expressions.
- String handling is the process of identifying and extracting sequences of characters from a string of characters.

STUDY / RESEARCH TASKS

1 Write code for a calculator app that allows the user to enter one or two numbers and then carry out all of the main arithmetic operations.
2 Write code for an app that allows the user to input two numbers and then carry out each of the relational operators returning an output of TRUE or FALSE.
3 Write code for an app that extracts the vowels from the alphabet.
4 Write code for an app that takes the numbers 1–10 and extracts them into odd and even numbers.
5 Research how Google uses Boolean operators to create accurate search results.
6 Is it possible to produce a completely random number?

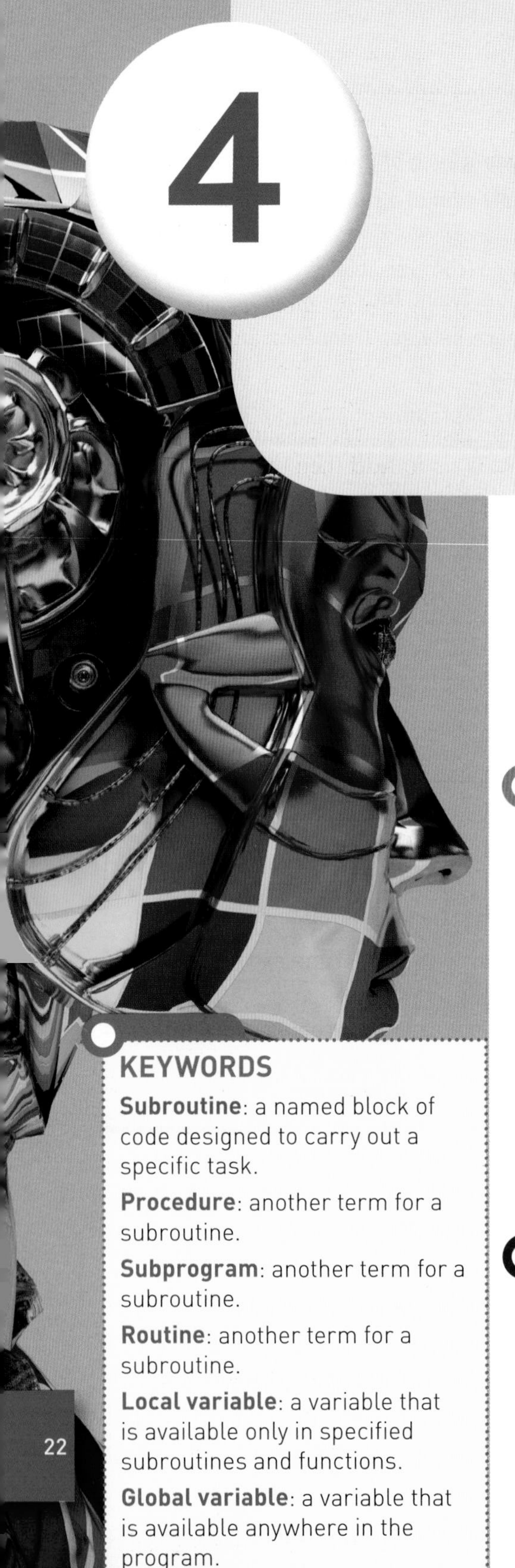

4 Subroutines, local and global variables

SPECIFICATION COVERAGE

3.1.1.9 Exception handling

3.1.1.10 Subroutines (procedures/functions)

3.1.1.11 Parameters of subroutines

3.1.1.12 Returning a value/values from a subroutine

3.1.1.13 Local variables in subroutines

3.1.1.14 Global variables in a programming language

LEARNING OBJECTIVES

In this chapter you will learn:

- what a subroutine is and how they are used to create programs
- how to create subroutines
- what a function is and how to create them
- what parameters and arguments are and how they are used within a function
- what local and global variables are.

A-level students will learn:

- what an exception is and how a program should deal with it.

KEYWORDS

Subroutine: a named block of code designed to carry out a specific task.

Procedure: another term for a subroutine.

Subprogram: another term for a subroutine.

Routine: another term for a subroutine.

Local variable: a variable that is available only in specified subroutines and functions.

Global variable: a variable that is available anywhere in the program.

INTRODUCTION

In programming a **subroutine** is a named block of programming code that performs a specific task. All programs therefore are made up of a series of subroutines. They provide structure to programs in the same way that chapters provide structure to a book. Subroutines are also called **procedures**, **subprograms** or **routines**.

Subroutines use variables that can either be local or global. **Local variables** are those that can only be used within that subroutine whereas **global variables** are accessible throughout the program.

Subroutines

A subroutine is self-contained and it carries out one or more related processes. These processes are sometimes called algorithms, which in turn are made up of lines of code. Subroutines must be given unique identifiers or names, which means that once they have been written they can be called using their name at any time while the program is being run.

KEYWORD

Event: something that happens when a program is being run.

For example you may want to write a program to maintain the contents of a file. You would need to write code to handle tasks such as adding a new record, amending existing details and deleting an old record. In this case you might have a subroutine to handle **events** that are generated from a main menu and then each of the three tasks has its own subroutine. For example if the variable **Selected** is set to **Add** then the procedure **AddRecord** would be called.

```
Subroutine MainMenu
  Input Selected
  If Selected = "Add" Then Subroutine AddRecord
  If Selected = "Amend" Then Subroutine AmendRecord
  If Selected = "Delete" Then Subroutine DeleteRecord
End Subroutine
:
Subroutine AddRecord
'Code to add a new record to a file
End Subroutine
:
Subroutine AmendRecord
'Code to locate and amend an existing record
End Subroutine
:
Subroutine DeleteRecord
'Code to delete an existing record
End Subroutine
```

Breaking up a program into manageable blocks like this has many benefits:

- They can be called at any time using the subroutine's unique name.
- They allow you to gain an overview about how the program is put together.
- You can use a top-down approach to develop the whole project.
- The program is easier to test and debug because each subroutine is self-contained.
- Very large projects can be developed by more than one programmer.

Visual Basic forces you to work with subroutines. In Visual Basic as soon as you try to write code that is connected to a control, Visual Basic creates a subroutine for you. Object-oriented programming takes this concept

KEYWORD

Module: a number of subroutines that form part of a program.

one stage further by putting all the code and the relevant data in the same **module** and the modules are put together to form the overall program. In this context a module is one part of a program that may contain several subroutines. See Chapter 6 for more details.

Functions

KEYWORD

Function: a subroutine that returns a value.

Functions are similar to subroutines but return a value. For example, most modern pocket calculators have a large range of functions. The most basic are probably the square and square root keys. The idea is that you enter a number, press the function key you want and the calculator gives you a result based on that number.

A function in a computer program performs much the same task as the buttons on a calculator. The user supplies the function with data and the function returns a value. For example you could create a function that calculates the volume of a cylinder – you supply the height and radius and the function returns the volume.

Figure 4.1 Function keys on a calculator

This process is not limited to numeric data; for example, you could create a function to count the number of times the letter 'h' occurs in a given block of text, or to check to see if a file has read/write or read-only access restrictions in place.

There are two benefits of using functions in a program:

- Some processes are very complex and involve many lines of code, but in the end they produce a single result. Including all those lines of complex code in the middle of your program will probably make it harder to understand, so instead you could put the code in a function and put the function itself somewhere else in the program, away from the main body of the program. This also means that if you want to alter the function it is easier to find. It also makes the main body of the code easier to work through.
- If you have to carry out the same process in lots of different places in the program, then instead of having to rewrite the same code over and over again, you would create the code once as a function and call it from the various places through the program. This has the benefit of keeping programs smaller, and if you need to alter the way the function works, you only have to alter one version of it.

KEYWORD

Functional programming: a programming paradigm that uses functions to create programs.

Functional programming is a method of programming that only uses functions. There is more on this in Chapters 46 and 47.

Parameters and arguments

KEYWORDS

Parameter: data being passed into a subroutine.

Argument: an item of data being passed to a subroutine.

Block interface: code that describes the data being passed from one subroutine to another.

In order for a subroutine or function to operate efficiently you need a way to control the data that it takes in. This is usually done by using **parameters** and arguments. A parameter works like a variable in that it identifies the data that you want a subroutine to take in and use. The **argument** is the actual value being passed to the subroutine.

The way that this is implemented varies depending on the programming language being used. As the subroutine being called is external to the current subroutine, there needs to be a mechanism for ensuring that the program knows how to handle the subroutine that has been called. It does this using a **block interface**, which is essentially a block of code that specifies the type and characteristics of the data being passed.

Local and global variables

As we have seen, it is highly likely that your program will be split up into lots of subroutines and functions. If you do this then you have to decide on the scope of any variables created. This means you have to construct your program in a way that either:

- limits the existence of the variable to the subroutine or function in which it was declared – a local variable, or
- allows the variable to be used anywhere in the program – a global variable.

The value of a variable is constantly changing throughout the program and as you have seen, values may be passed around between subroutines. If the subroutine changes the value stored in the variable, this may be passed back to the original subroutine or on to another subroutine. An important aspect of programming is keeping track of the state of variables and one of the main causes of program errors is when the value of a variable is changed within one subroutine, that then has an impact on another subroutine. This is known as a side effect.

It is good practice to use local variables wherever possible and using them has a number of advantages:

- You cannot inadvertently change the value being stored somewhere else in the program.
- You could use the same variable name in different sections, and each could be treated as a separate variable.
- You free up memory as each time a local variable is finished with, it is removed from memory.

You should only use a global variable where it needs to be available throughout the whole program. For example, you might store the password to a program as a global variable if you wanted to make a password accessible to different sections of your code.

When programming, different syntax is required to indicate whether the variable is local or global. For example in Visual Basic:

- Local variables are declared using the `Dim` statement:
  `Dim Age As Integer` declares a local variable called `Age`.
- Global variables are described as public:
  `Public Password As Text` declares a global variable called `Password`.

In Python all variables are assumed to be local when they are defined. If you want a variable to be global you must tell Python to make it global by using the `global` keyword, which actually declares functions not local variables.

- `global Password` creates a global variable called `Password`
- `Password = "password"` creates a local variable called `Password`.

In some programming languages if you declare a variable in your code and it is not inside a subroutine or function, then it is assumed to be global. Therefore you need to be careful when declaring variables. The following extract of code shows how you might set Password as a global variable and then other variables as local in Visual Basic:

```
Public Password As String
Private Sub CalculateMathsGrade
  Dim Score As Integer
  Dim Grade As String
  If Score > 50 then
    Grade = "Pass"
  Else
    Grade = "Fail"
End Sub
Private Sub CalculateEnglishGrade
  Dim Score As Integer
  Dim Grade As String
  If Score > 50 then
    Grade = "Pass"
  Else
    Grade = "Fail"
End sub
```

There are two subroutines: `CalculateMathsGrade` and `CalculateEnglishGrade.` The variable `Password` is a global variable and therefore is accessible from either of the two subroutines. The `Score` and `Grade` variables are local to each subroutine. This means that although they have the same names in both, they are actually different variables containing different values, a bit like having two people called John Smith with different characteristics.

Exception handling

There are many situations where a subroutine has to stop because of an exceptional circumstance that causes an error. This is not necessarily an unexpected event, just one that causes the current subroutine to stop. An example would be a division by 0 error, where the subroutine is expecting a number, but instead gets an undefined value caused by the division. When this happens, the subroutine has been 'thrown' an error, which it must deal with. If it is unable to 'catch' the error, the program could produce a fatal error, causing the program to stop running completely.

In the same way that subroutines are triggered by events, there need to be blocks of code that handle errors that are triggered whenever the error occurs. These are often referred to as catch blocks, which are specific blocks of code that are triggered in response to specific errors.

The normal procedure when this happens is:

- an error is thrown so the current subroutine stops or is paused
- the current state of the subroutine is saved
- the **exception handling** code (or catch block) is executed to take care of the error
- the normal subroutine can then be run again, picking up from where it was saved.

KEYWORD

Exception handling: the process of dealing with events that cause the current subroutine to stop.

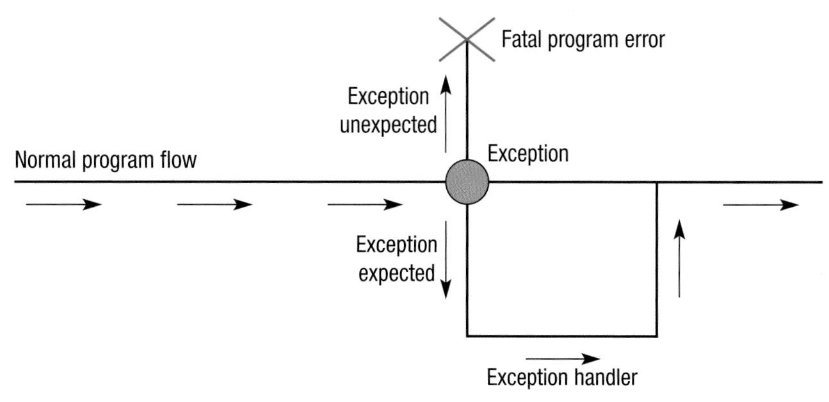

Figure 4.2 The exception handling process

Practice questions can be found at the end of the section on pages 46 and 47.

TASKS

1. Define the following terms and explain the difference between the two:
   a) subroutine **b)** function.
2. Give examples of in-built functions that you might find in a programming language.
3. Why is it good practice to construct programs using subroutines?
4. What is the advantage of using functions?
5. Define the following terms and explain the difference between the two:
   a) algorithm **b)** code.
6. What is a module and how does it differ from a subroutine?
7. Explain how parameters and arguments are used to pass values and variables into subroutines.
8. What is a block interface?
9. Why is it good practice to use local variables whenever possible?
10. Give an example of where you might use a global variable.
11. Explain how an exception handler works.

KEY POINTS

- Subroutines or procedures are a way of breaking up code into manageable blocks of code, each of which performs a specific task.
- Subroutines are likely to have other related subroutines.
- Breaking a program up into subroutines is beneficial for several reasons, mainly related to being easier to manage and maintain the program.
- A function is a type of subroutine that returns a set value, for example, the square root function.
- Parameters and arguments are the data that are passed into the function on which it performs its computations.
- Local variables can only be used within the subroutine in which they were created.
- Global variables can be used anywhere in a program.
- Exception handling is the way in which the program deals with events that may cause it to stop.

STUDY / RESEARCH TASKS

1. Write a program that calculates the square and square root of any given number. Try to include the following features:
   a) a subroutine
   b) a function
   c) a subroutine or function call
   d) a parameter or argument
   e) a local variable
   f) a global variable.
2. Extend your program to include other common mathematical functions.
3. Write a program that takes in a set of numbers and then calculates the mean, median and range using either in-built or user-defined functions.
4. Find examples of typical events that would require exception handling code.

5 Structured programming

SPECIFICATION COVERAGE

3.1.2.1 Structured programming

3.3.1.2 Design

LEARNING OBJECTIVES

In this chapter you will learn how to:

- use hierarchy/structure charts to plan the design for a program
- use flowcharts to describe a system
- write pseudo-code and test it by dry running
- use sensible naming conventions for program components
- write well-structured commented code.

INTRODUCTION

As we have seen in the first four chapters, there are lots of aspects to consider when creating a program including:

- working out the main processes
- identifying the data that is needed and how it will be stored
- working through the calculations that will be carried out on the data
- deciding what type of statements are needed
- organising the code into subroutines to create a working program.

Time and effort spent on designing a computer program are always well worth it, and good program design should result in a more efficient and error-free result. It will also make creating the code easier if you plan ahead.

This chapter looks specifically at the techniques that can be applied to **procedural** or **imperative programming languages**. These languages use sequences of instructions in the form of algorithms and subroutines as described in the previous chapter. A-level students need to be aware of other paradigms including object-oriented and functional programming languages. These are covered in Chapter 6 and Chapters 46 and 47 respectively.

KEYWORDS

Procedural programming languages: languages where the programmer specifies the steps that must be carried out in order to achieve a result.

Imperative programming languages: languages based on giving the computer commands or procedures to follow.

Hierarchy or structure charts

KEYWORDS

Hierarchy chart: a diagram that shows the design of a system from the top down.

Structure chart: similar to a hierarchy chart with the addition of showing how data are passed around the system.

Top-down approach: when designing systems it means that you start at the top of the process and work your way down into smaller and smaller sub-processes.

Hierarchy or **structure charts** use a **top-down approach** to explain how a program is put together. Starting from the program name, the programmer breaks the problem down into a series of steps.

Each step is then broken down into finer steps so that each successive layer of steps shows the subroutines that make up the program in more detail.

The overall hierarchy of a program might look like this:

- programs are made up of modules
- modules are made up of subroutines and functions
- subroutines and functions are made up of algorithms
- algorithms are made up of lines of code
- lines of code are made of up statements (instructions) and data.

We have come across all of these before apart from the module. In larger programs a module is a self-contained part of the program that can be worked on in isolation from all the other modules. This enables different programmers to work independently on different parts of large programs before they are put together at the end to create the program as a whole.

The text in a hierarchy chart at each level consists of only a few words – if you want more detail about what the process involves you need to move further down the diagram. The component parts for each section are organised from left to right to show how the system will work.

Figure 5.1 shows just part of a structure diagram of a program showing the program at the top, the modules beneath that and so on.

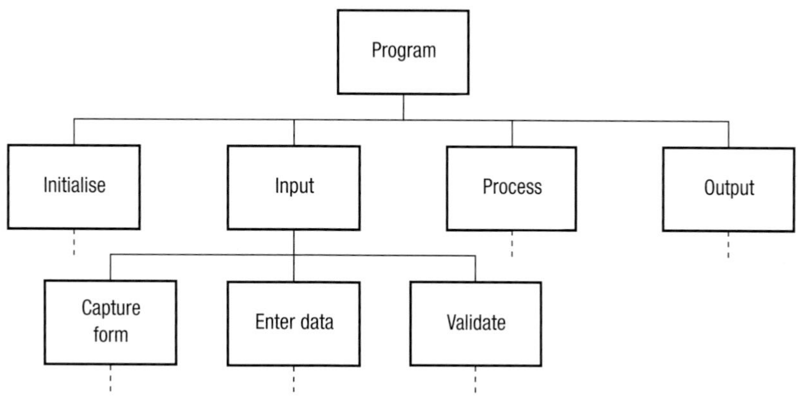

Figure 5.1 A simple hierarchy or structure chart

Flowcharts

KEYWORD

Flowchart: a diagram using standard symbols that describes a process or system.

A **flowchart** uses a set of recognised symbols to show how the components of a system or a process work together. Some of the more common symbols are shown in Figure 5.2.

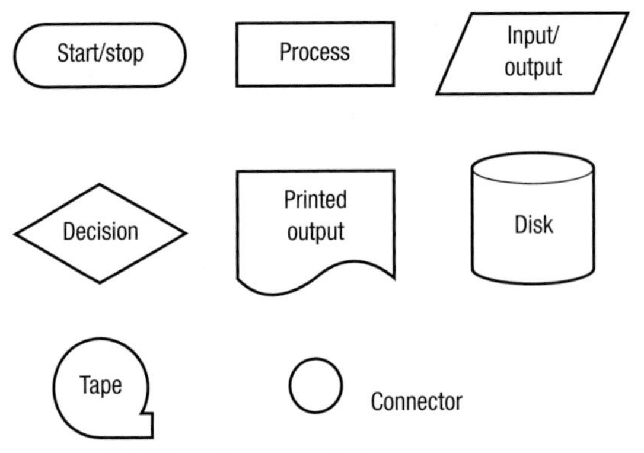

Figure 5.2 Flowchart symbols

KEYWORD

System flowchart: a diagram that shows individual processes within a system.

A **system flowchart** shows the tasks to be completed, the files that will be used and the hardware that will be needed but only as an overview. It is normally possible to create just one flowchart that shows the whole system, but this is not always a good idea as modern programs can be very large and cramming every process on to one flowchart might make it too complex to be of any real use. It might be more advantageous to create a separate systems flowchart for each section of the project.

The systems flowchart in Figure 5.3 shows the first few processes that are used when a person starts to use an ATM (Automated Teller Machine) at a bank.

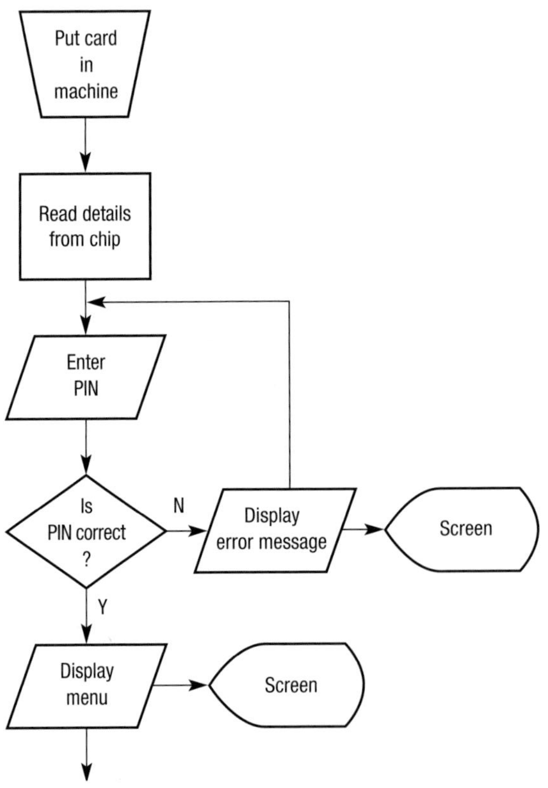

Figure 5.3 An example of system flowchart for an ATM

Pseudo-code

KEYWORD

Pseudo-code: a method of writing code that does not require knowledge of a particular programming language.

So far we have looked at diagrammatic ways of organising a program, but the code that a programmer creates does not use diagrams, it uses lines of code.

Pseudo-code is a way of writing code without having to worry too much about using the correct syntax or constructs. It consists of a series of commands that show the purpose of the program without getting bogged down with the intricacies of the chosen high-level language. The programmer will need to convert the pseudo-code into high-level code at a later date.

Pseudo-code is not a true programming language though it may well use some of the constructs and language of a high-level language. There is only one rule that needs to be adhered to if pseudo-code is to be of any use and that is that the language used needs to be consistent. Using the command 'Save to File' in one place then 'Write to File' in another will only make the job of converting to a true high-level language harder.

Pseudo-code can be used at many levels. The line:

```
Sort NameFile by surname
```

does exactly the same as these lines:

```
Repeat
    Compare adjacent values, swap if necessary
Until No more swaps are needed
```

It will be up to the programmer to decide how far down they need to break their pseudo-code before they can start to actually write the code.

Pseudo-code is very useful in that it allows a programmer to sort out the overall structure of the program as it will eventually appear. The fact that it can be used at many levels means the programmer does not have to work out all the fine detail from the start. The process of turning pseudo-code into programming code is covered in more detail in Chapter 17.

Naming conventions

KEYWORD

Naming conventions: the process of giving meaningful names to subroutines, functions, variables and other user-defined features in a program.

Adding variables to a program as you go along is a recipe for disaster and it shows a serious lack of planning. Before you start your actual code you should draw up a list of all the variables you intend to use, including details of the data types and whether they are going to be local or global variables.

These are some variables that are being declared so they can be used in a Visual Basic program:

```
Dim LoadFileName As String
Dim Counter As Integer
Dim AverageScore as Single
Dim RejectFlag As Boolean
```

Giving the variables, constants, modules, functions and subroutines in a program meaningful names is good practice. It makes a lot more sense to call a variable that stores the number of pupils in a group `GroupSize` than to call it `Size` or `C3`.

In the same way that programmers should sort out the variables, they should also draw up a list of the functions and subroutines they intend to use along with details of what each will do, what it will be called, and what parameters it will need to have assigned to it.

Code layout and comments

The final step to good program construction is to use the features of the programming language to make the code itself as programmer-friendly as possible. This might include adding suitable comments, especially to more complex or unusual sections of code, and using gaps and indents to show the overall structure of a program. Indenting loops can help to identify where a loop begins and ends. It also helps when you are trying to debug a program.

The following two sets of code do the same thing – they place the first 12 values from the two times table in an array.

This first example provides no support for the programmer at all.

```
For X = 1 To 12
W(X) = 2 * X
Next
```

This second example has made use of a number of features:

```
'routine to place multiples of 2 in array TwoTimes()
For Count = 1 To 12
  'counter counts from 1-12
  TwoTimes(Count) = 2 * Count
  'result in array TwoTimes
Next Count
'end loop
```

The helpful features are:

- comments to show the purpose of the algorithm itself
- comments to show the purpose of each line
- sensible variable names such as `Count` and `TwoTimes`
- the contents of the loop have been indented.

Dry runs and trace tables

> **KEYWORDS**
>
> **Dry run**: the process of stepping through each line of code to see what will happen before the program is run.
>
> **Trace table**: a method of recording the result of each step that takes place when dry running code.

No matter how careful a programmer is, even the simplest programs are likely to contain bugs. These come in a number of guises and some are trapped when the program is compiled, and others are trapped by the operating system. However some bugs can remain elusive and the programmer might have to resort to **dry running** the appropriate section of code.

Dry running is the process of following code through on paper. The variables that are used are written down in a **trace table**. Note that dry running can be done on pseudo-code or actual programming code. It is useful to dry run pseudo-code as any errors in the overall design of the algorithm can be identified before too much time has been spent programming it.

This is a simple example of a dry run:

```
For Counter = 1 To 3
   If StoreName(Counter) > StoreName(Counter + 1)_
   Then
      TempName ← StoreName(Counter)
      StoreName(Counter) ← StoreName(Counter + 1)
      StoreName(Counter + 1) ← TempName
   End If
Next Counter
```

The array called `StoreName` has four elements. The initial values for `StoreName` and the other variables are shown in the trace table below.

Counter	TempName	StoreName			
		1	2	3	4
0	<empty>	Kevin	Jane	Beth	Linda

The program is now dry run.
`Counter` is set to 1 and the contents of `StoreName(1)` are compared with the contents of `StoreName(2)`. Because "Kevin" is greater than "Jane" (alphabetically speaking) `TempName` takes the value "Kevin", `StoreName(1)` takes the value "Jane" and finally `StoreName(2)` takes the value held in `TempName` – "Kevin".

Counter	TempName	StoreName			
		1	2	3	4
		~~Kevin~~	~~Jane~~	Beth	Linda
1	Kevin	Jane	Kevin		

`Counter` now increments to 2 and `StoreName(2)` is compared to `StoreName(3)`. "Kevin" is greater than "Beth" so `TempName` becomes "Kevin". Even though `TempName` already contains "Kevin" it is important to realise that this is overwritten by the same name.

Counter	TempName	StoreName			
		1	2	3	4
		~~Kevin~~	~~Jane~~	~~Beth~~	Linda
~~1~~	~~Kevin~~	Jane	~~Kevin~~		
2	Kevin		Beth	Kevin	

`Counter` now increases to 3, and “Kevin” is compared to “Linda”. “Kevin” is less than “Linda” so the program jumps to the `End If` statement.

Counter	TempName	StoreName			
		1	2	3	4
		~~Kevin~~	~~Jane~~	~~Beth~~	Linda
~~1~~	~~Kevin~~	Jane	~~Kevin~~		
~~2~~	Kevin		Beth	Kevin	
3					

`Counter` has now reached the end value of the loop so the program moves on to whatever comes next.

You have probably realised by now that this algorithm is part of a simple sort routine. Whilst a program is being developed a programmer might also use techniques such as single stepping, where the program is executed one line a time. The programmer can see the values of the variables being used and may choose to insert breakpoints. A breakpoint stops the execution of a program either so the programmer can check the variables at that point or possibly just to show that a program has executed a particular section of code.

Practice questions can be found at the end of the section on pages 46 and 47.

KEY POINTS

- Hierarchy or structure charts use a top-down approach to explain how a program is put together.
- A flowchart uses a set of recognised symbols to show how the components of a system or a process work together.
- Pseudo-code is a way of writing code without having to worry too much about using the correct syntax or constructs.
- All the variables you intend to use including details of the data types and whether they are going to be local or global variables should be identified at the start.
- Good program construction is to use the features of the program to make the code itself as programmer-friendly as possible.
- Dry running is the process of following code through on paper. The variables that are used are written down in a trace table.

TASKS

1. Draw a systems flowchart that shows how the computer system at a supermarket handles the sale of goods at the POS (point of sale terminal).
2. The customers at a supermarket have the option of paying for their goods by credit or debit card. Draw a systems flowchart to show how this process works.
3. A programmer might choose to use both a flowchart and pseudo-code when developing a program. Describe the benefits and drawbacks of using each of these systems.
4. High-level languages support ‘user-defined variable names’. Explain what is meant by this term.
5. What steps can a programmer take to make the code they write easier for another programmer to follow?

STUDY / RESEARCH TASKS

1. Why is it considered bad practice to use the GoTo statement when programming?
2. An alternative programming paradigm to procedural languages is logic programming, for example Prolog. In what way does this paradigm differ and what are the implications for the way in which programs might be designed?
3. ‘All programs can be structured using only decisions, sequences and loops.’ Explain whether you think this is true.
4. In what way do Pascal, Algol and some other programming languages enforce structured design?

6 Object-oriented programming concepts

A level only

SPECIFICATION COVERAGE

3.1.2.1 Programming paradigms

3.1.2.3 Object-oriented programming

LEARNING OBJECTIVES

In this chapter you will learn:

- the key principles and methods of object-oriented programming (OOP) including encapsulation, inheritance, instantiation, polymorphism, abstraction and overriding
- that OOP programs are made up of classes and objects
- that classes are a blueprint containing properties and methods
- that objects are instances of a class containing the same properties and methods of the class
- how to create class diagrams.

INTRODUCTION

In Chapter 5 we looked at structure charts as a method of planning and organising programs. This is a top-down approach that breaks programs into modules, which in turn get broken down into subroutines and functions. Object-oriented programming can be thought of as an extension to this structured approach as it is entirely modular.

The key difference between procedural and object-oriented programming is that in procedural programming, the lines of code and the data the code operates on are stored separately. An object-oriented program puts all the data and the processes that can be carried out on that data together in one place called an object and allows restrictions to be placed on how the data can be manipulated by the code.

The examples in this chapter are given in Python, but all of the languages offered by AQA for the Paper 1 exam also support object-oriented programming.

Object-oriented programming can be described as being organised in a way that reflects the real world. For example, in real-life you may have an object, such as a bank. Inside that object there are various other objects such as customers and financial transactions. Inside each of those objects there are a number of data items and behaviours. For example, there are data about customers. These data are handled in a particular way and therefore have to be processed accordingly. For example, one process might be to add new customer data. Another process might be that money withdrawn needs to be deducted from the balance.

In object-oriented programming, a banking application would be created to mirror these real-life relationships. So there might be one object for customers and another for transactions. The customer object will then contain customer data and all the processes needed for that data.

There are a number of advantages to this approach:

- Programs are written in modules, which means that it is easy to amend programs as only the effected module needs editing.
- It is also easier to add new functionality to a program by adding a new module.
- Most programs are written by teams of programmers so the **modular design** approach allows groups of programmers to work independently on self-contained modules.
- Objects can inherit attributes and behaviours making code reusable throughout the program.
- Changes carried out to data are made within an object rather than in the program. This makes it less likely that changes made to code will inadvertently affect the results of another routine, which is a common cause of bugs in software programs.
- Libraries can be created enabling code to be reused easily.

KEYWORD

Modular design:a method of system design that breaks a whole system down into smaller units, or modules.

Encapsulation

The concept of keeping the data and the code within the same object is called **encapsulation**. The code is known as **methods**, which are the subroutines contained within an object that are designed to perform particular tasks on the data within the object. This is the main concept behind object-oriented languages meaning that the objects are self-contained. The implication of this is that there are far fewer side-effects than with procedural languages. A side-effect is where a subroutine changes the value of a variable which then has implications for other parts of the program.

In theory this cannot happen in object-oriented programming as the data can only be directly manipulated by the methods contained within the object. This concept is sometimes called information hiding, which means that the data are directly available in the object that actually needs to use it. Code outside of the class can only access data through the defined methods.

KEYWORDS

Encapsulation: the concept of putting properties, methods and data in one object.

Method: the code or routines contained within a class.

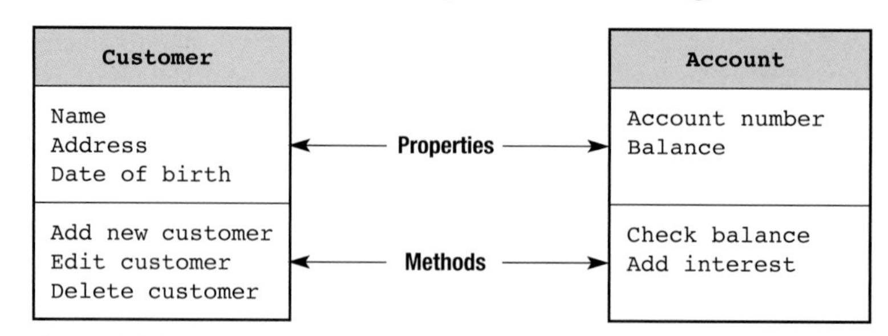

Figure 6.1 Classes containing properties and methods

Figure 6.1 shows two classes, `Customer` and `Account`, both containing their own **properties** and methods.

KEYWORD

Properties: the defining features of an object or class in terms of its data.

KEYWORDS

Class: defines the properties and methods of a group of similar objects.

Object: a specific instance of a class.

The two main building blocks of an object-oriented program are classes and objects:

- A **class** is a blueprint or master copy that defines a related group of things. It contains properties, which describe what the data are, and methods, which describe the way in which the data can behave. However, it does not store any data. Classes can be created based on other classes.
- **Objects** are created from a class. Every object is defined as an instance of a class so will have the same properties and methods from the class from which it is built. It will also contain the data on which the methods will be run.

In the banking example:

- A class might be called `Account` which defines the properties and methods of a bank account.
- All accounts have similar properties such as an `Account Number` and `CurrentBalance`.
- They also have the same methods – they can `GetCurrentBalance`, `AddInterest` and so on.
- Each individual type of account would be a subclass of the `Account` class and they would have all the properties and methods of the class `Account` in addition to properties and methods of their own. Subclasses are explained later in the chapter.
- For example, two subclasses based on the class `Account` may be `Current` and `Mortgage`. They both share the properties and methods from the `Account` class.
- `Current Account` will have its own specific properties, such as `Overdraft` and `PaymentMethods,` which are only a feature of current accounts. Similarly `Mortgage` will have its own specific properties, such as `EndDate`, which are unique to it.
- Objects are created from the class (or subclass) and represent a particular instance of that class. For example, an object created from the `Current` class would be one person's current account and will contain data about that specific account.
- Any number of objects can be created from a class.

When designing object-oriented programs, you can see how important it is to think carefully about the properties and methods of each object and how these might be organised into classes. Classes are fundamental to the design of object-oriented programs. Therefore they are stored where they can be reused either in the future, or by programmers working on other modules. The definitions of all classes are stored in a class library.

The code below shows a class called `Account` being created in Python. This is the base class:

```
class Account():

    def__init__(self, accountNumber, openingDate,
currentBalance, interestRate):

        self.accountNumber = accountNumber

        self.openingDate = openingDate

        self.currentBalance = currentBalance
```

```
        self.interestRate = interestRate
    def getAccountNumber(self):
        return self.accountNumber
    def getCurrentBalance(self):
        return self.currentBalance
    def addInterest(self):
        interest = self.currentBalance * self.
        interestRate
        self.currentBalance += interest
    def setInterestRate(self, interestRate):
        self.interestRate = interestRate
```

Inheritance

KEYWORD

Inheritance: the concept that properties and methods in one class can be shared with a subclass.

Inheritance in object-oriented languages acts in a similar way to the biological definition of inheritance. You start out with a set of characteristics and add to what already exists. You can add or change features and abilities but cannot delete them directly.

Taking the bank account example mentioned earlier you might start with a base class called **Account.** This will have properties and methods. For example the properties may be:

```
AccountNumber: String
DateOpened: Date
CurrentBalance: Currency
InterestRate: Real
```

When programmed, the properties become variables with an appropriate data type assigned.

The methods may include:

```
AddInterest
GetCurrentBalance
GetInterestRate
```

When programmed the methods become the subroutines (procedures or functions) required. In this case a procedure or function would be defined to calculate the amount of interest to add to the account.

The properties and methods defined in the base class are common to all types of account. For example, whether the account was a current account or mortgage account it would still have the same properties and methods. In addition, the other account types would have additional properties and methods so these can now be set up as subclasses along with any properties and methods that are unique to the subclass.

For example, the current account might have a new property **Overdraft**, which indicates whether the customer has an overdraft set up on their account. This is unique to current accounts so would not appear as a property on the mortgage account. A method to set the overdraft called **setOverdraft** could be defined in the subclass.

The mortgage account might have a property called **EndDate**, which is the date that the mortgage is paid off. This is not a property that you would use in a current account so needs to be set up in the **Mortgage** subclass. A method may be set up called **GetEndDate** to identify the date that the last payment needs to be made.

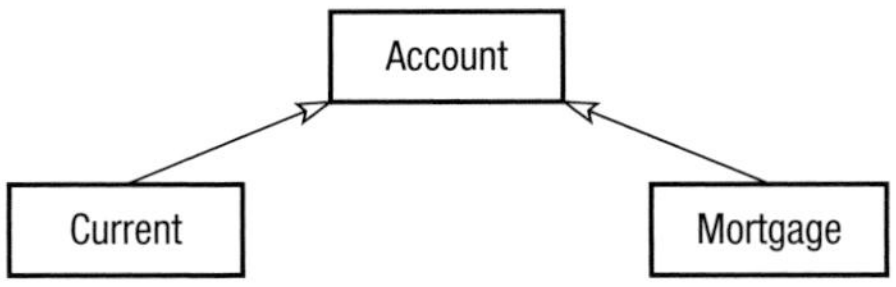

This relationship between the classes and subclasses can be shown as an inheritance diagram as in Figure 6.2. Note that the direction of arrows shows the path of inheritance.

Figure 6.2 An inheritance diagram for `Account`

Inheritance produces a hierarchical structure. In this scenario, **Account** could be described as a base class, super class or parent class, as it is the main class from which other classes **Current** and **Mortgage** are created. Classes that inherit from others are called subclasses, derived classes or child classes. This example has been simplified to include just two types of account, but the same principle can now be used to define further subclasses. For example, subclasses could be defined for savings accounts, trust fund accounts and so on.

```
class Current(Account):
    def __init__(self, accountNumber, openingDate,
    currentBalance, interestRate, paymentType,
    overdraft):
        Account.__init__(self, accountNumber,
        openingDate, currentBalance, interestRate)
        self.paymentType = paymentType
        self.overdraft = overdraft
    def setPaymentType(self, paymentType):
        self.paymentType = paymentType
    def setOverdraft(self, overdraft):
        self.overdraft = overdraft
    def getOverdraft(self):
        return self.overdraft
class Mortgage(Account):
    def __init__(self, accountNumber, openingDate,
    currentBalance, interestRate, endDate):
        Account.__init__(self, accountNumber,
        openingDate, currentBalance, interestRate)
        self.endDate = endDate
    def getEndDate(self):
        return self.endDate
    def setEndDate(self, endDate):
        self.endDate = endDate
```

Class diagrams for inheritance

> **KEYWORD**
>
> **Class diagrams**: a way of representing the relationship between classes.

Class diagrams are a standard method for representing classes, their properties and methods and the relationship between the classes. There are different ways of representing the relationships between the classes. This section deals with inheritance:

- They are hierarchical in structure with the base class at the top and the subclasses shown beneath.
- A subclass inherits the properties and methods of the base class.
- They use arrows to shows the direction of inheritance.
- Each class is represented with a box made up of three sections to include the class name, properties and methods.

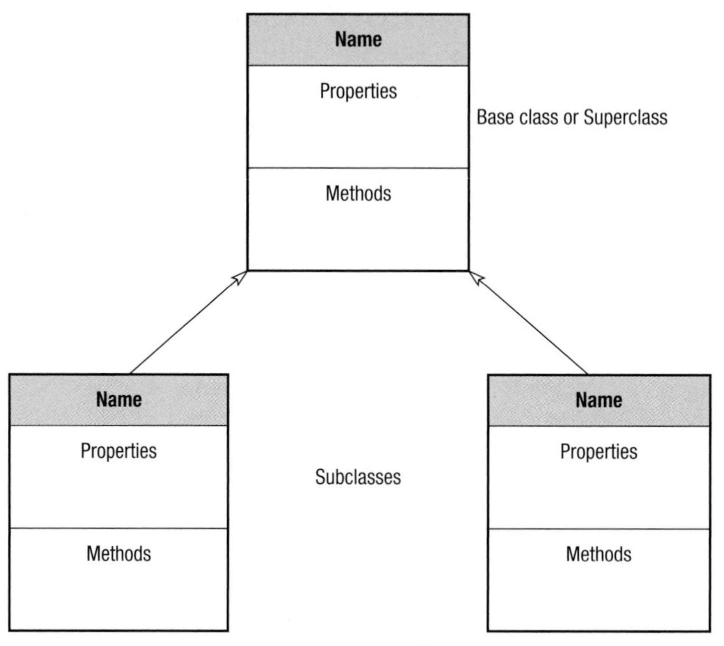

Figure 6.3 A basic class diagram

A class diagram for the account example might look like this:

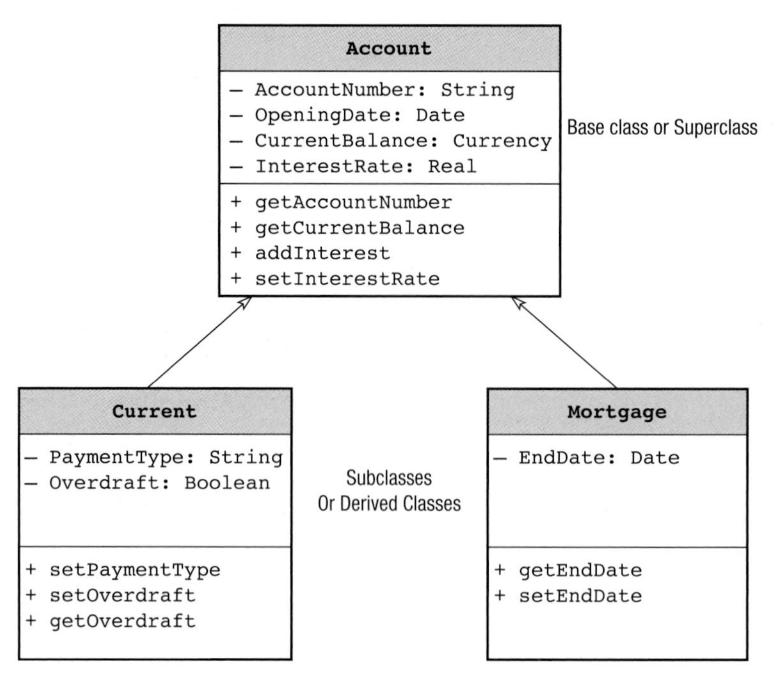

Figure 6.4 A class diagram for `Account`

The class diagram:

- Uses a + or – to indicate the visibility of the properties and methods to other classes. + means that the properties and methods are public to all classes. – means that the properties and methods are private and can only be used in that class.
- Uses a # to indicate that the properties and methods are protected so they can be used in that class and any of its subclasses.
- Uses arrows with the arrow pointing to the base class to show where the subclass is inheriting its properties and methods from.
- Defines the data types to be used for each variable.

Instantiation

KEYWORD

Instantiation: the process of creating an object from a class.

Instantiation is the process of creating an object from a class. In other words, you are creating an actual instance of an object using the properties and methods described in the class. With the `Account` example, an object could be created from the class `Account`, which is a specific customer's current account. Many different objects can be created from the same class so the programmer will need to go through the process of instantiation for every object needed in the program.

When programming, there is a subroutine called a constructor, which is called when an object is instantiated from a class to initialise the object.

```
new_account = Account(41344987, date.today(), 374.34,
0.032)
```

Polymorphism and overriding

KEYWORD

Polymorphism: the ability of different types of data to be manipulated with the same method.

The literal meaning of **polymorphism** is to take on many shapes. In object-oriented programming it describes a situation where a method inherited from a base class can be redefined and used in different ways depending on the data in the subclass that inherited it.

It is related to the hierarchy of classes and the way in which classes inherit properties from other classes. For example, there may be a common method that you want to carry out as part of your program. As this method is critical to the program, it could be defined in a base class, which is then inherited by other classes. However, the data contained in these new classes is perhaps of a different type. Rather than have to define a new method, polymorphism enables the original method to be redefined so that it will work with the new data.

For example, we could define a method that worked out the interest payment on an account. This method would be stored in the base class `Account`. It would then be inherited by the `Current` and `Mortgage` subclasses. The data needed to calculate interest might be different, for example, a current account may have a higher rate of interest, or use a different time period over which to calculate the payment. Each subclass would define a different method to calculate the interest, but these would have the same name as the calculate interest method in the base class.

KEYWORD

Overriding: where a method described in the subclass takes precedence over a method with the same name in the base class.

When the subclass implements the method it is called **overriding** because it overrides (or replaces) the method in the base class. In this example, the method for calculating interest is overridden for the `Current` object:

```
# new method added to the Current class overriding
  the addInterest method in the Account class
   def addInterest(self):
# if the account has an overdraft, interest is
   charged on the debt at 5%
   if self.overdraft:
           charges = self.currentBalance * 0.05
           self.currentBalance += charges
# otherwise interest is applied in the same way
   as the superclass (Account)
      else:
           Account.addInterest(self)
```

Abstract, virtual and static methods

Object-oriented languages handle objects in three different ways. You can think of the code inside an object as subroutines, which are a series of instructions that it will carry out on the data. Due to the nature of the relationships between objects and classes, it means that methods in one object may be used on data contained within another object.

In order to define the behaviour of the methods, they can be set up in three different ways:

- Static: the method can be used without an object of the class being instantiated.
- Virtual: the method is defined in the base class but can be overridden by the method in the subclass where it will be used. This is a feature of polymorphism.
- Abstract: the actual method is not supplied in the base class, which means that it must be provided in the subclass. In this case, the object is being used as an interface between the method and the data.

Which one you use depends on the nature of the methods and the data contained within the class and at what point you want the method to run.

Aggregation

Aggregation is a method of creating new objects that contain existing objects, based on the way in which objects are related. Object-oriented programming is concerned with recreating real-life objects as programming objects. In real life, objects are related to each other. For example, in a business you might have an object called `Workforce` and another called `Job Roles`. Under `Workforce` there may be further objects called `Manager` and `Employee`. All of these objects exist in their own right, but there is also relationship between them. For example:

- Managers and employees make up the workforce.
- Job roles can be taken on by managers or employees.
- Job roles define what the managers and employees do as part of the workforce.

Composition aggregation

KEYWORD

Composition aggregation: creating an object that contains other objects, and will cease to exist if the containing object is destroyed.

`Workforce` is made up of `Manager` and `Employee`. If you deleted `Workforce` you would by definition be deleting `Manager` and `Employee`. This is an example of **composition aggregation**, sometimes just called composition and it is where one object is composed (or made up of) two or more existing objects. You could say that `Workforce` is instantiated (created) when `Manager` and `Employee` are combined.

This can be shown as a class diagram as follows. Notice the shape of the arrow head, which indicates that `Manager` and `Employee` cannot exist unless `Workforce` does.

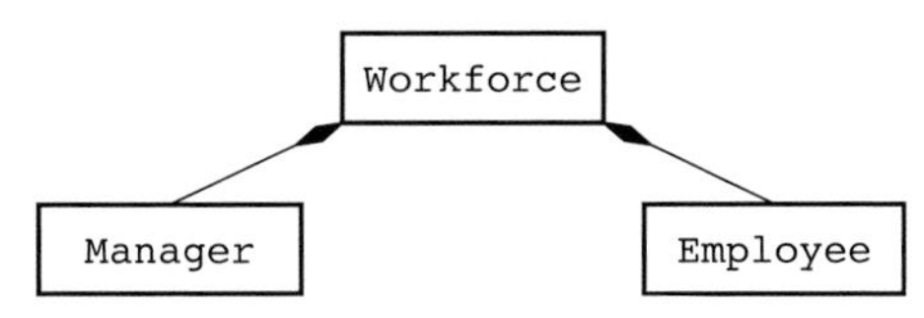

Figure 6.5 Class diagram showing composition aggregation

Association aggregation

KEYWORD

Association aggregation: creating an object that contains other objects, which can continue to exist even if the containing object is destroyed.

You could now extend the `Workforce` object to include a `Job Role` object. There is a relationship between `Manager`, `Employee` and `Job Role` in that the managers and employees all have specific job roles that they carry out. However, if you deleted `Job Role`, the `Manager` and `Employee` object would still be retained and still be usable objects as would the `Workforce` object. This is due to the nature of the relationship in the real world in that a job role is not fixed and therefore any manager or employee could be given any job role. This is an example of **association aggregation** where an object is made up of one or more objects but is not entirely dependent on those objects for its own existence.

This can be shown as a class diagram as follows. Notice the shape of the arrow head, which indicates that `Manager` and `Employee` can still exist even if `Job Role` does not.

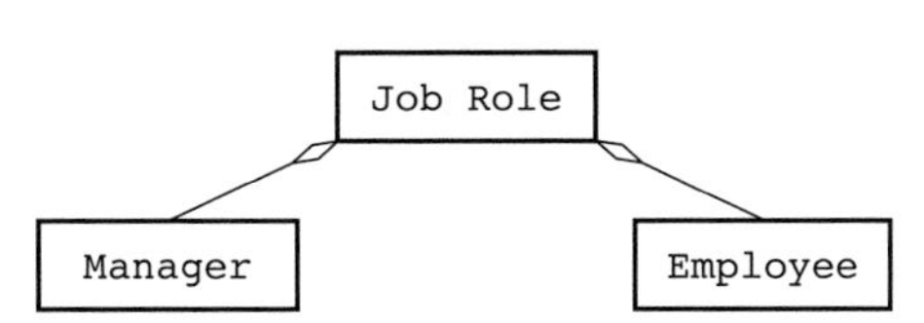

Figure 6.6 Class diagram showing association aggregation

Design principles

There are three key design principles that are recognised as producing the most elegant solution when programming using an object-oriented language.

- **Encapsulate what varies**: This is related to the concepts of encapsulation and information-hiding and the basic concept is that everything that varies should be put into its own class. This means that properties and methods are subdivided into as many classes as needed to reflect the real-life scenario.

For example, with our accounts program we could have an `Account` base class followed by `Current`, and `Mortgage` subclasses and we could stop at that. However, further analysis would suggest that you could create a subclass under `Current` for different types of current account such as `Standard`, `Premium`, `Student` and `Child` as these types of account may have properties and methods that are unique to them. You would continue with this process until you were sure that you had created a class for each unique set of properties and methods.

- **Favour composition over inheritance**: This principle refers to the way in which classes and objects are instantiated (created). As we have seen objects can be created from a base class and inherit its properties and methods. The alternative is to use aggregation or composition to combine existing objects to make new ones. This method is less error prone and enables simpler maintenance. For example, with reference to the `Workforce` example, rather than creating a new instance for `Workforce` we only need to create `Manager` and `Employee` and can then combine the two. Providing `Manager` and `Employee` are created correctly, then there should be no errors in `Workforce`. Similarly any changes made to `Manager` or `Employee` will be reflected in `Workforce`.
- **Program to interfaces, not implementation**: In object-oriented programming an interface defines methods to be used, which are then applied when classes are defined. In this sense an interface is an abstract type which is implemented when a class is created. When a class is created that adheres to the methods in the interface, it can be seen as an implementation of the interface. With our accounts example, being able to calculate interest and check the balance were two methods that all accounts must have regardless of the exact class they are in so these would be required by the interface. The way in which these operations are carried out would be defined within any class that wanted to use the interface. The interface ensures that they must be defined.
  Programs can then be written based on the interfaces rather than each individual implementation of a class. Using this methodology, if classes need to be added to or amended, this can be done with reference to the interface, meaning there will be little or no impact on the other classes in the program.

Practice questions can be found at the end of the section on pages 46 and 47.

TASKS

1. In what way does object-oriented programming reflect the way things work in real life?
2. Using a real-life example, define the following terms explaining the relationship between the three:
   a) class
   b) object
   c) inheritance.
3. Using the same real-life example, explain what properties and methods are.
4. Draw a class diagram for your real-life example.
5. How does encapsulation prevent side effects?
6. What are the two main ways to instantiate an object?

7 Explain the difference between static, abstract and virtual methods.

8 Explain the difference between composition aggregation and association aggregation.

9 What are polymorphism and overriding?

10 What is an interface in the context of object-oriented programming?

11 Explain the three main design principles for effective design.

STUDY / RESEARCH TASKS

1 For your real-life example above, use an object-oriented programming language to implement a solution. Include the following features:

- objects created using abstract, virtual and static methods
- inheritance
- aggregation
- polymorphism
- public, private and protected specifiers.

KEY POINTS

- Object-oriented programming can be described as being organised in a way that reflects the real-world.
- An object-oriented program puts all the data and the processes (methods) that can be carried out on that data in one place called an object.
- The concept of keeping the data and the code within the same object is called encapsulation.
- Class diagrams are a standard method for representing classes, their properties and methods and the relationship between the classes.
- Inheritance in object-oriented languages acts in a similar way to the biological definition of inheritance.
- Instantiation is the process of creating a real instance of a class, which is an object.
- Polymorphism describes a situation where a method inherited from a base class can be used in different ways depending on the data in the subclass that inherited it.
- When the subclass implements the method it is called overriding because it overrides (or replaces) the method in the subclass.
- Aggregation is a method of creating objects that contain other sorts of object.

Section One: Practice questions

1 The following code is part of a stock control system.

```
Dim Name As String
Dim Price As Real
Const VAT = 0.2
Type RecordDetails
    RecordType As String * 14
RecordCurrent As Integer
RecordRestock As Integer
End Type
```

a) Identify where each of the following have been used, and explain why the type of the variable chosen is appropriate:

i) a variable that is used to store a whole number

ii) a variable that is used to store a decimal number.

b) Why has a constant been used to store VAT?

c) Some computer languages support 'user-defined types'. Explain this term and give an example of a user-defined variable in the code.

2 A program has been written to analyse the results of a survey. For each of the following, name a suitable data type and give a reason for your choice:

a) the number of pets owned by a household

b) a telephone number such as 0122453322

c) whether a household's accommodation has central heating or not

d) the average number of children within a household.

3 It is considered poor design to define an Age field when storing personal details. Describe a better way of storing this data.

4 What values can a Boolean expression take?

5 The following section of pseudo-code is used to add and remove data in a queue.

```
'routine to add to a circular queue
'increment Rear pointer
     Rear ← Rear + 1
'Check to see if end of array has been reached
'If so go back to the start of the array
     If Rear = 9 Then Rear ← 0
'add data
     Put DataItem at position Rear in array
'routine to remove data from a circular queue
'remove data
     Take DataItem from position Front in array
'move Front on
     Front ← Front + 1
  'Check to see if the end of the array has been reached
  'If so go back to the start of the array
     If Front = 9 Then Front ← 0
```

a) State a line of code that has a comment in it.
b) State a line of code that is an assignment statement.
c) State a conditional statement that has been used.
d) An array contains the characters E, C and F, with the front pointer on E (at index 0 in the array) and the rear pointer on F (at index 2 in the array). Dry run the code above, showing what would happen if the characters A, D and G were added to the queue.
e) Why is it good practice to create programs in modules?

6 Write a program to implement the pseudo-code in question 5.

7 Explain what techniques programmers can use to assist with the design of a piece of software and how they can make their program code easy to follow.

8 Look at the following section of code and then answer the questions.

```
For Loop1 = 1 To NameCount - 1
 For Loop2 = 1 To NameCount - 1
  If NameStore(Loop2) > NameStore(Loop2 + 1) Then
    TempStore = NameStore(Loop2)
    NameStore(Loop2) = NameStore(Loop2 + 1)
    NameStore(Loop2 + 1) = TempStore
   End If
  Next
 Next
```

a) Name two different data types that are being used.
b) What is the purpose of the first line of code?
c) What is the purpose of the second line of code?
d) What is this algorithm doing?

9 Explain the difference between local and global variables.

10 An object-orientated programming language will be used to create a system related to animals.
a) Suggest suitable properties and methods for a base class.
b) Suggest two further subclasses that could be built from the base class.
c) Explain the difference between an object and a class.
d) Draw a class diagram to show your answers to parts **a** and **b**.
e) Give one example of inheritance in this example.
f) Explain how an object can be instantiated.
g) Give one example of where you may need to use overriding.

Section Two: Fundamentals of data structures

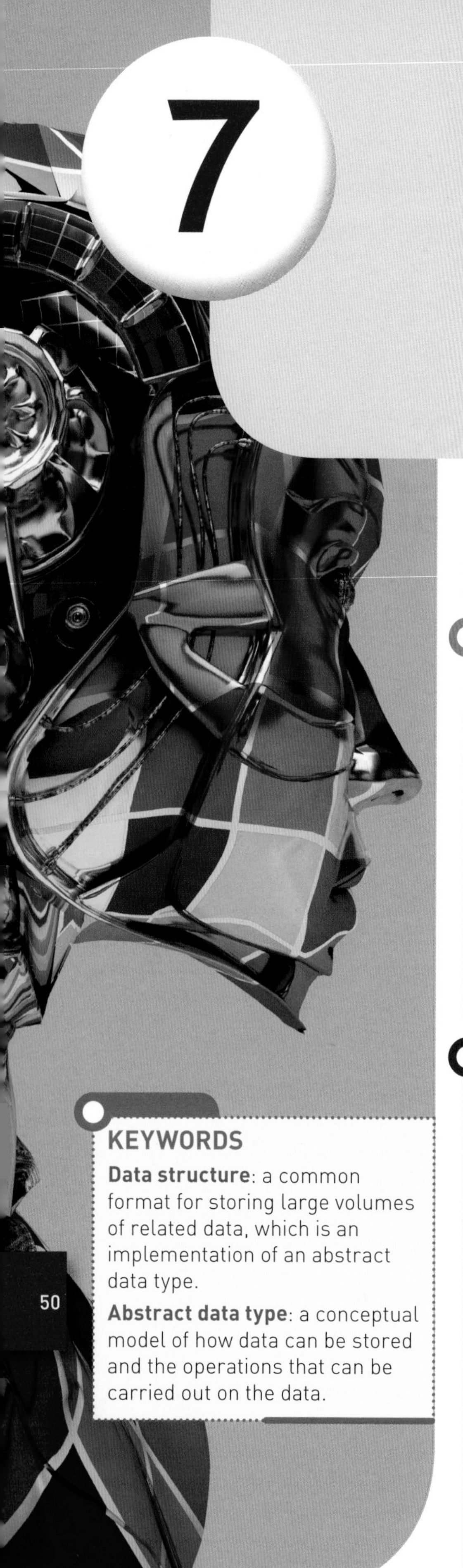

7 Data structures and abstract data types

SPECIFICATION COVERAGE

3.2.1.1 Data structures

3.2.1.2 Single- and multi-dimensional arrays (or equivalent)

3.2.1.3 Fields, records and files

3.2.1.4 Abstract data types/data structures

LEARNING OBJECTIVES

In this chapter you will learn:

- what data structures and abstract data types are
- what an array is and how to use one to store and access data
- what a file is and how it is used to store data
- the difference between text files and binary files
- how to read and write data to and from csv and binary files.

A-level students will learn:

- what static and dynamic data structures are
- the different characteristics of static and dynamic data structures.

INTRODUCTION

There is a difference between what is required for AS and A level here and this chapter has been split accordingly. AS-level students need only study to the end of the section on binary files. A-level students need to be aware of a much wider range of **data structures** and **abstract data types** and also understand how to implement them in a programming language:

- Chapters 7–11 explain the data structures required for A level, including coding on how to implement them.
- Chapters 12–16 cover some common algorithms required for A level, including how to implement the abstract data types described in Chapters 7–11.

In Chapter 1 we looked at the different ways individual items of data might be stored. For example, we looked at storing a person's age as an integer and their name as a string. These are known as data types. In this chapter we will look at the ways of storing larger volumes of data in formats that make it easy for programs and users to access and analyse. These are called data structures and abstract data types.

KEYWORDS

Data structure: a common format for storing large volumes of related data, which is an implementation of an abstract data type.

Abstract data type: a conceptual model of how data can be stored and the operations that can be carried out on the data.

Data structure and abstract data type

A data structure is any method used to store data in an organised and accessible format. Data structures normally contain data that are related and the way that the data are organised enables different programs to manipulate them in different ways. Different data structures tend to lend themselves to different types of applications. For example, a text **file** may be suitable for a database whereas a stack is suitable for handling exceptions.

> **KEYWORD**
> **File**: a collection of related data.

An abstract data type is a conceptual model of how the data are stored and the operations that can be performed upon them. The data structure is an implementation of this in a particular programming language.

Arrays

We came across the concept of an **array** in Chapter 1. An array is a list or table of data that has a variable name that identifies the list or table. Each item in the table is called an element. An array can have many dimensions but most arrays are either one-dimensional in which case they form a list or can be visualised as a two-dimensional table.

> **KEYWORD**
> **Array**: a set of related data items stored under a single identifier. Can work on one or more dimensions.

Lists and arrays are static data structures that are created by the programmer to store tables of data. In some programming languages programmers need to define just how big an array is going to be at the start of their program. This means that the size of the array and the amount of memory put aside for it does not change.

You might find that you want to store a sequence of data in some way. For example you might want to store the names of pupils in a class:

```
Name1 = "Derrick"
Name2 = "Peter"
Name3 = "Jill"
Name4 = "Lynn"
```

Carrying out any sort of work even on just these four names is going to be very cumbersome. Imagine how difficult this would be if you wanted to store 30 names or 3000 names. The best solution to this problem is to use an array. In the example above, we could call the array `StudentName`. Each element of the array can now be accessed using its position. For example, the third element in the array is Jill (assuming indexing starts at 1 and not 0). This would be shown as: `StudentName(3) = "Jill"`

Another example could be to set up a one-dimensional array called `DaysInMonth`. The third element would be set to 31 as that is the number of days in March. As this table contains just one row of data it could also be described as a list.

Element in DaysInMonth	1	2	3	4	5	6	7	8	9	10	11	12
Contents of that element	31	28	31	30	31	30	31	31	30	31	30	31

Figure 7.1 A one-dimensional array or list

An array has one or more dimensions – for example you might want to store the mock exam results of a group of pupils. The array then might be called **`Results`** and it would have two dimensions, one for the pupils and the other for the subjects and might look something like this:

	1	2	3	4	5	6	7
1	54	67	76	65	75	32	19
2	32	45	98	32	53	14	88
3	12	32	54	56	59	95	71
4	32	21	12	43	22	26	16
5	15	47	65	35	99	82	41

Figure 7.2 A two-dimensional array

You will note that the rows/columns are not labelled – it is up to the programmer to remember which axis refers to the pupil and which to the subject. In this diagram the 65 might represent the mark obtained by Hilary in the French exam. If the table were called **`Results`** then Hilary's French mark would be stored in **`Results(4, 1)`** where the 4 identifies the pupil and the 1 identifies the subject.

It is possible to work with multi-dimensional arrays. If you take the mock exam paper array further, you might decide to store the exam results for each exam paper. In this case the value in **`Results(4, 1, 2)`** could store the mark Hilary got in the second paper of the French exam.

In fact you can have many more dimensions than this – a four-dimensional array might store the marks gained for each question in each paper, so **`Results(4, 1, 2, 12)`** might store the mark Hilary was given for question 12 in paper 2 of the French mocks. As you add more and more dimensions to the array it becomes increasingly difficult to conceptualise.

Files

You will already be familiar with the concept of a file to store data. There are hundreds of different file types, all of which have their own structure depending on the specific use of the file. Some files are very specific in that they can only be used on certain applications. Many file formats however are portable, which means they can be used in a wide range of programs. Two common portable formats that can be used when programming are **text files** and **binary files**.

> **KEYWORDS**
>
> **Text file**: a file that contains human-readable characters.
>
> **Binary file**: stores data as sequences of 0s and 1s.
>
> **Record**: one line of a text file.
>
> **Field**: an item of data.

A text file is one that simply contains lines of text or other human-readable characters, where each line is usually referred to as a **record**. There may be different items of data stored and these are referred to as **fields**.

They may contain a header, which explains the content and structure of the file and an end of file (EOF) marker so that the programs using the file know when to stop. Common text file formats include txt used for non-formatted or plain text and csv (comma separated variables), both of which are used for transferring data between programs.

All files have an internal structure. For example, a csv file has fields that are split up using commas. Most text files are delimited like this in some way so that when the file is being used, the program knows where to look for each item of data in the file. The following examples show a typical structure where each row represents a record and the fields are separated either by tabs or commas:

A tab-delimited text (txt) file:

```
John Smith 22 Acacia Avenue LE11 1AA
Mary  Jones 1 High Street   LE12 5BD
Imran Siddiqi 12 Harrow Road LE13 1GG
Yin Li 24 Royal Road LE1 1AA
```

A comma separate variable (csv):

```
John,Smith,22 Acacia Avenue,LE11 1AA
Mary,Jones,1 High Street,LE12 5BD
Imran,Siddiqi,12 Harrow Road,LE13 1GG
Yin,Li,24 Royal Road,LE1 1AA
```

The two main actions you might want to carry out when working with text files are:

- to write data from the program into a text file
- to read data into the program from a text file.

The following extract of Visual Basic-based code shows how you would write data to a text file. In this case, data is being written from a two-dimensional array called `ArrayStore`:

```
FileOpen(1, "NewTable.csv", OpenMode.Output)
' look at each row/record in turn
For RecordCount = 1 to 30
 ' load first field from the next record into
 the temporary string
 RecordString = ArrayStore(recordcount,1)
 ' concatenate all the other fields
 For FieldCount = 2 To 4
    Recordstring = recordstring & "," & Arraystore
    (Recordcount,fieldcount)
 Next
 ' write 'record' to file
 Print(1, OutputString)
Next
FileClose(1)
```

The following extract of code shows how you would read data from a text file into a program using Visual Basic as an example:

```
'load csv file from folder
FileOpen(1, "C:\Users\NameList.csv", OpenMode.Input)
Do
```

```
 grdTableIn.Rows.Add(1)
 grdTableOut.Rows.Add(1)
 Input(1, DownLoadText)
 grdTableIn.Rows(RowCount).Cells(0).Value =
DownLoadText & RowCount
 RowCount = RowCount + 1
Loop Until EOF(1)
FileClose(1)
```

Binary files

A binary file is one that stores data as a series of 0s and 1s. Binary representation is one of the cornerstones of how computers work and is covered in detail in Chapter 25. At this stage it is important to understand that all program code and all of the data that you might use in a program including text, graphics and sound are all made up of 0s and 1s. These are usually organised into groups of 8 bits, called bytes.

Binary files contain binary codes and usually contain some header information that describes what these represent. As you can see from Figure 7.3, binary files are not easily readable by a human, but can quickly be interpreted by a program.

```
11101111 10111011 10111111 00111100 01101110 01101111 01100100 01100101
00100000 01101001 01100100 00111101 00100010 00110001 00110000 00110111
00110000 00100010 00100000 01110110 01100101 01110010 01110011 01101001
01101111 01101110 00111101 00100010 01100101 01100010 01100011 00110111
01100010 01100011 01100001 00110001 00101101 00110011 01100001 01100110
01100010 00101101 00110100 01100001 01100001 00110001 00101101 00111001
01100001 00110100 01100101 00101101 01100110 01100100 00110000 01100100
00110110 00110011 00110110 00110011 01100011 01100010 01100011 00110111
00100010 00100000 01110000 01100001 01110010 01100101 01101110 01110100
01001001 01000100 00111101 00100010 00101101 00110001 00100010 00100000
01101100 01100101 01110110 01100101 01101100 00111101 00100010 00110001
00100010 00100000 01110111 01110010 01101001 01110100 01100101 01110010
01001001 01000100 00111101 00100010 00110000 00100010 00100000 01100011
01110010 01100101 01100001 01110100 01101111 01110010 01001001 01000100
00111101 00100010 00110000 00100010 00100000 01101110 01101111 01100100
01100101 01010100 01111001 01110000 01100101 00111101 00100010 00110001
00110000 00110101 00110110 00100010 00100000 01110100 01100101 01101101
01110000 01101100 01100001 01110100 01100101 00111101 00100010 00110001
00110000 00110100 00110010 00100010 00100000 01110011 01101111 01110010
01110100 01001111 01110010 01100100 01100101 01110010 00111101 00100010
00110010 00100010 00100000 01100011 01110010 01100101 01100001 01110100
01100101 01000100 01100001 01110100 01100101 00111101 00100010 00110010
00110000 00110000 00110111 00101101 00110000 00110100 00101101 00110010
00110101 01010100 00110001 00111000 00111010 00110010 00111000 00111010
00110010 00110110 00100010 00100000 01110101 01110000 01100100 01100001
01110100 01100101 01000100 01100001 01110100 01100101 00111101 00100010
```

Figure 7.3 Output from a binary file

For example, the PNG image file is a binary file, can be used in a range of applications and requires less memory than some other image formats. Many program files (executables) are created as binary files so that they can be used on other platforms.

The two main actions you might want to carry out when working with binary files are:

- to write data from the program into a binary file
- to read data into the program from a binary file.

The following code shows how you would write data to a binary file:

```
OpenFile ("DemoFile.bin", Binary) For Output as # 1
Write 1, "Help"
Write 1,  True
Write 1,  3.123
Write 1,  5
Close #1
```

The following code shows how you would read data from a binary file into a program:

```
Dim TTData as Binary
Open "TTFile" For Binary As #2
  ReDim TTData(1 To LOF(2)) As Byte
  Get 2, 1, TTData
Close #2
```

Static and dynamic data structures

A level only

The way that data can be stored can be split into two broad categories – dynamic and static. This reflects the fact that sometimes the programmer will know how big a data structure will get and therefore how much memory is needed to store it. More often than not, the amount of data stored within a data structure will vary while the program is being run. Different data structures such as **queues** and **stacks** can be implemented either as static or dynamic structures.

- Static: A **static data structure** stores a set amount of data which is usually defined by the programmer. This is done by allocating a set amount of memory to the data structure. Accessing individual elements of data within a static structure is very quick as their memory location is fixed. However, the data structure will take up memory even if it doesn't need it. Records and some arrays are examples of static data structures.
- Dynamic: The word 'dynamic' means changeable. **Dynamic data structures** can use more or less memory as needed through the use of a **heap**. In basic terms, unused blocks of memory are placed on a heap, which are then usable within a program. A dynamic data structure is able to take more memory off the heap if it is needed and also put blocks of unused memory back onto the heap if it is not needed. This is a much more efficient use of resources and a more flexible solution as elements can be added and removed much more easily. Stacks, queues and binary trees are often implemented as dynamic structures.

The programmer will normally put a limit on the maximum amount of memory that any one data structure needs. However, it can lead to errors if elements are removed from empty structures or added to full ones. There is more on this in the following chapters.

KEYWORDS

Queue: a data structure where the first item added is the first item removed.

Stack: a data structure where the last item added is the first item removed.

Static data structure: a method of storing data where the amount of data stored (and memory used to store it) is fixed.

Dynamic data structure: a method of storing data where the amount of data stored (and memory used to store it) will vary as the program is being run.

Heap: a pool of unused memory that can be allocated to a dynamic data structure.

Static data structures	Dynamic data structures
Inefficient as memory is allocated that may not be needed.	Efficient as the amount of memory varies as needed.
Fast access to each element of data as the memory location is fixed when the program is written.	Slower access to each element as the memory location is allocated at run-time.
Memory addresses allocated will be contiguous so quicker to access.	Memory addresses allocated may be fragmented so slower to access.
Structures are a fixed size, making them more predictable to work with. For example, they can contain a header.	Structures vary in size so there needs to be a mechanism for knowing the size of the current structure.
The relationship between different elements of data does not change.	The relationship between different elements of data will change as the program is run.

Table 7.1 Comparison of static and dynamic data structures

Practice questions can be found at the end of the section on page 90.

KEY POINTS

- A data structure is any method used to store data in an organised and accessible format.
- An abstract data type is a conceptual model of how data are organised and the operations on them.
- An array is a data structure that contains data of the same type using a single identifier.
- A one-dimensional array is also known as a list.
- Arrays can be multi-dimensional.
- Files are used to store data.
- A text file is one that simply contains lines of text or other human-readable characters.
- A binary file is one that stores data as a series of 0s and 1s.
- Static data structures store a set amount of data which is usually defined by the programmer.
- Dynamic data structures can use more or less memory as needed through the use of a heap.

TASKS

1 How can you access each element in an array?
2 Explain how you could use an array to keep track of personal best times for the members of an athletic club. Your solution may require several dimensions.
3 Explain the terms file, record and field in relation to data structures.
4 What are the typical uses of text files?
5 What are the typical uses of binary files?
6 Identify two examples of:
   a) dynamic data structures
   b) static data structures.
7 Identify two advantages of using dynamic data structures.
8 Identify two disadvantages of using dynamic data structures.
9 Why are stack and queues considered to be dynamic data structures?

STUDY / RESEARCH TASKS

1 Write code that will:
   a) write data to a text file / read data from a text file
   b) write data to a binary file / read data from a binary file.
2 Write code that will:
   a) create an array
   b) read data from an array.
3 Find out how a stack is used to manage a memory heap.
4 PNG is a common binary format. Find out about other commonly used binary files.
5 Why might programmers create executable files as binary files?

8 Queues and stacks

A level only

SPECIFICATION COVERAGE

3.2.1.4 Abstract data types/data structures

3.2.2 Queues

3.2.3 Stacks

LEARNING OBJECTIVES

In this chapter you will learn:

- how stacks and queues work and what they are used for
- the difference between a circular queue, linear queue and priority queue
- how to write code to implement stacks and queues
- how to use nesting and recursion when implementing a stack.

INTRODUCTION

Queues and stacks are both examples of abstract data types that could be implemented as dynamic or static data structures. They are also abstract data types, which means that they do not normally exist as built-in data types but need to be created by the programmer using existing data types. For example, a stack might be built from an array. This chapter looks at the standard uses of stacks and queues and how to implement and work with them.

How stacks work

A **stack** is an example of a **LIFO** (last in first out) structure that means that the last item of data added is the first to be removed. A stack in a computer works in exactly the same way as a stack of books waiting to be marked or a stack of dishes waiting to be washed up – whichever item was added to the top of the stack last will be the first one to be dealt with.

However, unlike the washing up where items are literally taken off the stack as they are needed, the data in a computer stack is not actually removed. What happens is that a variable called the stack pointer keeps track of where the top of the stack is.

KEYWORDS

Stack: a LIFO structure where the last item of data added is the first to leave.

LIFO: last in first out refers to a data structure such as a stack where the last item of data entered is the first item of data to leave.

The process of adding a new item of data to the stack is called pushing and taking an item off the stack is called popping. A further action called peeking is used to identify the top of a stack. When an item is pushed onto the stack the stack pointer moves up and when an item is popped off the stack the pointer moves down, but a copy of the data is still left on the stack.

Here is a simplified example of a stack in use. Note that this stack can only store six data items.

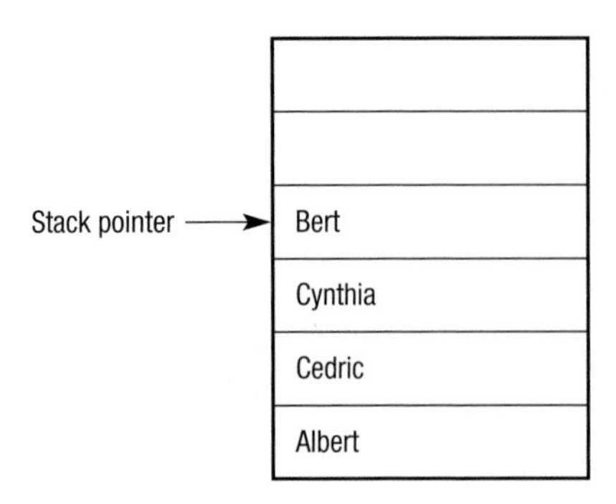

The stack pointer is used to show where the top of the stack is.

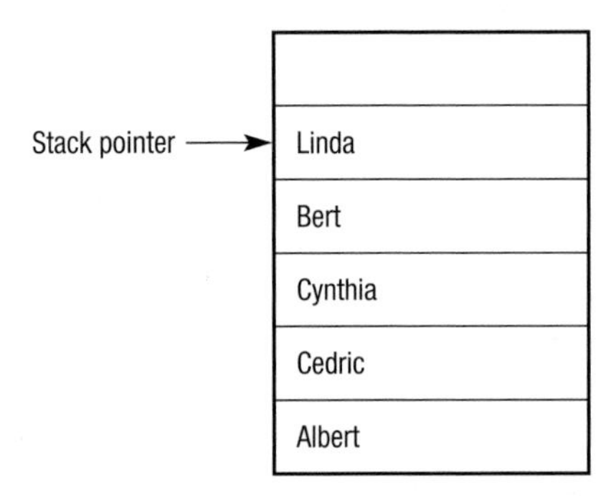

"Linda" has been pushed to the top of the stack so the pointer moves up.

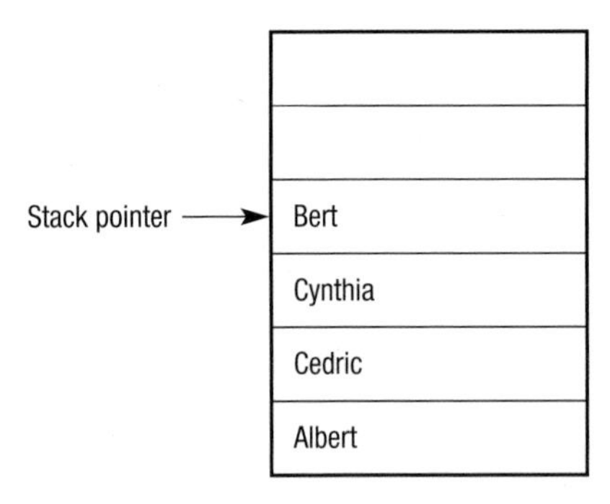

The stack is popped so the data at the **pointer** ("Linda") is read and the pointer moves down.

It is possible for the stack to need more memory than has been allocated to it. In the example given above, assuming that the stack had been set up as a static data structure, if the CPU tried to push on three more data items, the last one would have nowhere to go. In this case a stack overflow error would occur. Similarly if the stack was empty and the CPU tried to pop an item, this would result in a stack underflow as there is no data to be popped.

KEYWORD

Pointer: a data item that identifies a particular element in a data structure – normally the front or rear.

Implementing a stack

In the following two routines a single-dimension array called **StackArray** has been used to represent the stack. The variable **StackPointer** is being used to keep track of how far up or down the stack the pointer should be and **StackMaximum** stores the maximum number of values that can be stored on the stack.

```
'routine to push on to a stack
'check there is room on the stack
If StackPointer < StackMaximum Then
   'push on to the stack
   StackPointer ← StackPointer + 1
   StackArray(StackPointer) ← DataItem
Else
Error message "Data not saved – stack full"
End If
```

The error trap carries out an important task. The stack will only be allocated a limited number of memory locations, which in this case is kept in the variable **StackMaximum**. If the error routine was not there the stack would overflow – there would be too much data to store in it.

This routine shows how an item can be popped off a stack. Notice that the first line will trap an underflow error:

```
'Routine to pop off a stack
'check the stack is not empty
If StackPointer > 0 Then
   'pop off the stack
   DataItem ← StackArray(StackPointer)
   'decrease stack pointer
   StackPointer ← StackPointer – 1
Else
   Error message "There is no data to pop from the
stack"
End If
```

Uses of stacks

There are many uses for stacks. Due to their LIFO nature they can be used anywhere where you want the last data item in to be the first one out. A simple application would be to reverse the contents of a list as shown below:

1	2	3	4
Andrew	Jane	Mark	Wendy

The list above would go into the stack as follow:

Wendy
Mark
Jane
Andrew

If you now pull the names off the stack in order you would get:

1	2	3	4
Wendy	Mark	Jane	Andrew

Stack frames

> **KEYWORDS**
>
> **Stack frame**: a collection of data about a subroutine call.
>
> **Call stack**: a special type of stack used to store information about active subroutines and functions within a program.

Stacks can be used to store information about a running program. In this case it is known as a **stack frame** and a pointer will be used to identify the start point of the frame. This is used as a **call stack** when running programs as it can be used when a subroutine is running to call functions and pass parameters and arguments.

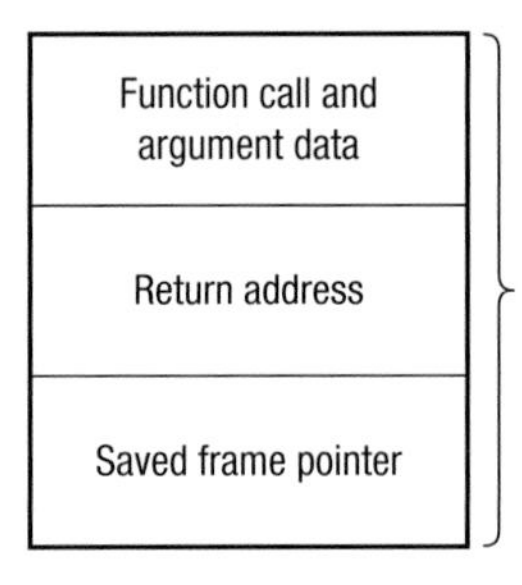

The function is called and data passed to it. The return address is placed on the stack so that when the function is finished, it will look at the return address so it knows where to go back to.

The subroutine is running using local variables. When a function is called, the current position is saved in the stack as a saved frame pointer.

> **KEYWORD**
>
> **Interrupt**: a signal sent by a device or program to the processor requesting its attention.

This is the same mechanism that is used for handling interrupts and exceptions in programs. **Interrupts** and exceptions are events where hardware or software demand the attention of the processor and cause the current program to stop. This could be something happening inside the program that is running or it could be an external event, such as a power failure, or a printer running out of paper.

When this happens, special blocks of code called interrupt handlers and exceptions handlers are loaded into memory and executed. Whilst the new demand is being dealt with, the details of the first program are stored on a stack. As soon as the interrupt or exception has been dealt with, the details are taken back off the stack and the first program can carry on wherever it left off.

Nesting and recursion

> **KEYWORDS**
>
> **Nesting**: the process of putting one statement inside another statement.
>
> **Recursion**: the process of a subroutine calling itself.

It is common practice to nest program constructs. For example, you might want to put one selection process inside another, or you might have a selection process being carried out inside an iterative loop. In this case the details of the successive nested loops are stored on the stack.

```
For HourCounter = 0 To 23
  For MinuteCounter = 0 To 59
    For SecondCounter = 0 to 59
     Output Hour , Minute , Second
```

```
        Next SecondCounter
      Next MinuteCounter
    Next HourCounter
```

This pseudo-clock won't keep very good time, but it does show how For/ Next loops can be nested inside each other.

Stacks also play a vital role in a process called recursion. This is where a subroutine calls itself in order to complete a task.

Queues

KEYWORDS

Queue: a FIFO structure where data leaves in the order it arrives.

FIFO: first in first out refers to a data structure such as a queue where the first item of data entered is the first item of data to leave.

A **queue** is called a **FIFO** (first in first out) structure. A queue in a computer acts just like people queuing for a bus – the first person in the queue is going to be the first person to get on the bus and the first item of data to be put into the queue in a computer will be the first to be taken out.

A common use of queues is when a peripheral such as a hard disk or a keyboard is sending data to the CPU. It might be that the CPU is not in a position to deal with the data straight away, so they are temporarily stored in a queue. Data being sent from the CPU might also need to be stored, and this is certainly the case when a document is sent to a printer. In this case your document is placed in the queue until the printer (which is a very slow device compared to the CPU) has the time to deal with the job. Queues that are used in this way are also known as buffers.

Here is a simplified example of how a queue is used. This queue can only store six data items.

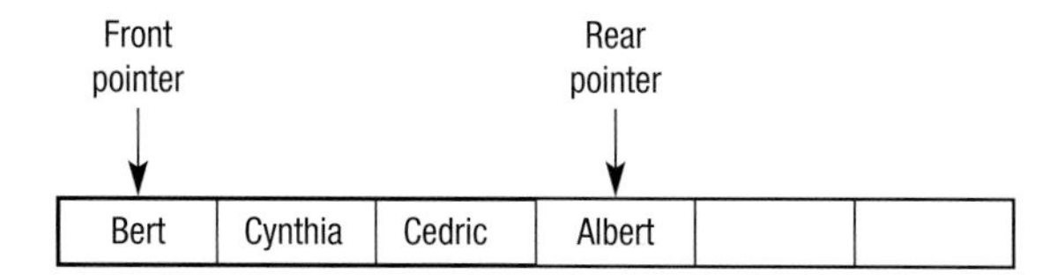

The queue has already been sent four data items, but none has yet been removed. The first item in the queue is "Bert" indicated by the front pointer. The last item in the queue is "Albert" indicated by the rear pointer.

When an item is added to the queue it is added at the end. If a new item ("Linda") is added, notice that the rear pointer has moved and now points to the new item "Linda". The front pointer has not moved.

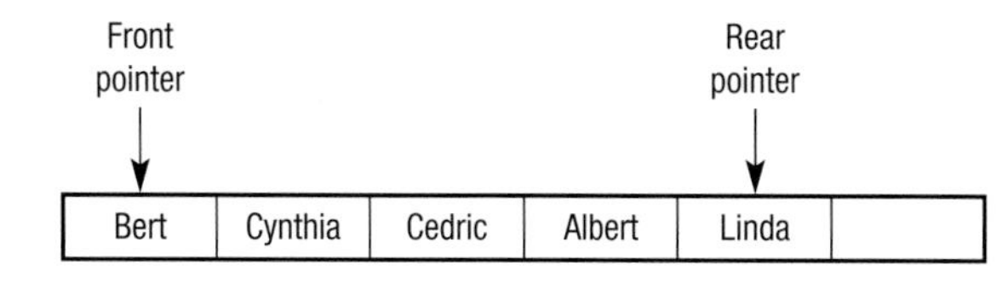

When a name is taken from the queue it is taken from the front. In this case "Bert" is removed from the queue and the pointer moves to the next item in the queue. The rear pointer does not move.

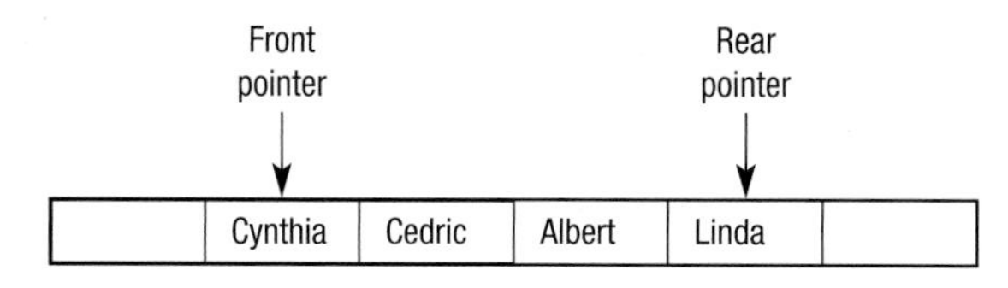

Linear, circular and priority queues

The examples above show a **linear queue**, that is, where you can envisage the data in a line. The first item in is the first item out. The maximum size of the queue is fixed in this case, although it could be dynamic. A typical method for storing data in a queue is to use a one-dimensional array.

A **circular queue** can be envisaged as a fixed-size ring where the back of the queue is connected to the front. This is often referred to as a circular buffer. As with a linear queue, it is the pointers that move rather than the data. However, with a circular queue the first items of data can be seen as being next to the last item of data in the queue.

A common implementation is for buffering, when items of data need to be stored temporarily while they are waiting to be passed to/from a device.

> **KEYWORDS**
>
> **Linear queue**: a FIFO structure organised as a line of data, such as a list.
>
> **Circular queue**: a FIFO data structure implemented as a ring where the front and rear pointers can wrap around from the end to the start of the array.

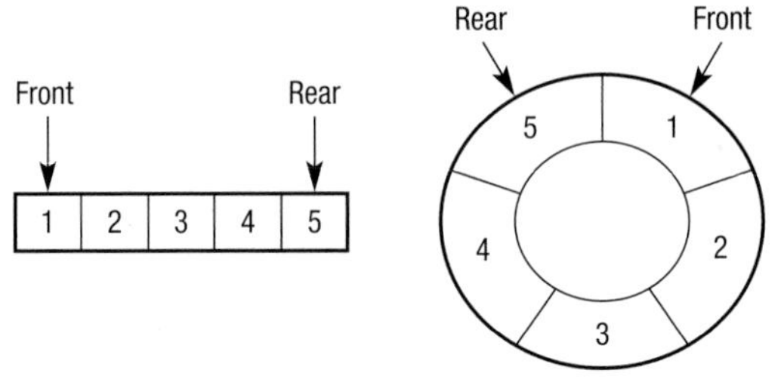

Figure 8.1 Linear and circular queues

A **priority queue** adds a further element to the queue which is the priority of each item. For example, if documents are being sent to print on a network printer then it might be possible for the user or systems manager to control the queue in some way. They may be able to force print jobs to the top of the queue or to put print jobs on hold whilst others are pushed through. This is known as a 'priority' queue and requires the programmer to assign priority ratings to different jobs. Higher priority jobs are effectively able to jump the queue. Where two jobs have the same priority, they will be handled according to their position in the queue.

> **KEYWORD**
>
> **Priority queue**: a variation of a FIFO structure where some data may leave out of sequence where it has a higher priority than other data items.

Implementing a linear queue

A queue is typically made up of a number of data items of the same type. Therefore, a common implementation is to use an array. To demonstrate the principle, this example shows a queue with a fixed size of nine elements. There are currently five items in the queue and FP shows the front pointer and RP shows the rear pointer.

FP				RP				
0	1	2	3	4	5	6	7	8
Alice	Belinda	Carly	Daphne	Erica				

Note that it is possible for the queue to become empty or full as data is added and removed, and that not every element has to have data in it. Therefore, when the queue is implemented we need to know:

- the name of the array
- the maximum size of the queue
- whether the queue is full or empty
- where the front of the queue is
- where the rear of the queue is.

Assuming that element 0 is the front of the queue and element 4 is the rear, when an item is removed, the queue will then look like this:

	FP			RP				
0	1	2	3	4	5	6	7	8
	Belinda	Carly	Daphne	Erica				

The front pointer has moved +1 so that the front is now pointing at element 1. The rear pointer does not change so remains on position 4.

Any item of data added to the queue is added to the rear. In this case it would be added in position 5 as this is the next available position. For example, if we add "Beth":

	FP				RP			
0	1	2	3	4	5	6	7	8
	Belinda	Carly	Daphne	Erica	Beth			

The front pointer is now on position 1 and the rear pointer is on position 5. Items can now be added and removed with the front and rear pointers moving accordingly. For example, if we removed the next two elements and added a new name "Jessica", the queue would look like this:

			FP			RP		
0	1	2	3	4	5	6	7	8
			Daphne	Erica	Beth	Jessica		

The front pointer will have moved forward to position 3 and the rear pointer will have moved to position 6.

Eventually, if data items keep being added, the rear pointer will reach the end of the array and there will be no more room in the array to add new elements, despite some earlier locations in the array being empty because elements have been removed from the front of the queue. The simplest way to deal with this is to always keep the front pointer pointing at index 0 in the array, and to move elements forward in the array each time an item is removed. This method is simple, but it can be quite time consuming to move the elements along in the array, especially if the queue is a long one. Therefore a more efficient method of dealing with this problem, known as a circular queue, is more common.

Implementing a circular queue

A circular queue works in a similar way to a linear queue except the front and rear pointers move when an item is added or removed, making more efficient use of memory. For example, in the linear queue above, items 1, 2 and 3 have all been removed. However, there is no way of adding items into those empty elements in the array as the front pointer has moved to element 3.

The circular queue makes use of the spaces that are freed up at the front of a queue after they have been removed. It does this by wrapping the rear pointer around the array starting again at element 0 once the queue becomes full. If we start with the same queue as before, the front pointer is 0 and the rear pointer is 4.

FP				**RP**				
0	1	2	3	4	5	6	7	8
Alice	Belinda	Carly	Daphne	Erica				

If two items are removed, the queue will then look like this:

		FP		RP				
0	1	2	3	4	5	6	7	8
		Carly	Daphne	Erica				

Four new items are now added to the queue: "Jane", "Davina", "Yvonne "and "Kelly". Notice that the rear pointer is now on 8.

		FP						RP
0	1	2	3	4	5	6	7	8
		Carly	Daphne	Erica	Jane	Davina	Yvonne	Kelly

As this is a circular queue, the rear pointer can now wrap back around to the beginning. If a further item is added, the rear pointer would move to position 0 as this free. To add "Maria":

RP		FP						
0	1	2	3	4	5	6	7	8
Maria		Carly	Daphne	Erica	Jane	Davina	Yvonne	Kelly

In the following pseudo-code the variable `Rear` is used to point to the end of the queue and `Front` is used to point to the start of the queue. The pseudo-code does not deal with the situations of the queue being either empty or full.

The code for adding a new item to a nine-element queue looks something like this:

```
'routine to add to a circular queue
'increment Rear pointer
Rear ← Rear + 1
'Check to see if end of array has been reached
'If so go back to the start of the array
If Rear = 9 Then Rear ← 0
'add data
Put DataItem at position Rear in array
```

The code for taking an item from the front of the queue might look like this:

```
'routine to remove data from a circular queue
'remove data
Take DataItem from position Front in array
'move Front on
Front ← Front + 1
'Check to see if the end of the array has been
reached
'If so go back to the start of the array
If Front = 9 Then Front ← 0
```

Implementing a priority queue

A priority queue can also be implemented using an array by assigning a value to each element to indicate the priority. Items of data with the highest priority are dealt with first. Where the priority is the same, then the items are dealt with on a FIFO basis like a normal queue.

There are two possible solutions using an array. One option is to use a standard queue where items are added in the usual way at the end of the queue. When items are removed, each element in the array is checked for its priority to identify the next item to be removed. Where this method is used, adding data is straightforward but removing it is more complex.

Starting with the same queue, this time a priority is included shown here in subscript and assuming that 1 is highest priority.

FP				RP				
0	1	2	3	4	5	6	7	8
Alice$_2$	Belinda$_1$	Carly$_2$	Daphne$_3$	Erica$_4$				

If an item is added, it is simply added with its priority at the end and the rear pointer is moved. If "Jane" were added with a priority of 1:

FP					RP			
0	1	2	3	4	5	6	7	8
Alice$_2$	Belinda$_1$	Carly$_2$	Daphne$_3$	Erica$_4$	Jane$_1$			

When data is removed, it is done so in order of priority. There are two items with a priority of 1. In this case, "Belinda" would be removed first as she is closest to the front of the queue. "Jane" would be the next item to be removed. In this example it shows how the principle of FIFO is still being used as "Belinda" entered the queue before "Jane".

An alternative is to maintain the queue in priority order, which means that when a new item is added, it is put into the correct position in the queue. Removing items can then be done in the usual way by taking the item at the front of the queue. Where this method is used, removing data is straightforward but adding it is more complex.

Working on the same list, this time the names would be in priority order. To remove the next item is just a case of removing the item at the front of the queue.

FP				RP				
0	1	2	3	4	5	6	7	8
Belinda$_1$	Alice$_2$	Carly$_2$	Daphne$_3$	Erica$_4$				

Therefore the first item to be removed would be "Belinda" as she has the highest priority:

	FP			RP				
0	1	2	3	4	5	6	7	8
	Alice$_2$	Carly$_2$	Daphne$_3$	Erica$_4$				

If a new item is added, it will be put into the correct position based on its priority. Where it has the same priority it will be added after the existing items of the same priority. For example, if "Yvonne" is added with a priority of 1:

	FP				RP			
0	1	2	3	4	5	6	7	8
	$Yvonne_1$	$Alice_2$	$Carly_2$	$Daphne_3$	$Erica_4$			

If "Kelly" is added with a priority of 2:

	FP					RP		
0	1	2	3	4	5	6	7	8
	$Yvonne_1$	$Alice_2$	$Carly_2$	$Kelly_2$	$Daphne_3$	$Erica_4$		

Practice questions can be found at the end of the section on page 90.

TASKS

1 What is meant by the terms pushing and popping?

2 The name "Robert" is pushed on to an empty stack. "Felicity", "Martin" and "Sam" are then pushed onto the same stack in that order. What data will be on the stack after the following operations? Pop one name, push on "Henry" then push "George", finally pop off one name.

3 Explain the purpose of the stack pointer.

4 A stack can be described as a LIFO and a queue as a FIFO. Use examples to explain the terms LIFO and FIFO.

5 Explain the difference between static and dynamic data structures with reference to stacks and queues.

6 Explain the difference between linear, priority and circular queues.

KEY POINTS

- Queues and stacks are dynamic data structures.
- A stack is an example of a LIFO (last in first out) structure which means that the last item of data added is the first to be removed.
- A queue is called a FIFO (first in first out) structure which means that the first item of data added is the first to be removed.
- There are three types of queues: linear, circular or priority.

STUDY / RESEARCH TASKS

1 Explain how a circular queue can be used to cope with a user entering data via a keyboard.

2 Write code that demonstrates the use of a flag to indicate if a user has pressed an invalid key.

3 Explore other methods of implementing a priority queue other than using an array, for example using linked lists or binary trees.

4 Write code to show how a pointer can be used to indicate the highest value held in a 20-element array.

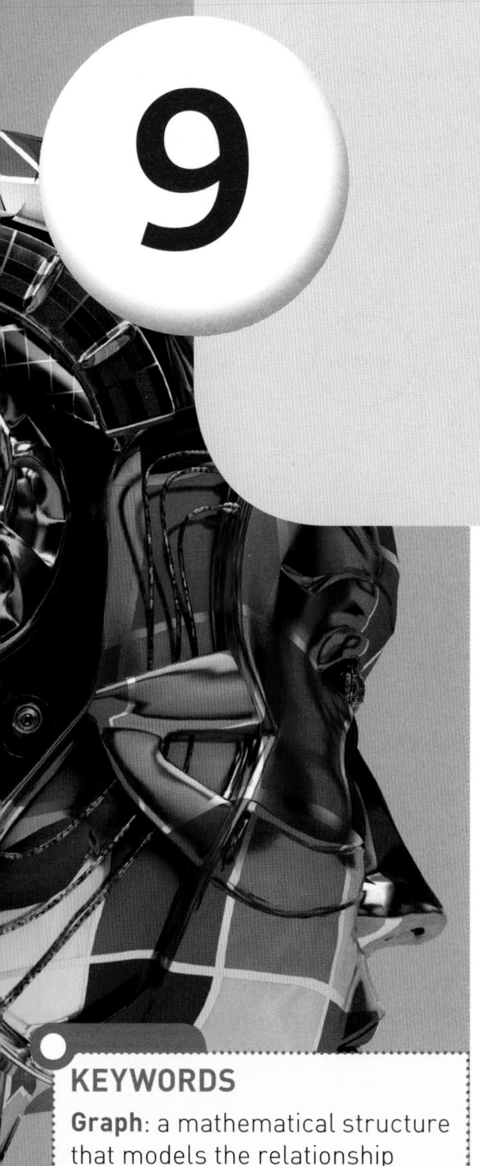

9 Graphs and trees

A level only

SPECIFICATION COVERAGE

3.2.4 Graphs

LEARNING OBJECTIVES

In this chapter you will learn:

- what graphs are and how they are constructed and used
- that graphs can be directed or undirected, weighted or unweighted
- how to use adjacency lists and matrices to represent graphs
- what trees are and how they are used
- how to create a binary tree.

INTRODUCTION

A **graph** is a mathematical structure that models the relationship between pairs of objects. The pairs of objects could be almost anything including places, people, websites, numbers, physical components, biological, chemical data or even concepts. The study of the use of graphs is called **graph theory** and it is useful in computing as it allows the programmer to model real-life situations that would otherwise be abstract.

KEYWORDS

Graph: a mathematical structure that models the relationship between pairs of objects.

Graph theory: the underlying mathematical principles behind the use of graphs.

Arc: a join or relationship between two nodes – also known as an edge.

Vertex/vertices: an object in a graph – also known as a node. (Vertices is the plural.)

Weighted graph: a graph that has a data value labelled on each edge.

To start with an example, a graph could be used to model the relationship between two places and how they are connected via a train line. In graph theory, objects are called nodes or **vertices** and each connection is called an edge or **arc**. In this simple example, we have two vertices, one for each town and one edge, which in this case will be the train connection between the two towns. A simple graph may look like this:

A **weighted graph** can be created by adding values to the edges. In this example, we might add the travel time between the two towns, so the weighted graph would look like this:

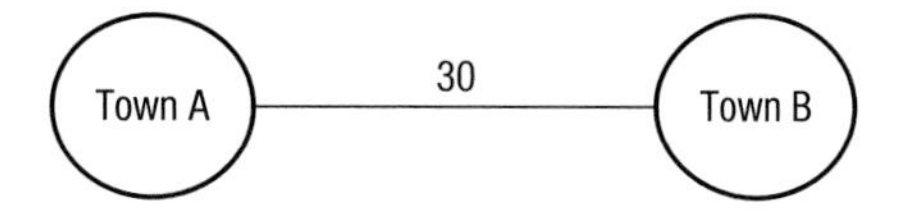

To extend this example, you might add in all of the towns on a particular network, with the travel time between each point. Figure 9.1 shows a graph that models the real-life situation so you can see that there is no direct connection between some of the towns, therefore there is no edge between some of the vertices.

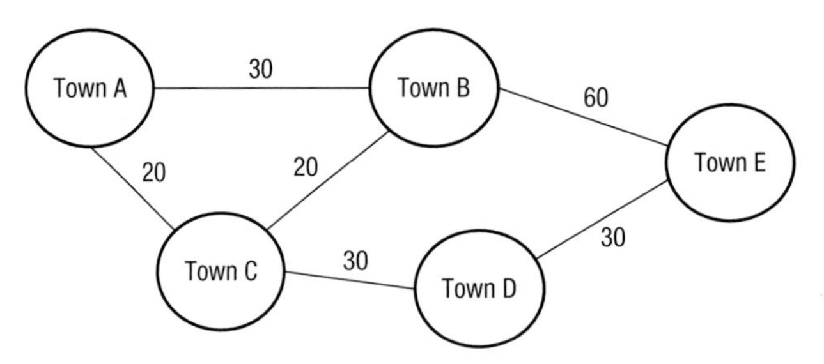

Figure 9.1 A graph structure to show journey times between towns

The graph now becomes quite useful as it could be used, for example, to find the quickest journey times between two towns. For example, to travel by train from Town A to Town E would be quicker via Town C and Town D than via Town B.

Graphs can also be directed or undirected, which refers to the direction of the relationship between two vertices.

> **KEYWORDS**
>
> **Undirected graph**: a graph where the relationship between vertices is two-way.
>
> **Directed graph**: a graph where the relationship between vertices is one-way.

An **undirected graph** is when the relationship between the vertices can be in either direction. In this example, the train will go in either direction between the towns, which means there is a two-way direction between the vertices in the graph.

A **directed graph** (also known as a digraph) is where there is a one-way relationship between the vertices. For example, we may produce a digraph to represent a real-life situation where we are creating a family tree. Figure 9.2 is a graph that shows parents and siblings.

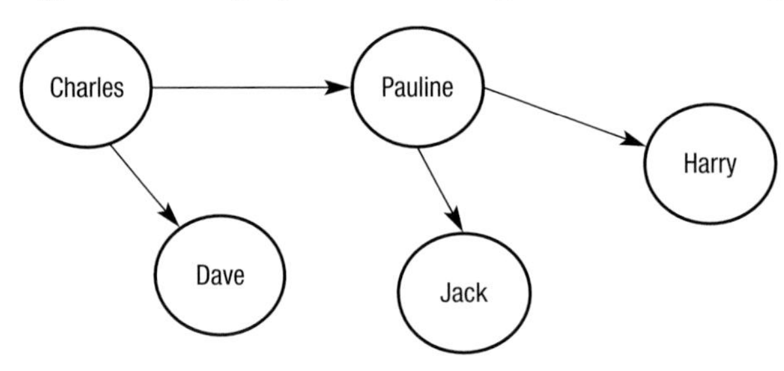

Figure 9.2 Graph to show parents and siblings

In this case, Charles is the parent of Dave and Pauline, Pauline is the parent of Jack and Harry. The arrows indicate that this is a one-way relationship.

Uses of graphs

Graphs have a wide range of uses in computing, as they are able to model complex real-life problems. There are a number of applications:

- Human networks: Human beings belong to numerous networks including family, work and social groups. For example, LinkedIn is an online network of business professionals and works on links between people. Once you create a profile, you link to other professionals that you know and they in turn link to all of the other people that they know. Each person is a vertex, and each edge is a relationship between one person and another.

- Transport networks: All transport works on the basis of a departure point, arrival point and route. The departure and arrival points form the vertices and the routes form the edges. There are several applications for graph theory including calculating quickest routes, planning timetables, scheduling and organising staff.
- The Internet and web: It is possible to 'map' the Internet or the World Wide Web using graph theory. In the case of the Internet, each connected device is a vertex with the physical connection forming the edge. With the web, each website is a node, with each linked site forming the edge. Figure 9.3 shows a map of the Internet.

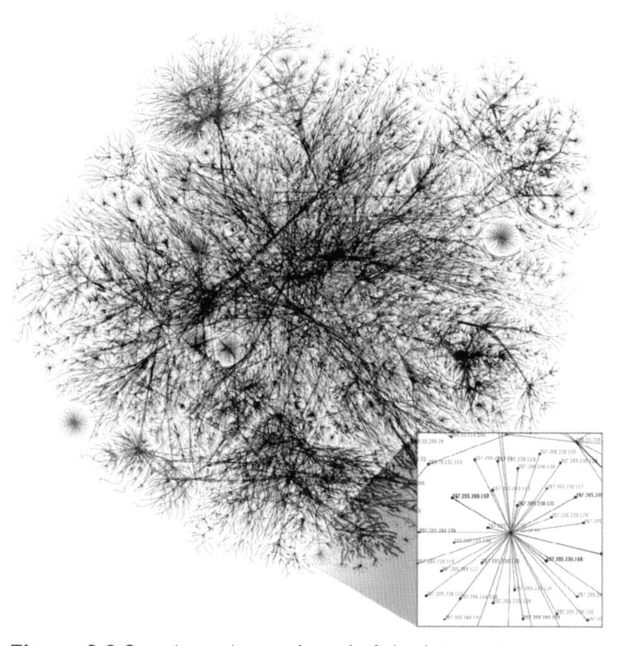

Figure 9.3 Graph to show a 'map' of the Internet

Notice that rather than forming a web shape, it looks more like a fireworks display. Each of the concentrations of colour is where there is an ISP as all the data is being routed through their servers.

- Computer science: **Latency** is a key factor in communication networks. It refers to the time delay that occurs when data is being transmitted. Graph theory could be used to calculate the quickest path to send data around a microprocessor where each vertex is a processor component and the edges are the buses that carry the data.
- Medical research: Understanding how diseases spread is critical to their prevention. For example, if studying the spread of a flu virus, each case of flu could be a node, or more likely, each location where there has been an outbreak would be a vertex. The edges would be the distance between locations. A weighted graph could be used to analyse the extent of outbreaks in particular locations and how much that then spreads between vertices.
- Project management: Any kind of large-scale project can be modelled using a graph. For example, this might be an engineering, construction or IT project. In this case the vertices would be each of the actions needed to complete the project and the edges would be the relationships and dependencies that exist between the tasks.
- Game theory: This is used in wars and conflicts to try to understand the causes of conflict and predict the likely actions that people might take for different strategies. For example, in a battle, the vertices could be the actions that one group might take, with the edges being the outcomes.

KEYWORD

Latency: the time delay that occurs when transmitting data between devices.

Graph theory is also an important concept in relation to Dijkstra's algorithm. This calculates the shortest path between nodes. It has been used for applications such as working out shortest distances between cities and calculating shortest distances between vertices in computer networks. This is covered in detail in Chapter 13.

Adjacency list

KEYWORD

Adjacency list: a data structure that stores a list of nodes with their adjacent nodes.

A graph is an example of an abstract data type. So far we have considered the graph in graphical form, but we need to represent it in a way that can be stored and manipulated by the computer. The first method is to use a list, called an **adjacency list**.

Adjacent means 'next to', so the idea of the adjacency list is to store the value of the vertices along with the vertices that they are next to. There are three basic formats for the list depending on whether the graph is direct or undirected and whether it is weighted.

Undirected graph

Vertex	Adjacent vertices
A	B,C
B	A,C,E
C	A,B,D
D	C,E
E	B,D

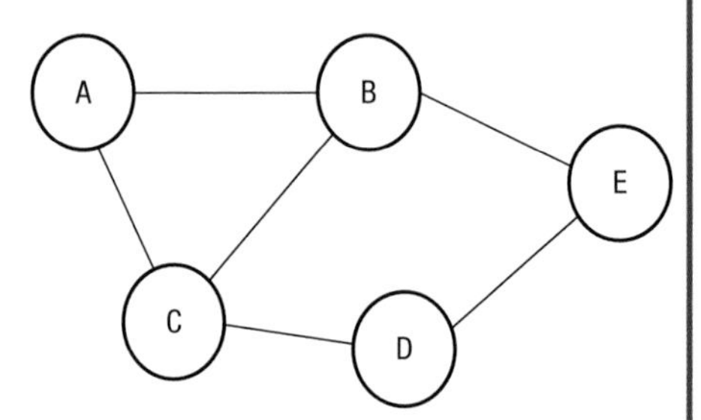

The list shows each vertex and each vertex that it is adjacent to. All adjacencies are shown as this is a two-way relationship.

Directed graph

Vertex	Adjacent vertices
A	B,D
B	E
C	
D	C,E
E	

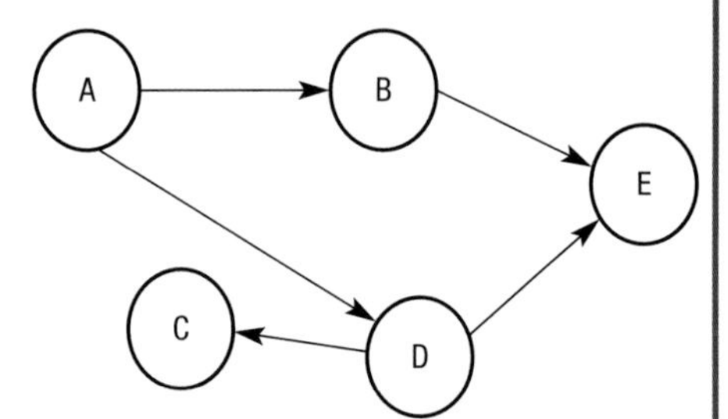

The list only shows the one-way relationship between the vertices. For example, D is connected to C but C is not connected to D.

Weighted graph

Vertex	Adjacent vertices
A	B,20,C,30
B	A,20,C,30,E,25
C	A,30,B,30,D,35
D	C,35,E,40
E	B,25,D,40

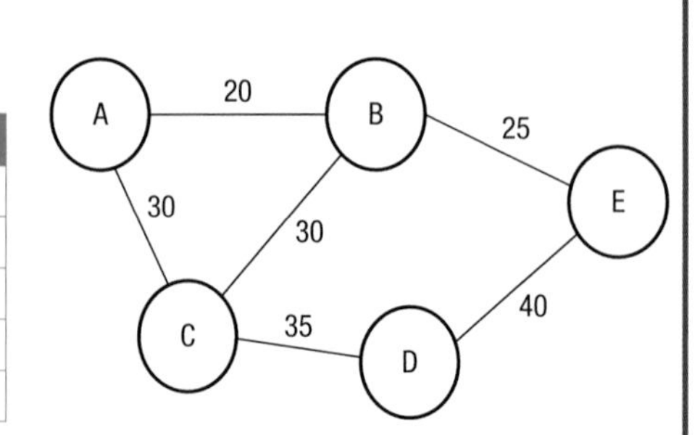

The list shows the value of each edge after each adjacent vertex. For example, A is adjacent to B with a weighted value of 20, A is adjacent to C with a weight value for 30 and so on. Notice that this example is an undirected weighted graph.

> **KEYWORD**
>
> **Adjacency matrix**: a data structure set up as a two-dimensional array or grid that shows whether there is an edge between each pair of nodes.

Adjacency matrix

The second method for storing the data is to use an **adjacency matrix**. This method uses a two-dimensional array or grid populated with 1s and 0s.

Undirected graph

	A	B	C	D	E
A	0	1	1	0	0
B	1	0	1	0	1
C	1	1	0	1	0
D	0	0	1	0	1
E	0	1	0	1	0

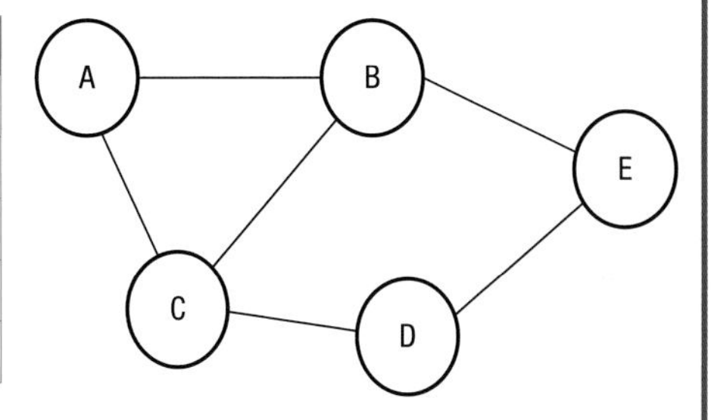

This works by putting a 1 in each cell where there is an edge and a 0 in each cell where there is not an edge. For example, A is adjacent to B so there will be a 1 in the grid where A and B intersect in the matrix. A is not adjacent to D so there will be a 0 in the grid where A and D intersect in the matrix.

Directed graph

	A	B	C	D	E
A	0	1	0	1	0
B	0	0	0	0	1
C	0	0	0	0	0
D	0	0	1	0	1
E	0	0	0	0	0

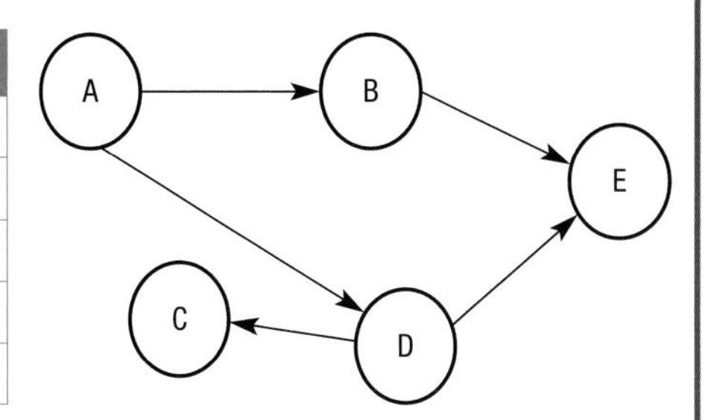

In this case, you read the matrix row by row, inserting a 1 where there is a one-way relationship between two vertices and 0 where this isn't. For example, A has a one-way relationship to B so there is a 1 in the cell where A and B intersect in the matrix. B does not have a one-way relationship to A, so there is a 0 in the cell where B and A intersect in the matrix.

Weighted graph

	A	B	C	D	E
A	∞	20	30	∞	∞
B	20	∞	30	∞	25
C	30	30	∞	35	∞
D	∞	∞	35	∞	40
E	∞	25	∞	40	∞

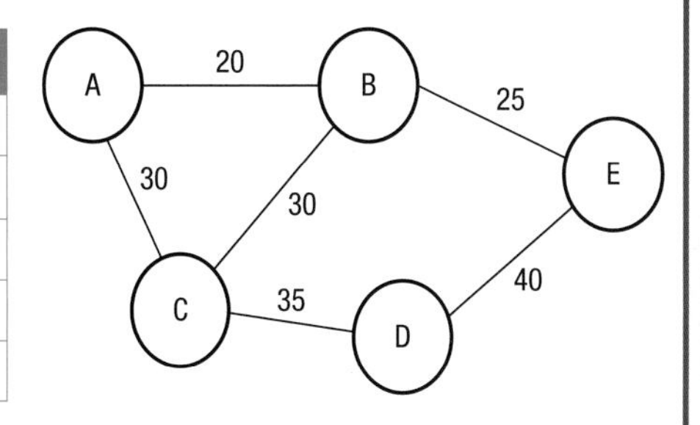

In this case, you follow the same process for an undirected graph, but this time you input the weighted value rather than a 1. Instead of the 0, the infinity sign is used.

Adjacency list vs adjacency matrix

When deciding on which implementation to use it usually comes down to two factors: speed and memory. Speed refers to how quickly the program will be able to access the data structure and produce a result. Memory refers to the amount of memory that each implementation will use. Bear in mind that the graph structure is likely to be used with very large datasets, making these issues critical. If you consider the simple examples above, five vertices will produce 25 data items that need storing. 100 vertices would produce 10000 and 1000 vertices would produce a million data items. If a single byte was used to store each data item that would create a 1 MB file. If you envisage thousands or even millions of vertices, the file sizes can get very large.

Table 9.1 shows the main factors.

Table 9.1 Comparison of adjacency list and adjacency matrix

Adjacency list	Adjacency matrix
Only stores data where there is an adjacency (edge) so requires less memory.	Stores a value for every combination of node adjacencies so requires more memory.
The list has to be parsed to identify whether particular adjacencies exist, which increases the time taken to process the data.	Adjacencies can be identified more quickly as every combination is already stored. Effectively the matrix forms a look-up table of each node to every other node.
Where there are not that many edges (few adjacencies), this method would be more suitable for the reasons stated above. This is known as a sparse graph.	Where there are many edges (lots of adjacencies), this method would be more suitable for the reasons stated above. This is known as a dense graph.

There is more information on working out time and size complexity of different algorithms in Chapter 25.

Trees

A **tree** is an abstract data structure that is very similar to a graph in that it has **nodes** and **edges**. It is called a tree because it is visualised as a hierarchical structure (like a family tree) with branches. Trees can have a root node, with all the other nodes branching away from the root.

The key difference with a tree compared to a graph is that it is connected and undirected and can contain no cycles or loops. For example, A goes to B and C, but you could not go from A to B to C or from A to C to B and back to A.

A tree could be visualised as follows:

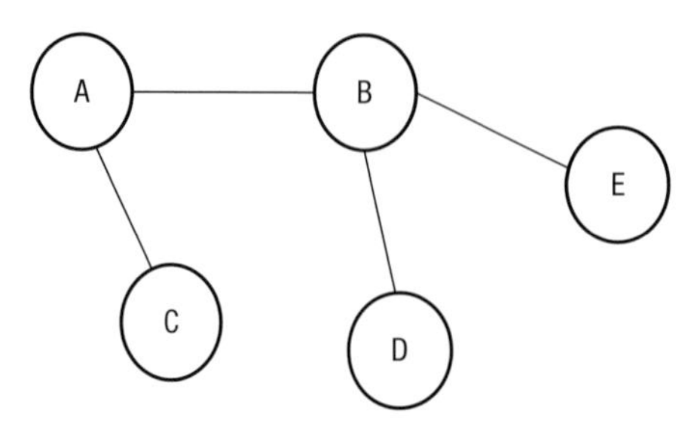

Figure 9.4 A tree structure

KEYWORDS

Tree: a data structure similar to a graph, with no loops.

Node: an object in a graph – also known as a vertex.

Edge: a join of relationship between nodes – also known as an arc.

KEYWORDS

Root: the starting node in a rooted tree structure from which all other nodes branch off.

Parent: a type of node in a tree, where there are further nodes below it.

Child: a node in a tree that has nodes above it in the hierarchy.

Leaf: a node that does not have any other nodes beneath it.

In this example, there are five nodes and four edges:

- A is the **root** node as all the other nodes branch away from it.
- A is also a **parent** node as it has two **child** nodes B and C.
- B is also a parent node and has two child nodes, D and E.
- C, D and E have no child nodes. These are sometimes called **leaf** nodes.
- You can see that there are no cycles. For example, A has an edge with C forming a single path. It would not be possible for example to go from A to C to D and back to A.

Trees have a number of uses. They:

- can be used to store data that has an inherent hierarchical structure. For example, an operating system may use a tree for directories, files and folders in its file management system
- are dynamic, which means that it is easy to add and delete nodes
- are easy to search and sort using standard traversal algorithms. There is more on this in Chapter 16
- can be used to process the syntax of statements in natural and programming languages so are commonly used when compiling programming code.

Binary search trees

A common implementation of a tree is the binary search tree. This is a directed and rooted tree, which can have no more than two branches off each node and is commonly used to store data that are input in a random order. The nature of the structure means that data are automatically sorted as they are entered and that it can be 'traversed' in order to search for and extract data from it.

The first item of data to be used is stored in the root node. The next (and any subsequent) data item is dealt with by the following routine:

- If the value of the new data item is less than the value in the current node then branch left, otherwise branch right.
- Keep repeating this process until you come to an 'empty' branch, then put the new value in the node at the end of the branch.

KEYWORD

Binary tree: a tree where each node can only have up to two child nodes attached to it.

This sounds awkward but look at the diagram below and try to follow through how the name "Fred" has been added to the **binary tree**.

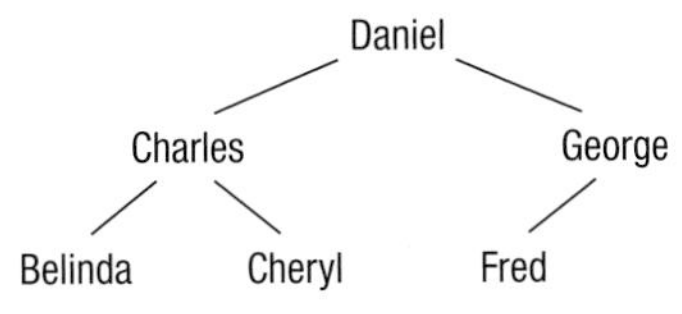

Figure 9.5 An example of a binary tree

Daniel is the root node. Belinda, Cheryl and Fred can be classed as leaf nodes because they have no nodes below them. Charles can be described as the parent and Cheryl the child.

Implementing a binary tree

The code for creating a binary tree needs three arrays. The first (called `Node` in the example below) stores the data itself. The second (`Left` in the example) stores which node the left branch from a node moves to and the third (`Right`) copes with branches to the right.

The data to add to the tree is stored in the variable `AddItem` and the root node has already been set up with the name "Jim". The algorithm adds further data items to the binary tree:

```
'Find next gap in the Node array
NodeCount ← 1
While Node(NodeCount) is not empty
  NodeCount ← NodeCount + 1
End While
'NodeCount stores the next blank
Node(NodeCount) ← AddItem
'start at the root node
PresentNode ← 1
While Node(PresentNode) is not empty do
  'Branch Left or Right?
  If AddItem < Node(PresentNode) Then
    'If Left branch is empty then assign NodeCount
    If Left(PresentNode) = 0 Then
      Left(PresentNode) ← NodeCount
    End If
    PresentNode ← Left(PresentNode)
  Else
    'If Right branch is empty then assign NodeCount
    If Right(PresentNode) = 0 Then
      Right(PresentNode) ← NodeCount
    End If
    PresentNode ← Right(PresentNo)
  End If
End While
```

If the root starts with the name "Jim", the arrays should look like this after you have added the names Kevin, Alice and Belinda to the tree.

	Node ()	Left()	Right()
1	Jim	3	2
2	Kevin	0	0
3	Alice	0	4
4	Belinda	0	0

The binary tree this represents will look like this.

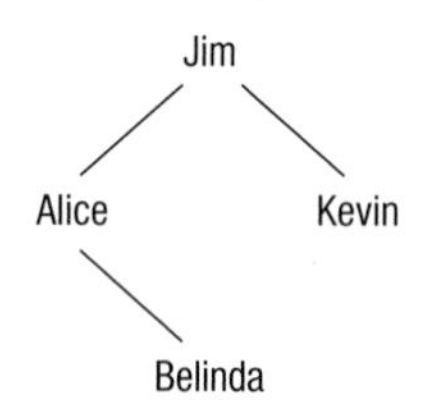

Figure 9.6 The resultant binary tree

Practice questions can be found at the end of the section on page 90.

TASKS

1 Create an adjacency list and matrix from the following graphs.

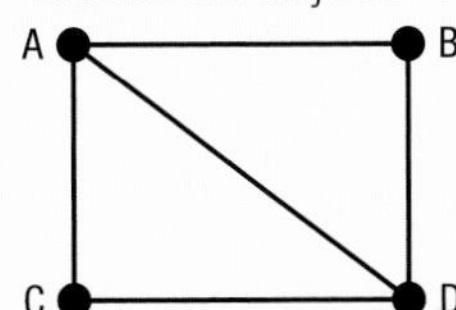

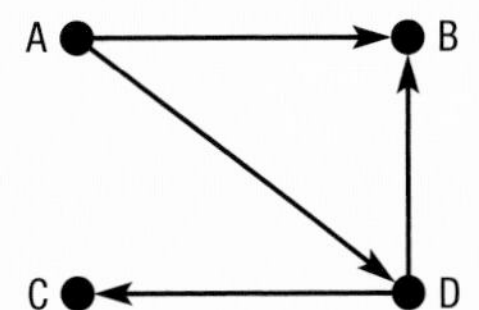

2 Draw an undirected graph from the following adjacency matrix.

	A	B	C	D
A	0	1	1	1
B	1	0	1	1
C	1	1	0	0
D	1	1	0	0

3 Draw a directed graph from the following adjacency list.

Vertex	Adjacent vertices
A	B
B	D
C	D
D	A

4 Draw a weighted undirected graph from the following data adjacency list.

Vertex	Adjacent vertices
A	B,5,C,3
B	A,5,D,2
C	A,3
D	B,2

5 Draw a weighted undirected graph from the following data adjacency matrix.

	A	B	C	D
A	∞	10	20	∞
B	10	∞	∞	30
C	20	∞	∞	20
D	∞	30	20	∞

6 Explain where it might be more appropriate to use an adjacency list compared to an adjacency matrix.

7 Draw a binary tree from the following array:

Sequence	Vertex	Left	Right
1	E	2	0
2	D	3	0
3	A	0	4
4	B	0	5
5	C	0	0

8 Represent the following binary tree using arrays:

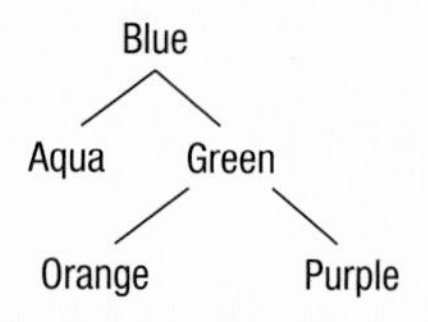

Show what would happen to the tree and the array if the following two items of data were added: Yellow, Magenta.

STUDY / RESEARCH TASKS

1 Explain how the following well-known abstract problems were solved using graph theory:

a) bridges in Konigsberg
b) the travelling salesperson problem
c) the four-colour theorem (for colouring maps)
d) six degrees of separation (in social networks).

2 'No two web pages are separated by more than 19 clicks.' How could graph theory help you work out whether this is true or not?

3 Explain how graph theory could help computer security experts understand how worms spread.

4 How might graph theory be applied to the problem of timetabling in a school or college?

5 How might a tree be used to create routing algorithms that define how data is sent around networks?

6 What are red–black B trees? How do they differ from binary trees?

KEY POINTS

- Graphs are a data structure made up of vertices (nodes) and edges, which are the connections between the vertices.
- Graphs can be used to analyse the connections and relationships between data items and are a useful tool for modelling real-life situations.
- Graphs can be directed or undirected, meaning that there may be a one-way or two-way connection between each vertex.
- Graphs can be weighted, meaning that a value can be applied to the edges between nodes.
- An adjacency list or matrix can be used to identify which vertices are connected to which others and whether there is any weight associated with the edge.
- A tree structure is a connected, undirected graph that contains no cycles.
- A binary tree structure is a special type of tree where each vertex can have no more than two children.

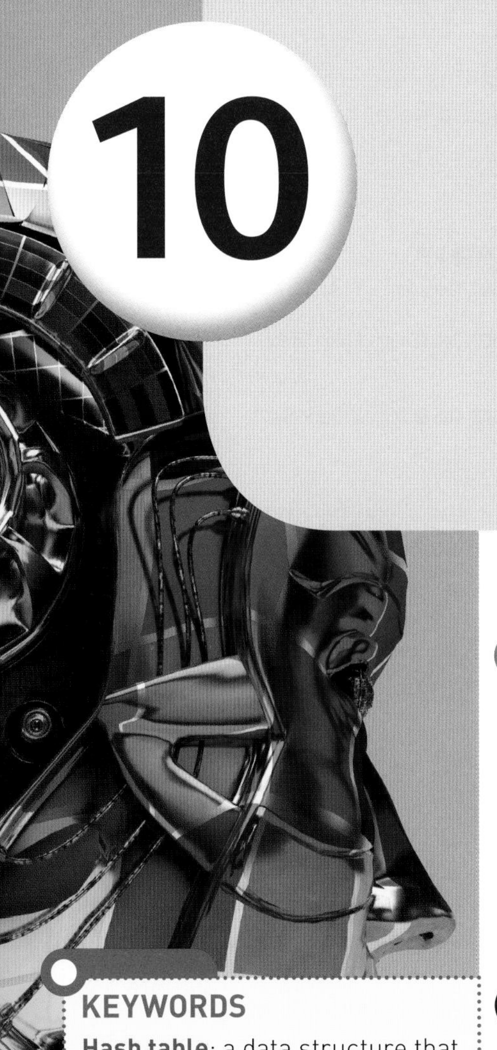

10

Hash tables and dictionaries

A level only

SPECIFICATION COVERAGE

3.2.6 Hash tables

3.2.7 Dictionaries

LEARNING OBJECTIVES

In this chapter you will learn:

- what hash tables are used for and how to create them
- how to create a hashing algorithm
- how to avoid collisions in hashing algorithms
- what a dictionary data structure is and how to construct one.

KEYWORDS

Hash table: a data structure that stores key/value pairs based on an index calculated from an algorithm.

Key/value pair: the key and its associated data.

INTRODUCTION

Hash tables and data dictionaries are both data structures made up of two parts. They can be viewed as two-dimensional arrays, or tables with one dimension being the data and the other being the key that identifies the location of the data in the table. Each **key/value** combination is unique within the data structure.

Hash tables

KEYWORD

Hashing algorithm: code that creates a unique index from given items of key data.

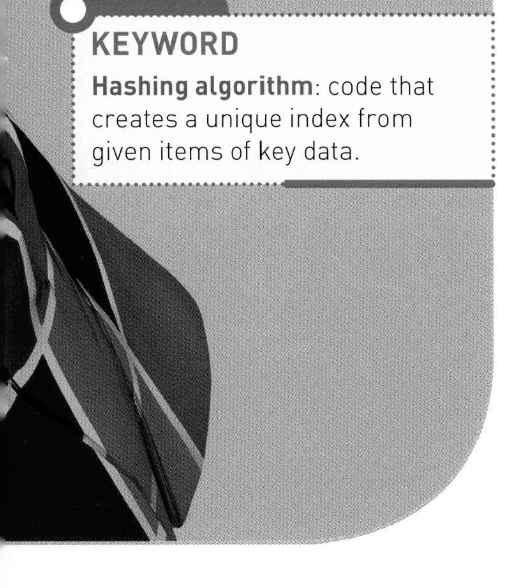

A hash table is a data structure made up of two parts, a table (or array) of data, and a key, which identifies the location of the data within the table. A **hashing algorithm** is carried out on the key, which then acts as an index to the specific location of that data item within the array. You could think of it as a look-up table that uses a key/value combination.

When the data need to be retrieved, for example, if a search is carried out on the data, the same hashing algorithm is used on the key being searched to calculate the index and therefore retrieve the data in one step. This is a very efficient search technique and it is why hashing tables are used extensively in applications where large datasets need to be accessed or where speed of access is critical.

As an example, a customer database stored as an array may contain records of millions of customers including CustomerID, Name, Address and so on. A hashing algorithm could be applied to the CustomerID field, which can be used as the key in this case. This would generate an index for each customer, which would point to the location of the record in the array.

This could be visualised as follows:

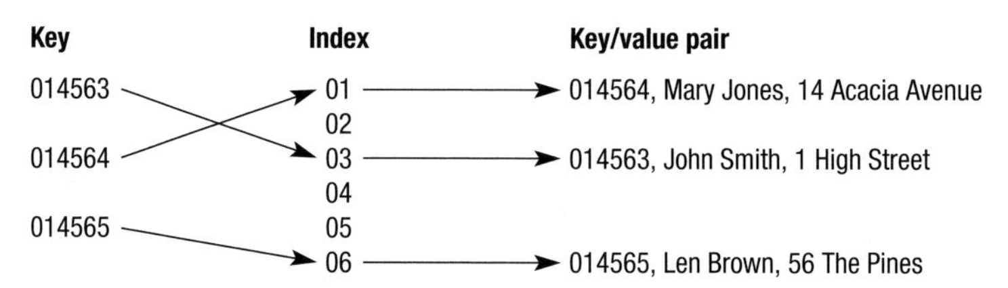

Figure 10.1 Visualisation of a hash table

The array into which the data are being stored can be envisaged as a series of slots, each of which has a unique index. The index can then be used to access all of the data stored in the record. Note that the key/value pair is the key and all of the data stored in relation to that key. In this case it would be a customer record.

Uses of hashing algorithms

Hashing algorithms have many applications:

- Databases: Used to create indices for databases enabling quick storage and retrieval of data.
- Memory addressing: Used to generate memory addresses where data will be stored. It is particularly useful for cache memory, where data is placed temporarily allowing the user fast access to programs and data stored in the **cache**.
- Operating systems: As an example of memory addressing, some operating systems use hashing tables to store and locate the executable files of all its programs and utilities.
- Encryption: Used to encrypt data, hence the term 'encryption key'. In this case the algorithm must be complex so that if data is intercepted it is not possible to reverse-engineer it.
- Checksums: A value calculated from a set of data that is going to be transmitted. On receipt the algorithm is run again on the data and the two results are compared as a way of checking whether the data has been corrupted during transmission.
- Programming: Used to index keywords and identifiers as the compiler will need access to these quickly as it compiles the program.

KEYWORD

Cache: a high-speed temporary area of memory.

Hashing algorithms

To generate the index, you need a suitable algorithm. To start with we will look at a very simple example to show the concept. You might have an array with six elements used to store six values. We could calculate the index using an algorithm that adds the numbers (digits) in the key together and then performs a modulo 6 sum on the result, as there are six slots in our hash table.

- For the first data item the value of the key might be 25463.
- Add the numbers (digits) together 2 + 5 + 4 + 6 + 3 = 20.
- Perform modulo 6 calculation so divide by 6 = 3 r 2.
- Therefore the Index = 2.
- The data is placed in slot 2.

0	
1	34255
2	25463
3	
4	
5	

- The second data item might have a key with the value 34255.
- Add the numbers (digits) together 3 + 4 + 2 + 5 + 5 = 19.
- Perform modulo 6 calculation so divide by 6 = 3 r 1.
- Therefore the Index = 1.
- The data is placed in slot 1.

This process then continues for every key. You can see from this how the index is created from the data in the key. The real benefit of using an algorithm is that it is used to store the data in the first place and then used to locate the data when it is needed. The indices therefore are created and recreated when they are needed.

Choosing a hashing algorithm

The basic example above demonstrates a few features and associated problems when creating a suitable algorithm:

- A numeric value needs to be generated from the data in order to perform the calculation. For non-numeric keys such as text and other characters, the ASCII or Unicode value of each character is normally used.
- It must generate unique indices. For example, if the next item of data was 43525, the algorithm would generate the index of 1 again. There is already data stored in this location so this has created a **collision**. It is theoretically possible to create a perfect hashing algorithm that avoids collisions, but in practice, they are unavoidable. A good algorithm will create as few as possible and needs a mechanism to cope with collisions as they occur.
- It needs to create a uniform spread of indices. For example, if you were storing millions of items of data into millions of slots the algorithm needs to provide an even spread of index values from the data and avoid **clustering**. This cuts down the possibility of collisions.
- There must be enough slots to store the volume of data. For example, if a database is going to store 1 million records, the algorithm must be capable of producing at least 1 million indices. In fact it would need more than this to avoid collisions as the table fills up. Hash tables have a **load factor** which is the number of keys divided by the number of slots. A high load factor means that it will become increasingly difficult for the algorithm to produce a unique result.
- It has to balance the issues of speed with complexity. For example, an algorithm for a database needs to calculate the index very quickly. An algorithm for encryption needs to be very complex, but may not need to calculate the index quickly.

KEYWORDS

Collision: when a hashing algorithm produces the same index for two or more different keys.

Clustering: when a hashing algorithm produces indices that are not randomly distributed.

Load factor: the ratio of how many indices are available to how many there are in total.

Creating suitable algorithms is sometimes described as a 'black art' as there is no universally accepted method for doing it and the design of the algorithm depends to a large extent on the application.

Collisions

One of the main features of a hashing algorithm is that it must produce a unique **index**. Where a collision occurs, there must be some way of handling it so that a unique index can be assigned to the key.

There are two main methods:

- **Chaining**: In this case, if a collision occurs, a list is created in that slot and the key/value pair becomes elements of the list. If another collision occurs, that key/value pair becomes the next element in the list and so on. Figure 10.2 shows the concept.

KEYWORDS

Index: the location where values will be stored, calculated from the key.

Chaining: a technique for generating a unique index when there is a collision by adding the key/value to a list stored at the same index.

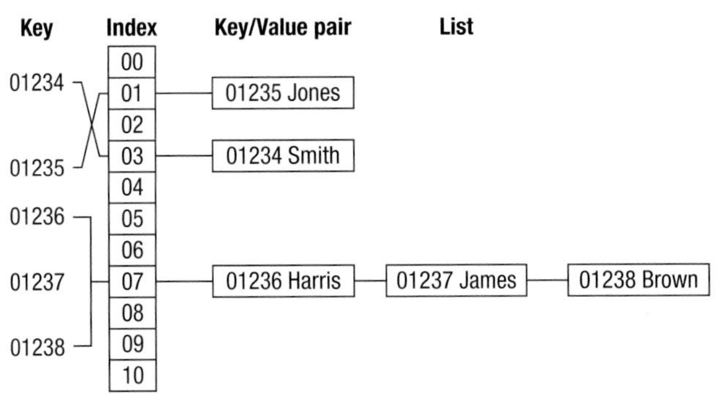

Figure 10.2 Chaining of key/value pairs when there is a collision

Where the index is unique, the key/value pairs work in the normal way. Where two or more keys generate the same index, a list is formed. It is called chaining as the additional key/value pairs get chained together inside a list. Each key/value pair is uniquely identified by its position within the list. In this example the keys 01236, 01237 and 01238 all produced the same index so their key/values have been chained together.

- **Rehashing**: In this case, if a collision occurs, the same algorithm is run again, or an alternative algorithm is run until a unique key is created. This normally uses a technique called probing, which means that the algorithm probes or searches for an empty slot. It could do this by simply looking for the next available slot to the index where there was a clash.

KEYWORD

Rehashing: the process of running the hashing algorithm again when a collision occurs.

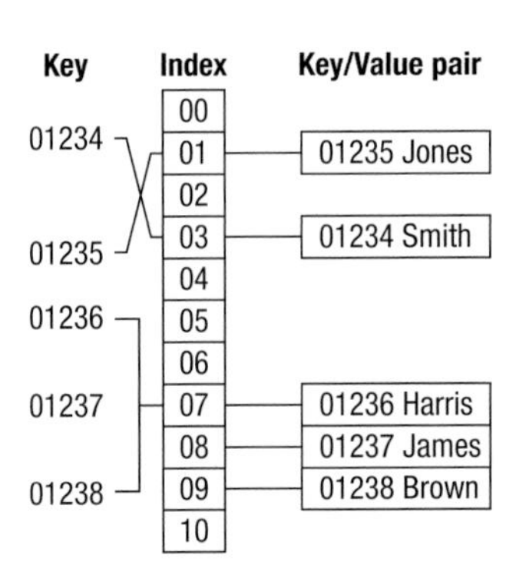

Figure 10.3 Probing as a result of a collision

Figure 10.3 shows a simple linear probe where the next available slot is used. This is not very sophisticated because if the hashing algorithm is leading to clustering as in this example, the results are still going to be clustered in around the same slots. A more sophisticated method is to apply another hashing function to the index where the clash occurred, in order to generate another one.

The following extract of code shows a hashing algorithm that creates a key using the day and date of birth multiplied by 28 with modulo 100 applied to the result. Notice that it also rehashes if there is a collision:

```
Private Sub cmdFindRecord_Click(ByVal sender
As System.Object, ByVal e As System.EventArgs)
Handles cmdFindRecord.Click
```

```
Dim FindRecord As String
Dim FindDay As Integer
Dim FindMonth As Integer
Dim HashKey As Integer
' calculate hash key
FindRecord = txtFindRecord.Text
FindDay = Val(Mid(FindRecord, 1, 2))
FindMonth = Val(Mid(findRecord, 4, 2))
HashKey = (28 * (FindMonth - 1) + FindDay) Mod 100
txtHashKey.Text = "(28 * (" & FindMonth & " - 1) +
" & FindDay & ") Mod 100 = " & HashKey
If grdTable.Rows(HashKey).Cells(3).Value =
txtFindRecord.Text Then
  ' record found using first hash key
  grdTable.Rows(HashKey).DefaultCellStyle.
  BackColor = Color.Red
  MsgBox("Match found using hashing algorithm")
  grdTable.Rows(HashKey).DefaultCellStyle.
  BackColor = Color.White
Else
  Do
    HashKey = HashKey + 1
    If grdTable.Rows(HashKey).Cells(3).Value =
    txtFindRecord.Text Then
      ' found using rehashing
      grdTable.Rows(HashKey).DefaultCellStyle.
      BackColor = Color.Red
      MsgBox("Collision occured. Match found
      using linear re-hashing")
      grdTable.Rows(HashKey).DefaultCellStyle.
      BackColor = Color.White
      Exit Sub
    End If
  Loop Until grdTable.Rows(HashKey).Cells(3).Value
  = "" Or HashKey = 100
  ' record not found
  MsgBox("No match for that date")
End If
End Sub
End Class
```

KEYWORDS

Dictionary (data structure): a data structure that maps keys to data.

Associative array: a two-dimensional structure containing key/value pairs of data.

Dictionaries

A **dictionary** is an abstract data type that maps keys to data. It is called an **associative array** in that it deals with two sets of data that are associated with each other. The dictionary data structure is analogous with a real-life dictionary in that there is a word (the key) associated with a definition (the data). This is similar to a hash table in that it has key/value pairs.

In the same way that a real-life dictionary is accessed randomly, a dictionary data structure also requires random access. The common procedures that you would need to carry out on a dictionary would be to add, retrieve and delete data. Unlike a real-life dictionary however, the data inside a dictionary data structure is unordered.

We could use the customer database example again here. Each record has a key, which might be the `CustomerID`. This key links to all of the data that is stored about the customer. At any time we may want to retrieve, add or delete a customer record. Dictionary data structures are often used to implement databases due to the fact that there will be inherent associations within the data and that they need to be searched and sorted as a matter of routine in order to retrieve data.

In simple terms the dictionary data structure can be envisaged as a two-dimensional array:

Table 10.1 A two-dimensional array

Key e.g. CustomerID	Associated data
01234	James Cochran, 12 Harbour Mews, Leicester
01235	Mary Abbot, 56 Eagle Street, Manchester
01236	Keith Fletcher, 3 Yarborough Road, Leeds
01237	Hussain Khan, 68 Lemon Street, Derby
01238	David Lui, 87 Threddle Lane, Northampton
01239	Rachel Young, The Forest Lodge, Kettering

As you can see from this example, dictionaries and hash tables are very similar and in fact a hash table is one way of implementing a dictionary. Dictionaries can also be created using binary trees (see Chapter 9).

Some programming languages such as Visual Basic and Python have a dictionary data type built in. For example, in Python, it is possible to use a dictionary constructor to build a dictionary directly from the key/value pairs. Visual Basic has a dictionary object which allows key/value pairs to be added directly.

The following Visual Basic-based pseudo-code shows the implementation of a dictionary data structure using the data dictionary type. To add an item:

```
Dim Dict As Dictionary (Of String, Integer)
Dictionary.Add ("Anne", 10)
Dictionary.Add ("Dave", 52)
Dictionary.Add ("Ethel", 17)
```

The data in speech marks is the key and the integer is the value assigned to it in the format (`"key", value`) where key is a string and value is an integer. This code simply adds three names to the dictionary.

To retrieve an item from the dictionary the key/value pair need to be identified:

```
For Each pair As KeyValuePair(Of String, Integer)
In dict
    MsgBox(pair.Key & "  -  " & pair.Value)
Next
```

This code would display the list of all the names in the dictionary in a message box.

To delete an item:

```
Dictionary.Remove ("Anne")
```

This code would delete the item identified by the value input, in this case, "Anne".

Practice questions can be found at the end of the section on page 90.

TASKS

1. What is a hash table data structure?
2. Suggest a suitable hashing algorithm that could be used to store the names of everyone in your class. Write code to implement your solution.
3. Identify three possible applications for hashing algorithms.
4. What are the main features of a good hashing algorithm?
5. What can you do to minimise the likelihood of collisions when creating a hash table?
6. What is load factor?
7. What is clustering and how is it caused?
8. Explain in detail how a hashing algorithm can deal with collisions.
9. What is a dictionary data structure and how does it differ from a hash table?
10. What are the main actions that you might want to carry out on data stored in a dictionary?

STUDY / RESEARCH TASKS

1. How does private/public key encryption use hashing algorithms to encrypt data?
2. How do hashing algorithms written for encryption vary from those written for indexing databases?
3. Is it possible to write a perfect hashing function?
4. Research 'Google Sparse Hash'.

KEY POINTS

- A hash table is a data structure made up of two parts, a table (or array) of data, and a key, which identifies the location of the data within the table.
- A hashing algorithm is carried out on the key, which then acts as an index to the specific location of that data item within the array.
- Hashing algorithms must create a range of keys sufficient to assign unique values to each item of data.
- Collisions occur when the hashing algorithm generates the same key from two different items of data.
- Chaining or rehashing must be carried out in the event of a collision.
- A dictionary is an abstract data type that maps keys to data.
- Dictionaries and hash tables are similar data structures.

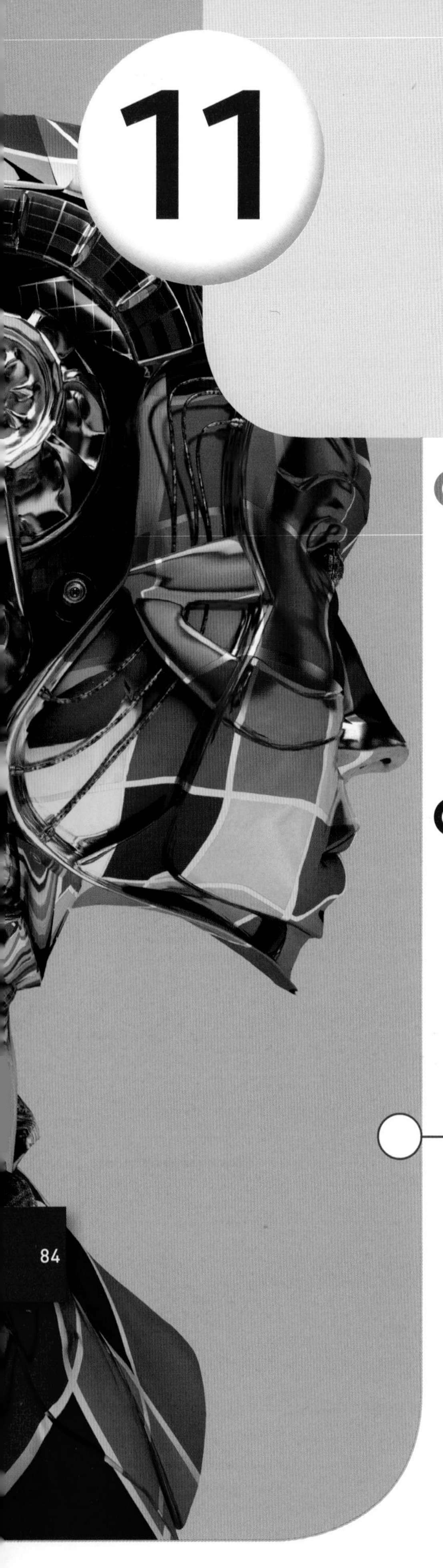

11 Vectors

A level only

SPECIFICATION COVERAGE

3.2.8 Vectors

LEARNING OBJECTIVES

In this chapter you will learn:

- what vectors are used for
- how to represent vectors as arrays, dictionaries and lists
- how to represent vectors as functions
- how to represent vectors as arrows. How to combine vectors using addition, multiplication and convex combination
- how to apply vectors.

INTRODUCTION

Vectors can be represented and applied in various ways both mathematically and geometrically. They are used in different ways in computing, for example:

- as a data structure
- as a method for mapping one value to another
- as a method of defining geometric shapes.

In this chapter we will look at all three interpretations of vectors.

Representing vectors as a data structure

When programming, vectors can be implemented as values stored in a list. For example, the first six values of the Fibonacci sequence could be represented as:

```
fibonacci[0] = 0; fibonacci[1] = 1; fibonacci[2] = 1;
fibonacci[3] = 2; fibonacci[4] = 3; fibonacci[5] = 5;
```

This representation could also be described as a one-dimensional array where each item of data is an element in the array, which can be accessed by its location:

Index	0	1	2	3	4	5
Data	0	1	1	2	3	5

A dictionary is a data structure that maps a key to a value. As we have seen, we can create sets of real numbers that can then be applied over the vectors.

The dictionary structure allows us to call an index, which is then used as a look-up to the real values.

```
{0: Value 1, 1: Value 2, 2: Value 3, 3: Value 4...}
```

The start of the Fibonacci sequence vector could be represented in a dictionary as:

```
{0:0, 1:1, 2:1, 3:2, 4:3, 5:5}
```

Representing vectors as a function

A function is a mathematical construct that maps an input to an output. For example, the function $f(x) = x^2$, simply maps the value of x to its square. A vector can be represented as a function. For example:

F = the function to create the vector

S = the complete set of values that the function can be applied to

R = the potential outputs of the function.

Therefore $F: S \rightarrow R$

Note that all of the output values must be drawn from R, which is being treated as a single field from which the function takes its values.

Representing vectors as arrows

Geometrically, vectors can be represented as arrows as shown in Figure 11.1.

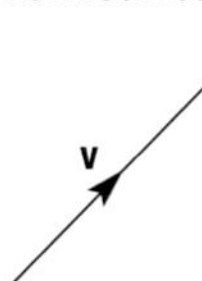

Figure 11.1 A vector represented as an arrow with magnitude and direction

KEYWORDS

Magnitude: one of the two components of a vector – refers to its size.

Direction: one of the two components of a vector.

Components: the values within a vector.

The two dimensions of size (or **magnitude**) and **direction** are shown. The direction of the arrow is shown by the arrow head and v represents the size. The start of the arrow is called the tail and the top of the arrow, the head. To quantify the size and direction of the arrow, think of it plotted on x and y axes:

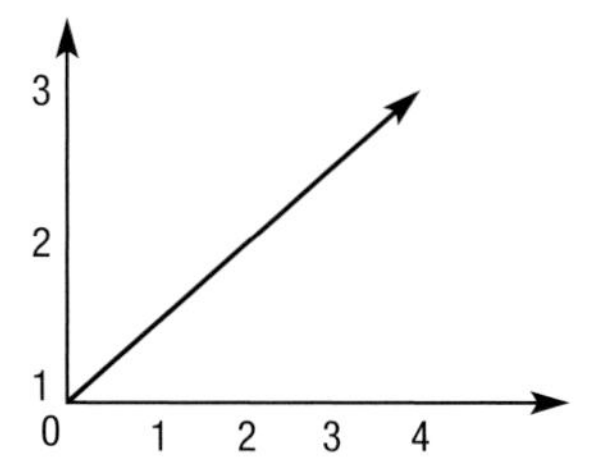

Figure 11.2 A vector visualised as an arrow with a measurement

The arrow can be represented as a vector **A** in the format $\mathbf{A} = (x,y)$. x and y are called the **components** of the vector and in this case would be the distance from (0, 0) on an x and y axis. Therefore, this vector is described as $\mathbf{A} = (4, 3)$ often shown in the format $\mathbf{A} = \binom{4}{3}$ to differentiate them from a coordinate pair used to plot points on a graph.

Already, you can see how useful vectors can be. With reference to vector graphics for example, it would now be possible to resize an image simply by changing the component values in the vector.

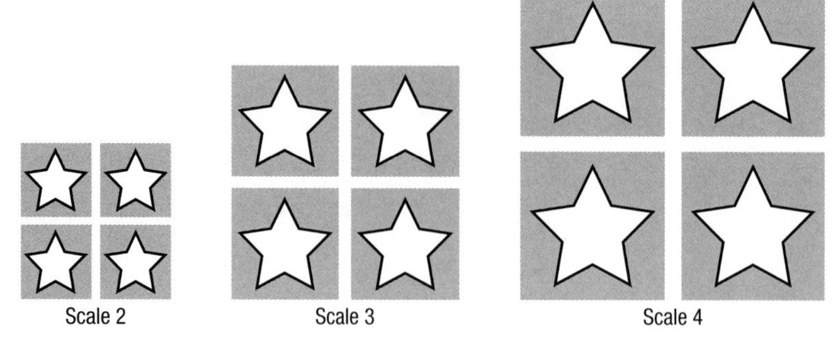

Figure 11.3 The effect of scaling a vector

Three-dimensional objects can be represented using the same method with the addition of a further component, the z axis.

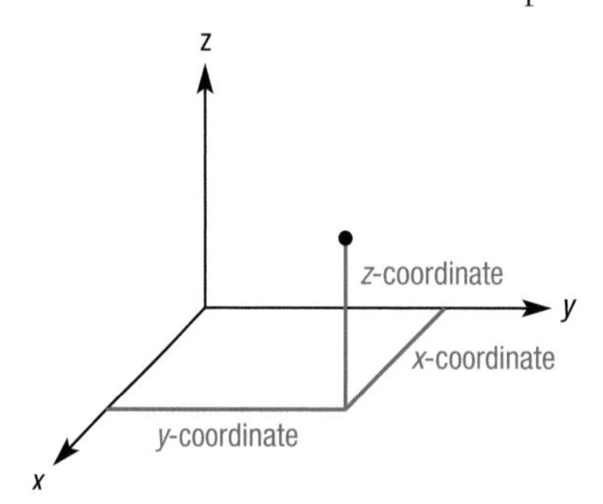

Figure 11.4 A visualisation of a vector in three dimensions

In this example, the vector could be represented as $\mathbf{A} = (x, y, z)$.

Vector addition

It is possible to add vectors together, which has the effect of translating or displacing the vector. Geometrically, this could be visualised as joining the tail of one to the head of another.

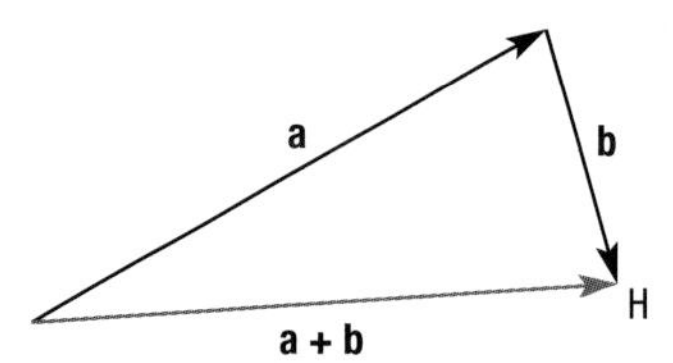

Figure 11.5 Adding vectors

Notice that a new point H has been created which may be used as the head of a new vector.

The sum of two vectors **A** and **B** can be represented as follows:

A = (A1, A2, A3)

B = (B1, B2, B3)

A + **B** = (A1 + B1, A2 + B2, A3 + B3)

Note that the two vectors must have the same dimension, which in this case is three components. For example, if:

A = (2, 3, 4) and **B** = (3, 5, 10) then **A** + **B** = (5, 8, 14)

Scalar–vector multiplication

KEYWORD

Scalar: a real value used to multiply a vector to scale the vector.

It is also possible to multiply vectors by a number, which has the effect of scaling. The number is called a **scalar** as it represents the amount by which you want change the scale of the vector. An analogy would be changing the scale of a map. If you zoom in you are changing the scale. In the case of a vector if you scale it by a factor of two, it will have twice the magnitude. The direction however, will not change as a result of scaling. You can envisage this geometrically as shown in Figure 11.6.

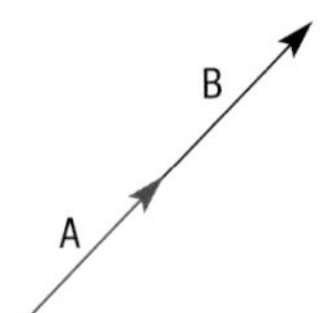

Figure 11.6 Scaling a vector

The original vector **A** = (3, 2). Multiply this by the scalar 2 results in vector **B** = (6, 4). Notice that the tail position and direction do not change.

Dot product

KEYWORD

Dot product: multiplying two vectors together to produce a number.

Dot product is the process of combining two vectors to calculate a single number. It is calculated in the following format:

$\mathbf{A} \cdot \mathbf{B} = A_x B_x + A_y B_y$

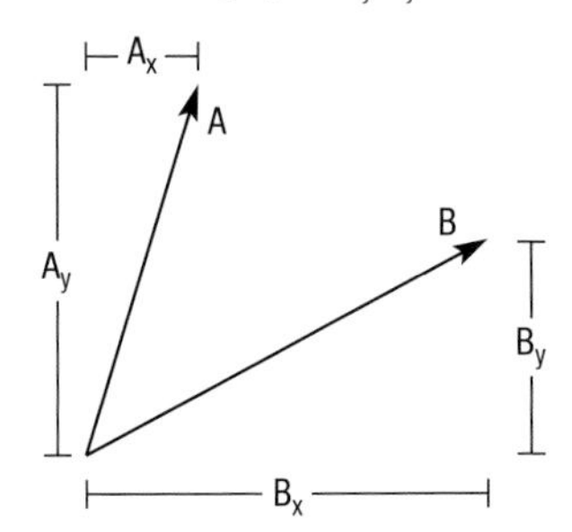

Figure 11.7 The dot product of two vectors

In this example, **A** = (3, 5) and **B** = (7, 2)

Therefore the dot product is $3 \times 7 + 5 \times 2 = 31$

This would also work in three dimensions by including *z* in the components. For example, two vectors with the coordinates **A** = (5, 3, 2) and **B** = (2, 7, 4) would result in a dot product of:

$5 \times 2 + 3 \times 7 + 2 \times 4 = 10 + 21 + 8 = 39$

Convex combinations

KEYWORDS

Convex combination: a method of multiplying vectors that produces a resulting vector within the convex hull.

Vector space: a collection of elements that can be formed by adding or multiplying vectors together.

When two vectors are combined to create a third, a relationship exists between the three vectors. In Figure 11.8 you can see that the new vector **c** has been created at right angles to the other vectors.

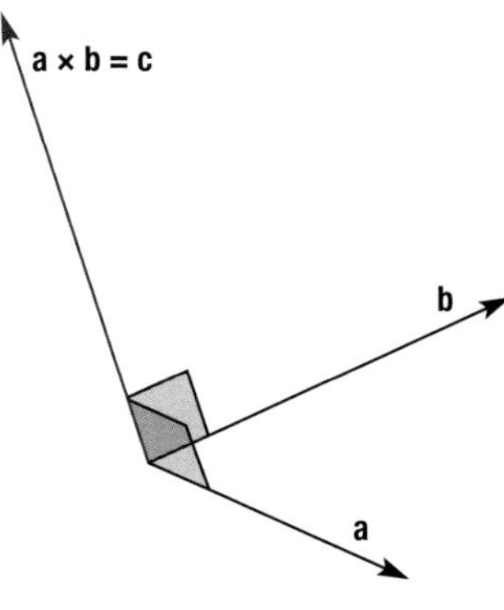

Figure 11.8 Combination of vectors

When these combinations are created, they have to be done according to certain mathematical principles. For example, a **convex combination** of vectors is one where the new vector must be within the **vector space** of the two vectors from which it is made.

This could be visualised as follows shown in Figure 11.9.

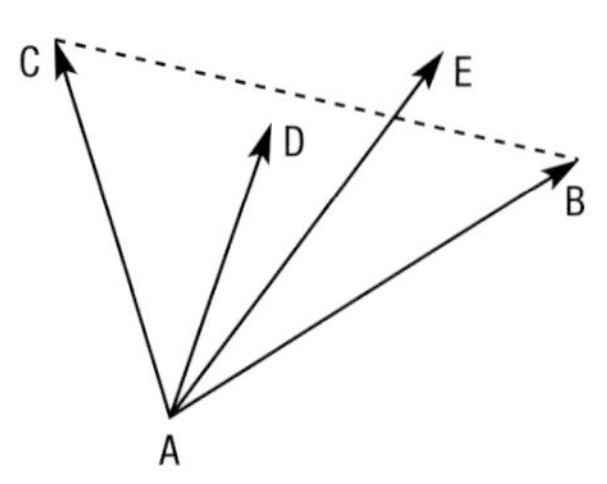

Figure 11.9 Convex combination of vectors

KEYWORD

Convex hull: a spatial representation of the vector space between two vectors.

Vector AD is created by combining vectors AC and AB. Notice an imaginary line between points B and C. The new vector must fall within the vector space defined by the points A, B and C in the diagram. This is a visual representation of what is called a **convex hull** that represents all of the points that make up the vector space. Notice point E, which represents the head for another vector. This falls outside the convex hull and is therefore not a convex combination.

Mathematically, to perform a convex combination, you will be multiplying one vector either by a scalar, or by another vector. This could be represented as:

$\mathbf{D} = \alpha AB + \beta AC$

where AB and AC are the two vectors

α and β represent the real number that each vector will be multiplied by.

α and β must both be greater than or equal to 0 and $\alpha + \beta$ must equal 1. D will then fall within the vector space.

Uses of dot product

Given two vectors **u** and **v**, it is possible to generate parity using the bitwise AND and XOR operations.

For example, where **u** = [1, 1, 1, 1] and **v** = [1, 0, 1, 1], the dot product would be **u**.**v** = 1. This is calculated by performing arithmetic over GF(2) where GF has two elements 0 and 1. This calculation works out the parity bit for even parity. The first vector will always be [1, 1, 1, 1] and in this example the second vector is [1, 0, 1, 1]. As you can see, we would expect the parity bit to be a 1 as the vector **v** currently has an odd number of 1s.

The calculation would work as follows:

u.**v** = [1, 1, 1, 1].[1, 0, 1, 1]

= 1 AND 1 XOR 1 AND 0 XOR 1 AND 1 XOR 1 AND 1

= 1 XOR 0 XOR 1 XOR 1

= 1

Arithmetic over GF(2) can be summarised in two small tables. Multiplication can be achieved by bitwise AND operation:

×	0	1
0	0	0
1	0	1

Addition can be achieved by bitwise XOR operation:

+	0	1
0	0	1
1	1	0

Subtraction is identical to addition, −1 = 1 and −0 = 0.

Practice questions can be found at the end of the section on page 90.

TASKS

1 Show how a simple vector could be represented as:
 a) a list
 b) a function
 c) an arrow.

2 Explain how a dictionary data structure can be used to represent a vector.

3 Use an example to show how you can add two vectors together and what effect this has on the vector.

4 Use an example to show how you can multiply a vector by a scalar and what effect this has on the vector.

5 Use an example to show the dot product of two vectors.

6 What is a convex combination of vectors?

KEY POINTS

- Vectors can be represented as a list of numbers, as a function and as a geometric point in space.
- Vectors can be created as a one-dimensional array or dictionary.
- Vectors can be combined using addition, multiplication and convex combination.
- Addition of vectors has the effect of translation or displacement.
- Multiplication of vectors by a scalar has the effect of scaling.
- Dot product can be used to generate parity.

STUDY / RESEARCH TASKS

1 Research how vectors are used to create computer games.

2 Explain how the length of a vector (envisaged as an arrow) is determined from its coordinates.

3 Write code to perform:
 a) vector addition
 b) multiplication of a vector by a scalar
 c) dot product calculation.

4 Research other methods of combining vectors including conical combination and affine combination.

5 Vector space has a number of axioms. Look into these and explain why they are essential in defining vector space. For example, associativity of addition, distributivity of scalar multiplication.

Section Two: Practice questions

1 The following data needs to be stored and accessed:

C, D, B, A, F, G

a) Describe how this data would be added to and then removed from a stack.

b) Describe how this data would be added to and then removed from a queue.

c) Show how these data items would be added to a binary search tree.

d) Assuming the data has been added to a binary search tree, write out the sequence of values that would be output from the tree following:

i) post-order traversal

ii) pre-order traversal

iii) in-order traversal.

2 The following adjacency list represents a graph.

Node	Adjacent nodes
A	B, 20, C, 30, D, 10
B	A, 20, D, 20
C	A, 30, D, 30
D	A, 10, B, 20, C, 30

a) Draw the graph.

b) Create an adjacency matrix for the graph.

c) Explain why this graph is not a tree.

3 Vector A is stored in an array A = {4, 5} and vector B is stored in an array B = {6, 12}.

a) What is an array?

b) What is a vector?

c) Calculate the result of adding the two vectors together, showing your working.

d) Calculate the dot product of these two vectors, showing your working.

4 Look at the following section of code that generates a hash value and then answer the questions.

```
Dim FindRecord As String
Dim FindDay As Integer
Dim FindMonth As Integer
Dim HashKey As Integer
FindRecord = txtFindRecord.Text
FindDay = Val(Mid(FindRecord, 1, 2))
FindMonth = Val(Mid(findRecord, 4, 2))
HashKey = (28 * (FindMonth - 1) + FindDay) Mod 100
```

a) What is the purpose of a hash value?

b) Explain whether or not you think that this is an effective algorithm, justifying your view.

c) Suggest an alternative hashing algorithm for generating the hash value.

Section Three: Fundamentals of algorithms

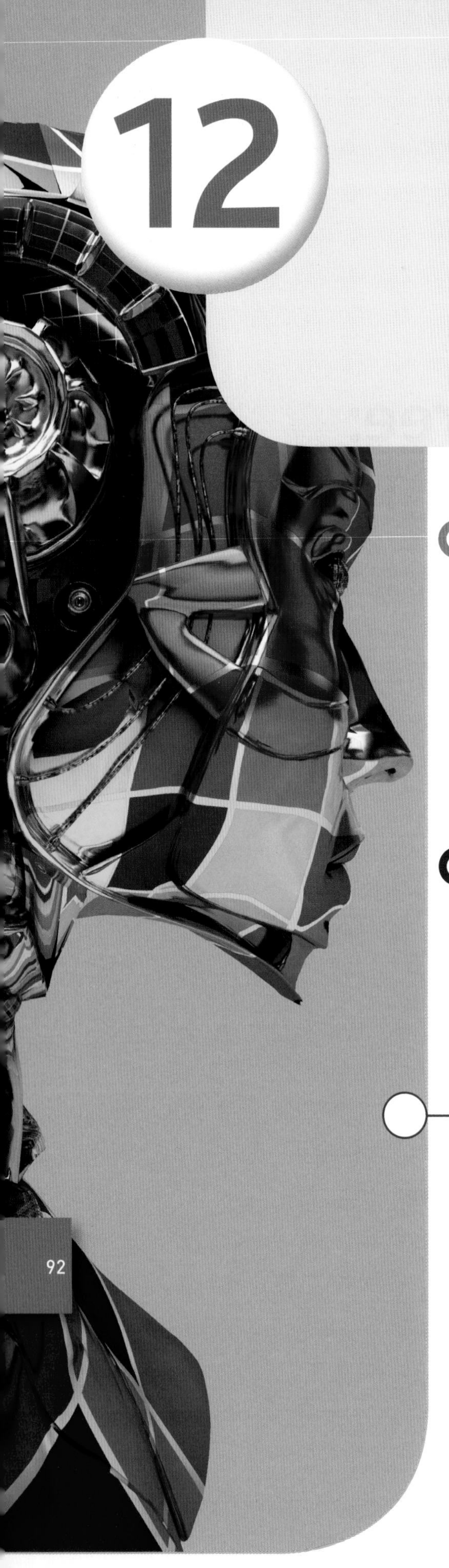

12 Graph and tree traversal

A level only

SPECIFICATION COVERAGE

3.3.1 Graph traversal

3.3.2 Tree traversal

LEARNING OBJECTIVES

In this chapter you will:

- consolidate your learning on graphs and trees
- learn how to implement and traverse a graph breadth first and depth first
- learn how to implement and traverse a binary tree pre-order, in-order and post-order
- learn how to apply stacks and queues and use recursion.

INTRODUCTION

In Chapter 9 we looked at the graph and tree data types and how they can be used. In this chapter you will learn how to implement and traverse the structures. The word 'traversing' means 'to move across' and that is what you do when you traverse a graph or a tree – you move across it, visiting nodes as you go.

Implementing a graph

As we saw in Chapter 9, graphs can be implemented using adjacency lists or matrices, which represent every vertex (node) and the edges (or connections) between the vertices.

Node	Adjacent nodes
A	B
B	A, C, E
C	B, D
D	C, E
E	B, D

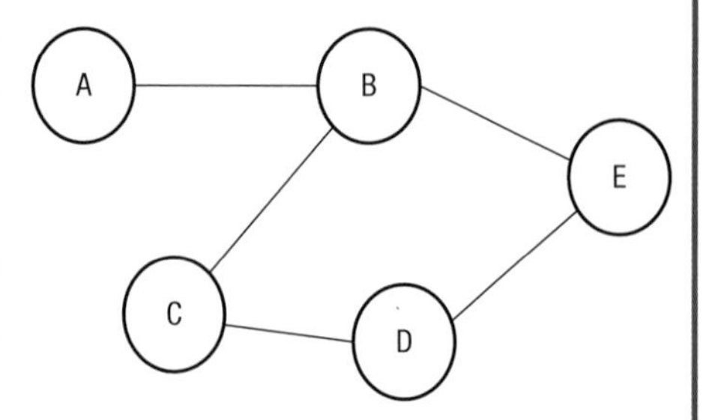

Figure 12.1 Adjacency list and corresponding graph

KEYWORDS

Implementation: creating code to produce a programmed solution.

Array: a set of data items of the same type grouped together with the same identifier.

Edge: a connection between two nodes in a graph or tree structure.

Graph: a data type made up of nodes and edges.

One possible **implementation** is to store this in an **array** showing each vertex and whether there is an **edge** between vertices. For example, the graph above could be represented by the following two-dimensional array:

Table 12.1 A two-dimensional array representing a graph

	A	B	C	D	E
A	0	1	0	0	0
B	1	0	1	0	1
C	0	1	0	1	0
D	0	0	1	0	1
E	0	1	0	1	0

A 1 represents an edge between the two vertices and a 0 means there is no edge. This approach can be used to represent any unweighted, undirected **graph**.

Traversing a graph

KEYWORDS

Depth first: a method for traversing a graph that starts at a chosen node and explores as far as possible along each branch away from the starting node before backtracking.

Breadth first: a method for traversing a graph that explores nodes closest to the starting node first before progressively exploring nodes that are further away.

Queue: a data structure where the first item added is the first item removed.

Node: an element of a graph or tree.

There are two ways of traversing the graph: depth first or breadth first.

- **Depth first** is a method that explores as far into a graph as possible before backtracking to visit unvisited nodes. It is often implemented using a recursive algorithm, which is explained later in the chapter.
- **Breadth first** is a method for traversing a graph that visits the nodes closest to a starting point first. A **queue** is used to keep track of the **nodes** to visit.

Using the graph in Figure 12.1 as an example, depth first works as follows:

Table 12.2 Depth first traversal

Explanation	Current node	Visited nodes
Select the node to start from (A).	A	
Mark node A as visited. Choose a node that is connected to A (B) and recursively call the search routine to explore from this node.	A	A
Mark node B as visited. Choose a node that is connected to B and has not been visited (C) and recursively call the search routine to explore from this node.	B	A B
Mark node C as visited. Choose a node that is connected to C and has not been visited (D) and recursively call the search routine to explore from this node.	C	A B C
Mark node D as visited. Choose a node that is connected to D and has not been visited (E) and recursively call the search routine to explore from this node.	D	A B C D
Mark node E as visited. All nodes connected to E have already been visited, so unwind recursion. There are no nodes left to visit during this unwinding, so the algorithm will terminate.	E	A B C D E

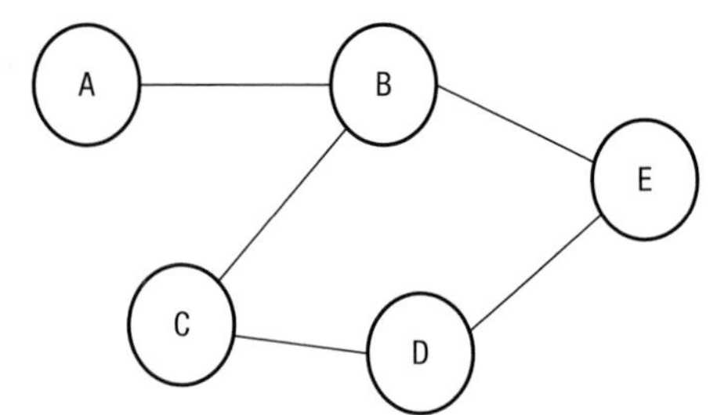

Figure 12.2 Adjacency list and corresponding graph

Using the graph in Figure 12.2 as an example, breadth first works by visiting the starting node and then all of the nodes attached to it in order. It then moves to the next closest nodes to repeat the process as follows:

Table 12.3 Breadth first traversal

Explanation	Contents of queue
Add the node to start exploring from (A) to the queue.	A
Add all nodes that are adjacent to node at front of queue (A) and not already full explored to queue (B).	A B
Remove A from queue as fully explored.	B
Add all nodes that are adjacent to B and not already fully explored to queue (C, E).	B C E
Remove B from queue as fully explored.	C E
Add all nodes that are adjacent to C and not already fully explored to queue (D).	C E D
Remove C from queue as fully explored.	E D
Add all nodes that are adjacent to E and not already fully explored to queue (none).	E D
Remove E from queue as fully explored.	E
Add all nodes that are adjacent to D and not already fully explored to queue (none).	D
Remove D from queue as fully explored. Queue empty so algorithm terminates.	

The following code shows a Visual Basic implementation of a grid with eight nodes. The code uses a CSV file to read in the adjacencies. Two procedures have been created to traverse the tree depth first and breadth first. The code is commented to explain how each subroutine works:

```
Public Class frmGraph
  Public RouteMatrix(8, 8) As Boolean
  Public NodeMatrix(8) As String
  Public VisitedMatrix(8) As Boolean
  Public NodeCount As Integer
  Public DataRow As String
  Public ThisNode As Integer
  Private Sub frmGraph_Load(ByVal sender As System.
  Object, ByVal e As System.EventArgs) Handles MyBase.
  Load
    ' count nodes
    NodeCount = 8
    ' populate arrays
    FileOpen(1, "C:\NodeTable.csv", OpenMode.Input)
    DataRow = LineInput(1)
    For Loop1 = 1 To NodeCount
      dgvRoutes.Rows.Add()
```

```
    Input(1, DataRow)
    NodeMatrix(Loop1) = DataRow
    VisitedMatrix(Loop1) = False
    dgvRoutes.Rows(Loop1 - 1).HeaderCell.Value =
    DataRow
    For Loop2 = 1 To NodeCount
      Input(1, DataRow)
      RouteMatrix(Loop1, Loop2) = DataRow
      dgvRoutes.Rows(Loop1 - 1).Cells(Loop2 - 1).
      Value = DataRow
    Next
  Next
End Sub
Private Sub btnDepthFirst_Click(ByVal sender
As System.Object, ByVal e As System.EventArgs)
Handles btnDepthFirst.Click
  Array.Clear(VisitedMatrix, 0, VisitedMatrix.
  Length)
  txtDepthOut.Text = ""
  ' start with node 'A'
  Depth(1)
End Sub
Private Sub Depth(ByVal ThisNode)
  txtDepthOut.Text = txtDepthOut.Text &
  NodeMatrix(ThisNode) & vbCrLf
  VisitedMatrix(ThisNode) = True
  ' Look at each node, check for route
  For Loop1 = 1 To NodeCount
    ' check for route
    If RouteMatrix(ThisNode, Loop1) = True Then
      'check node is unvisited
      If VisitedMatrix(Loop1) = False Then
        Depth(Loop1)
      End If
    End If
  Next
End Sub
```

```
Private Sub btnBreadth_Click(ByVal sender As
System.Object, ByVal e As System.EventArgs) Handles
btnBreadth.Click
  ' reset visited array
  Array.Clear(VisitedMatrix, 0, VisitedMatrix.
  Length)
  txtBreadthOut.Text = ""
  ' initialize
  Dim Queue(30) As Integer
  Dim QueueHead As Integer
  Dim QueueNext As Integer
  QueueHead = 1
  QueueNext = 1
  ThisNode = 1
  Queue(QueueNext) = ThisNode
  VisitedMatrix(1) = True
  QueueNext = QueueNext + 1
  Do
    ' take next item in queue
    ThisNode = Queue(QueueHead)
    QueueHead = QueueHead + 1
    txtBreadthOut.Text = txtBreadthOut.Text &
    NodeMatrix(ThisNode) & vbCrLf
    For Loop1 = 1 To NodeCount
      ' is node connected?
      If RouteMatrix(ThisNode, Loop1) = True Then
        ' has node been visited
        If VisitedMatrix(Loop1) = False Then
          ' add node reference to queue
          VisitedMatrix(Loop1) = True
          Queue(QueueNext) = Loop1
          QueueNext = QueueNext + 1
        End If
      End If
    Next
  Loop Until QueueHead = QueueNext
 End Sub
End Class
```

KEYWORDS

Binary tree: a structure where each node can only have up to two child nodes attached to it.

Pre-order: a method of traversing a tree by visiting the root, traversing the left subtree and traversing the right subtree.

In-order: a method of traversing a tree by traversing the left subtree, visiting the root and traversing the right subtree.

Post-order: a method of traversing a tree by traversing the left subtree, traversing the right subtree and then visiting the root.

Traversal: the process of reading data from a tree or graph by visiting all of the nodes.

Traversing a binary tree

Implementing a **binary tree** was explained in Chapter 9. In this section, you will learn how to traverse a tree. The process of traversing a binary tree extracts all the data from the tree in some sort of order. There are three ways of traversing a binary tree – **pre-order**, **in-order** and **post-order**.

To traverse a binary tree you start at the root node and move left, right or visit depending on the type of **traversal** you are using. Moving left or right entails 'looking' to see if there is a node in that direction and moving if there is. `Visit` entails extracting the data at that node.

Traversing the binary tree in Figure 12.2 gives the following results:

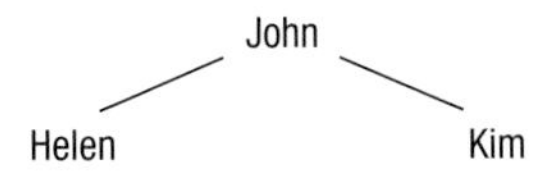

Figure 12.2

Table 12.4 Binary tree traversals

Pre-order	Visit, Left, Right	John, Helen, Kim
In-order	Left, Visit, Right	Helen, John, Kim
Post-order	Left, Right, Visit	Helen, Kim, John

Note that pre/in/post tells you when you do the visit stage.

This algorithm written in pseudo-code carries out an **in-order traversal**:

```
Set current node as root
Traverse
End
Define Procedure Traverse
  If there is a node to the left then
    Move left to child node
    Traverse
  End If
  Visit
  If there is a node to the right then
    Move right to child node
    Traverse
  End If
End Procedure
```

Here is how this algorithm traverses the binary tree in Figure 12.3.

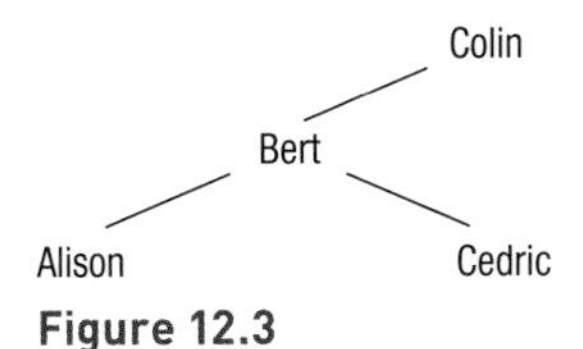

Figure 12.3

1. The root node is set as the current node ("Colin").
2. The procedure `Traverse` is called for the first time.
3. There is a node to the left of the current node so move to the node to the left so that we are now on the node containing "Bert".
4. The procedure `Traverse` is called for the second time. The details of the first call of `Traverse` are pushed on to the stack.
5. There is a node ("Alison") to the left of "Bert" so move left. Current node now becomes "Alison".

6 **`Traverse`** is called again. This time there is nothing to the left of the current node.
7 Visit the node – the term 'visit' is deliberately vague. It might mean 'print it out' or it might mean 'enter the person's date of birth' or any other process you want to carry out on each node.
8 Now we need to check if there is a node to the right of "Alison" but there is not, so move back up the branch to "Bert".
9 This call of **`Traverse`** has now been completed so the details of the previous call can be popped off the stack.
10 We jumped out of the second call of **`Traverse`** after the first question so we now visit the node "Bert".
11 Now we look to the right of "Bert". There is a node there ("Cedric") so we go to that node.
12 **`Traverse`** is called again so the details of the previous call of **`Traverse`** are placed on the stack.
13 Now we are at "Cedric" we look left, then visit then look right.
14 This call is now finished so back up the branch to "Bert".
15 We have now finished the call to "Bert" so that call of **`Traverse`** is also complete, so it's back up to "Colin".
16 Visit "Colin".
17 Try to go right, but there is nothing to go to.
18 Finish the first call to **`Traverse`**.

If you have followed this process through you should find that you have visited the nodes in alphabetical order – Alison, Bert, Cedric and finally Colin.

This looks and sounds like a very complex process, but in fact it is a very elegant solution to the problem. You must remember that although we have only traversed a tree with four nodes, the process would be exactly the same for a tree with 400 nodes. The only limitation is the number of calls of **`Traverse`** that the stack can handle.

Traversing a binary tree in pre-order follows the same routine but in this case you visit the root node as soon as you get to it. Traversing the tree given above in pre-order would result in Colin, Bert, Alison and Cedric. The only detail you would need to make the code carry out a pre-order traversal would be to move the visit to before the first **`If`** statement like this.

```
'Pre order traversal
Visit
If there is a node to the left then
  Move left to child node
  Traverse
End If
If there is a node to the right then
  Move right to child node
  Traverse
End If
```

A post-order traversal would result in the list Alison, Cedric, Bert and Colin. In this case you visit the node after you have tried to go both left and right from the node.

An interesting feature of all this is that no matter how you set out the four nodes, an in-order traversal will always produce a sorted list, but pre-order and post-order produce a different set of data if the data are rearranged. Typical uses of each traversal are:

- Pre-order: This can be used with expression trees to evaluate the expression using prefix notation. Evaluating an expression simply means that values are to be placed into the expression to produce a result. Prefix means that the operators in the expression are evaluated before the values.
- In-order: This is the equivalent of a **binary search** of the tree, which is explained in more detail in the next chapter.
- Post-order: This will produce Reverse Polish Notation (RPN) and this is covered in detail in Chapter 18. A post-order algorithm can also be used to empty the contents of a tree.

KEYWORD

Binary search: a technique for searching data that works by splitting datasets in half repeatedly until the search data is found.

Recursion

KEYWORD

Recursion: a technique where a function can call itself in order to complete a task.

Recursion is the process of calling a function from within itself. The concept is very elegant, but trying to understand how it works is rather more difficult. The algorithm described above that traverses a binary tree uses recursion. Each time a call is made the current state of the procedure must be stored on the stack.

The process to traverse a binary tree in order goes like this:

```
Define Procedure Traverse
  If there is a node to the left Then
    Go Left
    Traverse
  End If
  Visit
  If there is a node to the Right Then
    Go Right
    Traverse
  End If
End Procedure
```

After the procedure **`Traverse`** has been called for the first time the program will check to see if there is a node to the left. If there is it goes left then calls the procedure **`Traverse.`** This means that **`Traverse`** has been called from inside the procedure **`Traverse`**, and if the next node also has a node to its left then **`Traverse`** will be called from inside itself again.

Recursion has a base case and general cases. The base case is also known as the terminating case and defines the point at which the recursive function will stop calling itself. In the example above, the terminating case is when there are no more nodes left to visit in the tree. The general cases are all of the inputs which require the function to call itself. In the example above, **`Traverse`** will continue to call itself if there is a node either on the right or the left of the current node.

Practice questions can be found at the end of the section on page 132.

TASKS

1. Draw a binary tree for the following data: Rose, Jasmine, George, Naomi, Trevor and Stanley.
   a) List the nodes that will be visited in order to find the node that stores George.
   b) Traverse the tree in pre-order and write down the value at each node when you visit it.
   c) Repeat this process for a post-order traversal.
   d) Repeat this process for an in-order traversal.
2. Explain the term recursion and give an example where it might be used.
3. Write pseudo-code to show how the following graph could be traversed: depth first and breadth first.

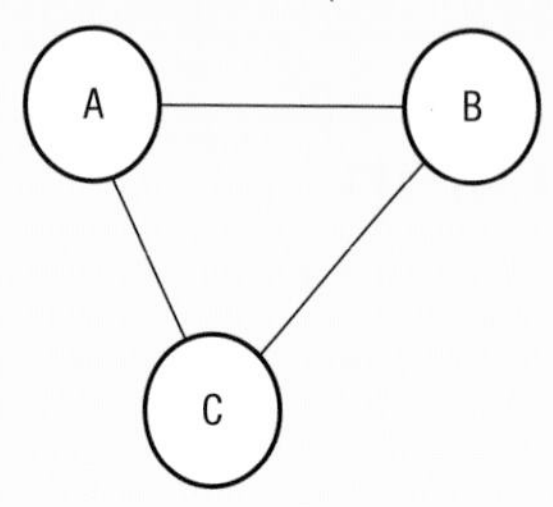

KEY POINTS

- Graphs can be represented using an adjacency list or matrix.
- Traversal is the process of visiting the vertices (nodes) in different orders to generate different results.
- Graphs can be traversed depth first or breadth first.
- Breadth first traversal finds the shortest path between vertices on unweighted graphs.
- Binary trees can be traversed in-order, post-order to pre-order, to create different outputs.
- Post-order traversal of a binary tree can be used to create Reverse Polish Notation.
- Recursion is where a function calls itself.

STUDY / RESEARCH TASKS

1. Write pseudo-code to traverse a binary tree for the following data: Rose, Jasmine, George, Naomi, Trevor and Stanley.
   a) List the nodes that will be visited in order to find the node that stores George.
   b) Traverse the tree in pre-order and write down the value at each node when you visit it.
   c) Repeat this process for a post-order traversal.
   d) Repeat this process for an in-order traversal.
2. Write code to implement and traverse the following graph.

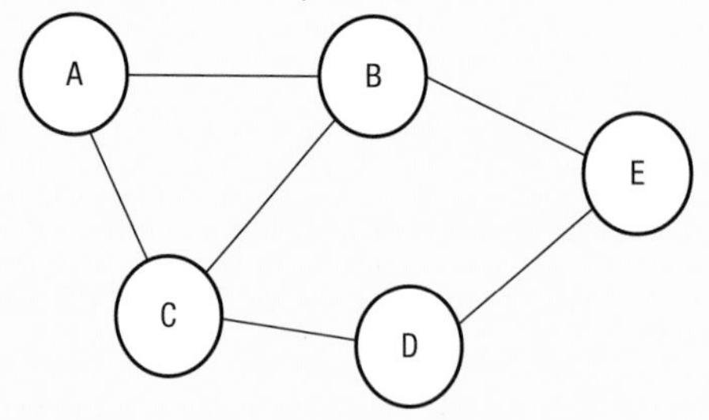

3. Turn your pseudo-code from question 1 above into an application using a high-level language of your choice.

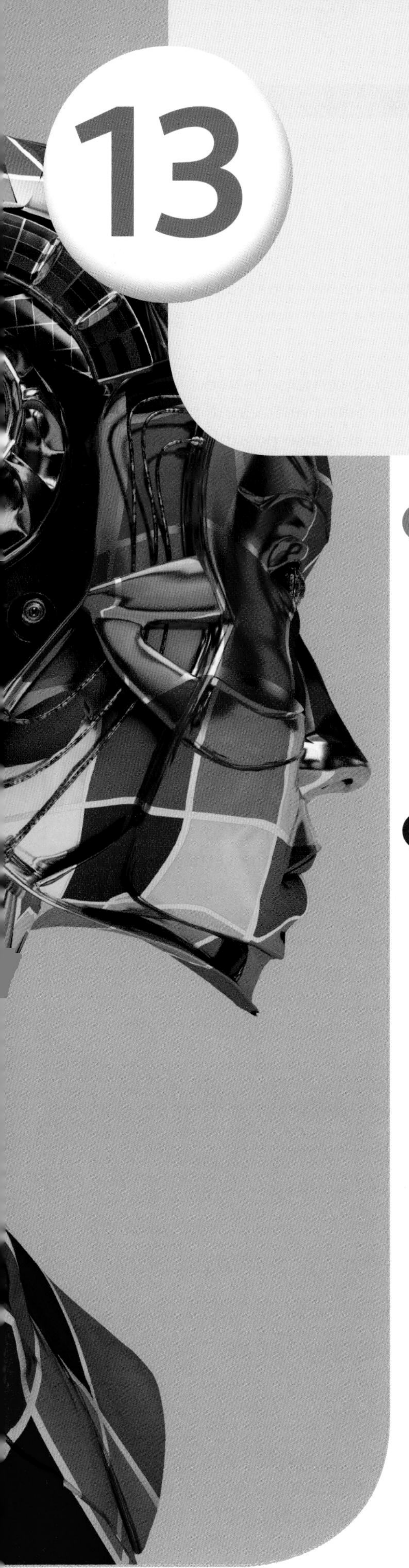

13 Dijkstra's shortest path algorithm

A level only

SPECIFICATION COVERAGE

3.3.6 Optimisation algorithms

LEARNING OBJECTIVES

In this chapter you will:

- consolidate your learning about graphs
- learn what Dijkstra's shortest path algorithm is and how it can be applied
- learn how to trace Dijkstra's shortest path algorithm
- learn how to implement Dijkstra's shortest path algorithm.

INTRODUCTION

Dijkstra's shortest path algorithm calculates the shortest distance between two vertices (nodes) on a graph data type. The algorithm can be used to find the shortest path from one node to all other nodes in a graph. It was devised by Dutch computer scientist Edsger Dijkstra and published in 1959. To understand the algorithm you must have a good understanding of the graph data type that we looked at in Chapter 10.

As an example, it can be used to solve problems like working out the shortest distance between two towns.

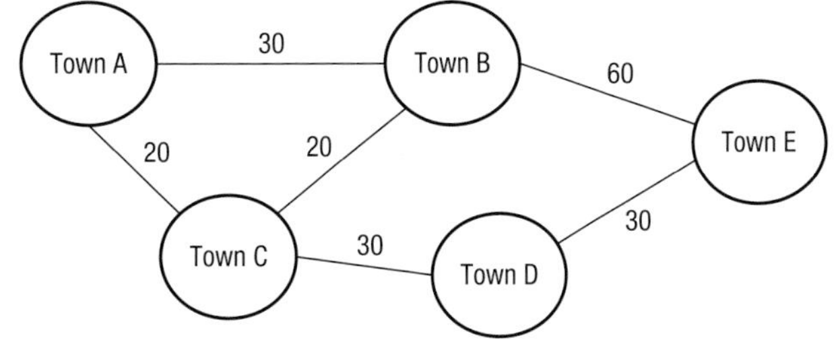

Figure 13.1 Graph to show journey time between towns

Consider the problem we looked at in Chapter 10 of working out the shortest distance between Town A and Town E. On a simple graph like this we could simply use trial and error to find the result. For example:

Possible routes	Distance	Total distance
A to B to E	30 + 60	90
A to C to B to E	20 + 20 + 60	100
A to C to D to E	20 + 30 + 30	80
A to B to C to D to E	30 + 20 + 30 + 30	110

The table shows all of the possible routes that we could take that do not involve circuits, and also shows the shortest path, which is to go from A to C to D and then to E.

As a quick reminder, graphs are made up of vertices (or nodes) and edges, which are the connections between them. Some vertices are not connected and therefore have no path between them. It is also possible to have weighted graphs as with the example above, where there is a value attached to each edge.

KEYWORD

Single source: in Dijkstra's algorithm it means that the shortest path is calculated from a single starting point.

Dijkstra's algorithm works by working out the shortest path between a **single source** (the starting vertex) and every other vertex. As a result it also produces the shortest path between the starting vertex and a destination vertex, as in the example above. It only works on weighted graphs with positive values on the edges.

Below are examples of some of the common applications that will require a shortest path algorithm. Dijkstra's algorithm is likely to be the basis of all of these.

- Geographic information systems (GIS) such as satellite navigation systems and mapping software where the vertices are geographical locations and the edges show distance or drive-time.
- Telephone and computer network planning where the vertices are the individual devices on the network and the edges could either be physical distance or a measurement of network capability, such as bandwidth.
- Network routing/packet switching: where the vertices are network devices and the edges are the connections between them. The algorithm can be used to send data packets along the most efficient route. In fact there is a routing protocol for TCP/IP networks called OSPF, which stands for open shortest path first.
- Logistics and scheduling: wherever there is a large network of vehicles, for example, delivery vehicles, buses or aeroplanes the algorithm can be used to calculate the optimum routes.

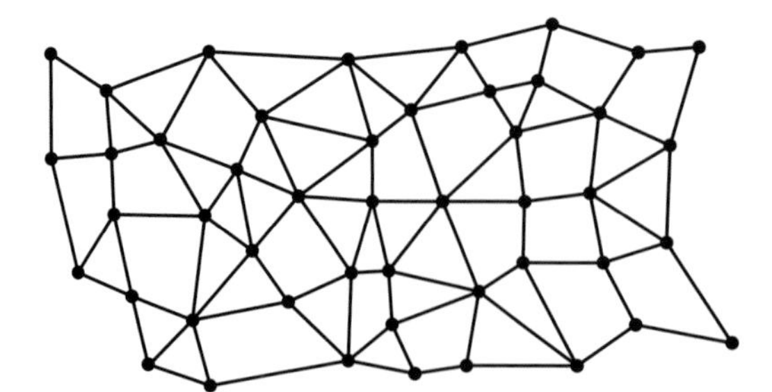

Figure 13.2 A graph with multiple vertices and edges

Figure 13.2 is a visualisation of the problem above showing any number of vertices and edges. As you can see, you could have a very large number of possible paths, making the trial and error system impractical.

KEYWORD

Shortest path: the shortest distance between two vertices based on the weighting of the edges.

Tracing Dijkstra's algorithm

The algorithm works as follows using Figure 13.3 as an example and assumes that we are looking for the **shortest path** between vertex A and G rather than the shortest path from A to every node.

1. Start from the first vertex (in this case A).
2. Work out the weight (also known as the cost) for each edge between that vertex and other connected vertices, e.g. A to B is 2 and A to C is 5.
3. Move on to the next nearest vertex and repeat the process. In this case it would be B. This time you need to add the two weights together to get a total weight between two points. For example:
   - A to B to C would be 6.
   - A to B to F would be 9.
4. We now have two options for getting from A to C:
   - We could go from A to C direct with a weight of 5.
   - We could go from A to B to C with a weight of 6.
5. As we are finding the shortest path, we now know that the quickest route from A to C is to go direct from A to C. We need to retain this information and ignore other routes that are longer.
6. Now repeat the process until all vertices have been visited and you get to the destination vertex, which in this case is G.

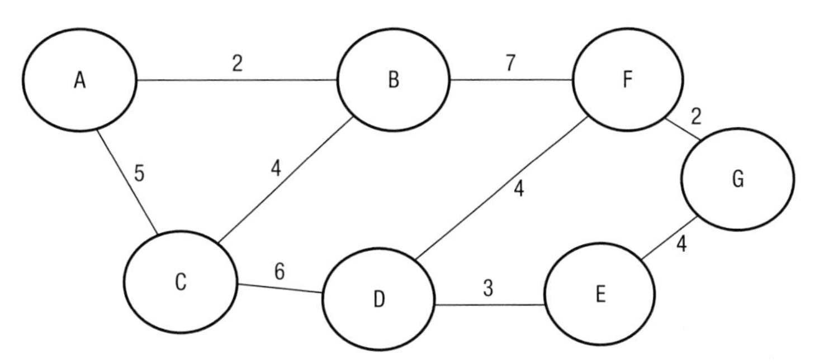

Figure 13.3 A map for tracing Dijkstra's algorithm

The calculations become a little more complicated as you need to keep an accumulated total of the weights between all sets of connected vertices, and then choose the shortest one. Table 13.1 traces each stage of the algorithm and we will work through the table a step at a time.

Table 13.1

Step	Vertex	A	B	C	D	E	F	G
1	A	0_A	2_A	5_A	∞_A	∞_A	∞_A	∞_A
2	B	0_A	2_A	5_A	∞_B	∞_B	9_B	∞_B
3	C	0_A	2_A	5_A	11_C	∞_C	9_B	∞_C
4	F	0_A	2_A	5_A	11_C	∞_F	9_B	11_F
5		0_A	2_A	5_A	11_C	∞_F	9_B	11_F

Step 1

- Place A in the first column and complete the distance between it and the other vertices.
- Notice that A to A is shown as 0. A to B is 2, A to C is 5. These are all shown with the subscript A to show clearly which vertex is being used. This becomes important later on.
- Notice that where there is no edge, the value is shown as infinite.
- We have now finished with the vertex A as there are no other edges.

Step 2

- Now move onto the next nearest vertex to A, which is B as it has the lowest value in the row above. Notice that the same value is placed in the table for B as in the row above. This is because we already know that this is the shortest path from A to B. In this case it is 2.
- The subscript A shows us clearly that the shortest path came from vertex A.
- The next path is B to C. This would be 4. However, we need to add on the 2, which is the shortest path that it took to get from A to B in the first place. This would give us a result of 6. However, this is higher than the path we already have between A and C, so we do not include it. Instead we keep the 5 from the row above. In other words, going from A to C direct is a shorter path than going from A to B to C.
- As you move through the rows you always keep the lowest value from the row above as this is the shortest path to that point.

Step 3

- Now move onto the next nearest vertex to B, which is C.
- Notice we can complete the table for the vertices that we have already visited and finished with in red. This makes it clear that the vertices have been dealt with and that we do not need to calculate them again.
- The next edge is between C and D. It has a weight of 6, but we have to add the shortest weight that it took to get to C in the first place, which you can see from the row above is 5. Therefore we put 11 in the table for the distance from C to D with a subscript C to show which vertex we came from.
- C is not connected to any other vertex that has not already been explored, so a standard way of showing this is to put the infinity sign in the table against the other vertices.
- Notice that we had a connection between A and F (via B) of only 9, so this stays in the column. This is because A to B to F is shorter than A to C to D to F.

Step 4

- Now move on to the next nearest vertex, which is F (with a weight of 9).
- Complete the table in red up to that point as before to show that we have finished with those vertices.
- Notice that we have been able to skip D and E as we already know that these will not produce the shortest path as the distance to D is equal to the shortest distance found to G so far. The algorithm will however have to calculate these distances first before it can carry out the next step.
- F connects to G in 2, but you have to add on the shortest path to this point, which is 9 making a total of 11.

Step 5

- There are no more edges to be compared so this final step simply lists the final values.
- Reading off the last row of the table you can see that the shortest path between A and G is 11 and looking at the subscript letters you can see that the route is A, B, F, G.

You can check this by looking at alternative routes and working out the total weight. The two other possible paths in this example are:

- A, C, D, E, G with a total weight of 18
- A, C, B, F, G with a total weight of 18.

Implementing Dijkstra's shortest path algorithm

The values for a weighted graph with eight vertices could be represented as a two-dimensional array as follows:

Table 13.2 A two-dimensional array containing details for a graph

	A	B	C	D	E	F	G
A	0	4	3	7	0	0	0
B	4	0	0	1	0	5	0
C	3	0	0	3	5	0	0
D	7	1	3	0	2	2	7
E	0	0	5	2	0	0	2
F	0	5	0	2	0	0	5
G	0	0	0	7	2	5	0

This would produce the following graph:

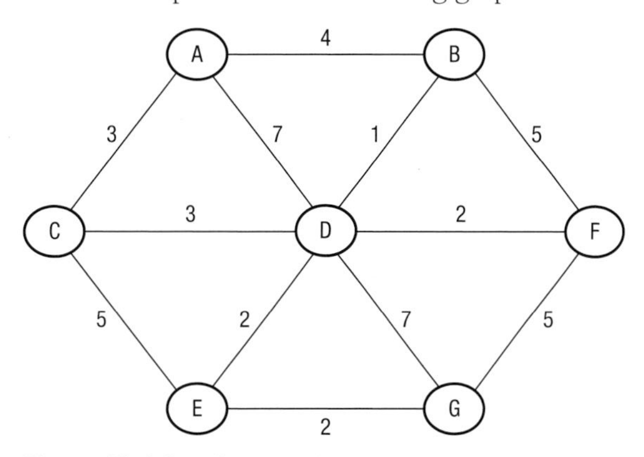

Figure 13.4 Graph created from the two-dimensional array in Table 13.2

This code reads in the data from an array stored in a csv file. It uses recursion to visit each node and mark it as visited, recording the shortest path between each. This means that it is able to produce the shortest path between any two vertices visited as well as provide a shortest path between the starting vertex A and the destination vertex G.

```
Public MinDist(8) As Integer
Public NodeFixed(8) As Boolean
Public Route(8) As String
Public ThisNode As Integer
Public ThisMin As Integer
Public ThisRoute As String
Public DistToThisNode As Integer
Public NodeCount As Integer
' generix 'load data from file routine'
Private Sub frmGraph_Load(ByVal sender As System.
Object, ByVal e As System.EventArgs) Handles
MyBase.Load
  Dim DataRow As String
  ' count nodes
  NodeCount = 7
  ' populate arrays
    FileOpen(1, "C:\NodeTable.csv", OpenMode.
    Input)
  DataRow = LineInput(1)
  For Loop1 = 1 To NodeCount
    dgvRoutes.Rows.Add()
    dgvLowestValue.Rows.Add()
    dgvLowestValue.Rows(Loop1 - 1).Cells(0).Value =
    999
    MinDist(Loop1) = 999
    Input(1, DataRow)
    NodeName(Loop1) = DataRow
    dgvRoutes.Rows(Loop1 - 1).HeaderCell.Value =
    DataRow
    dgvLowestValue.Rows(Loop1 - 1).HeaderCell.Value =
    DataRow
    For Loop2 = 1 To NodeCount
      Input(1, DataRow)
      RouteMatrix(Loop1, Loop2) = DataRow
      dgvRoutes.Rows(Loop1 - 1).Cells(Loop2 -
      1).Value = DataRow
    Next
  Next
End Sub
```

```
Private Sub btnFind_Click(ByVal sender As System.
Object, ByVal e As System.EventArgs) Handles
btnFind.Click
  ' reset tracking table
  For Loop1 = 1 To NodeCount
    dgvLowestValue.Rows(Loop1 - 1).Cells(0).Value =
    999
    dgvLowestValue.Rows(Loop1 - 1).Cells(1).Value =
    False
    Route(Loop1) = "A"
    NodeFixed(Loop1) = False
    MinDist(Loop1) = 999
  Next
  ' start recursive process
  MinDist(1) = 0
  CurrentNode(1)
End Sub
Private Sub CurrentNode(ByVal ThisNode)
  ' theoretically all distances start as infinity
  ' but infinity is not a concept a computer can
  ' cope with so I have used a value of 999
  DistToThisNode = MinDist(ThisNode)
  NodeFixed(ThisNode) = True
  ThisRoute = "" & Route(ThisNode)
  ' calculate distance using this node
  ' check all the nodes
  For Loop1 = 1 To NodeCount
    ' has node been fixed?
    If NodeFixed(Loop1) = False Then
      ' is node connected to ThisNode?
      If RouteMatrix(ThisNode, Loop1) <> 0 Then
        ' is potential distance shorter?
        If MinDist(Loop1) > DistToThisNode +
        RouteMatrix(ThisNode, Loop1) Then
          MinDist(Loop1) = DistToThisNode +
          RouteMatrix(ThisNode, Loop1)
          Route(Loop1) = ThisRoute &
          NodeName(Loop1)
          ' update display
```

```
                dgvLowestValue.Rows(Loop1 - 1).Cells(0).
                Value = MinDist(Loop1)
                dgvLowestValue.Rows(Loop1 - 1).Cells(1).
                Value = NodeFixed(Loop1)
                dgvLowestValue.Rows(Loop1 - 1).Cells(2).
                Value = Route(Loop1)
              End If
            End If
          End If
        Next
        'find shortest distance leading to am unfixed node
        ThisMin = 999
        For Loop1 = 1 To NodeCount
          ' update display to show progress
          dgvLowestValue.Rows(Loop1 - 1).Cells(1).Value =
          NodeFixed(Loop1)
          ' is node fixed?
          If NodeFixed(Loop1) = False Then
            ' is this the shortest unfixed node?
            If MinDist(Loop1) < ThisMin Then
              ' then record which node it leads to
              ThisNode = Loop1
              ThisMin = MinDist(Loop1)
            End If
          End If
        Next
        MsgBox("Current node is " & NodeName(ThisNode) &
        vbCrLf & "click to move on", 0, "")
        ' if ThisMin is still 999 then all nodes are fixed
        If ThisMin <> 999 Then
          CurrentNode(ThisNode)
        End If
      End Sub
    End Class
```

Figure 13.5 is a screenshot from the program that shows the process being tracked after a vertex has been visited.

Dijkstra's Algorithm

Route Table

	A	B	C	D	E	F	G
A	0	4	3	7	0	0	0
B	4	0	0	1	0	5	0
C	3	0	0	3	5	0	0
D	7	1	3	0	2	2	7
E	0	0	5	2	0	0	2
F	0	5	0	2	0	0	5
G	0	0	0	7	2	5	0
*							

Find Route

Theoretically all distances start as infinity but infinity is not a concept a computer can cope with so a value of 999 has been used instead.

Tracking Progress

	Lowest	locked	Route
A	999	True	
B	4	True	AB
C	3	True	AC
D	5	True	ABD
E	7	True	ABDE
F	7	True	ABDF
G	9	True	ABDEG
*			

Figure 13.5

Practice questions can be found at the end of the section on page 132.

TASKS

1 Draw the graph that would be produced from the following array.

	A	B	C	D	E
A	0	3	2	0	0
B	3	0	0	3	0
C	2	4	0	4	2
D	0	3	4	0	4
E	0	0	2	4	0

2 Write the array that would be needed to create the following graph.

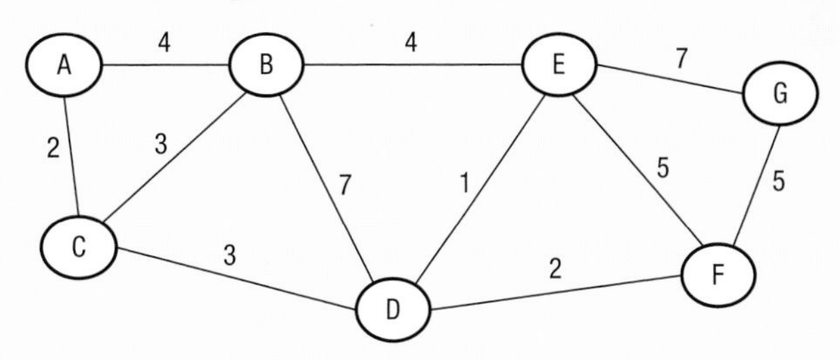

3 Using this graph, trace Dijkstra's algorithm to show the shortest path between A and G.

STUDY / RESEARCH TASKS

1 Explain why Dijkstra's shortest path algorithm would not solve 'the travelling salesman problem'.

2 What is meant by a greedy algorithm?

3 Explain the time complexity of Dijkstra's algorithm using Big O notation. (You may need to refer to Chapter 22 for details on Big O notation.)

4 Research the open shortest path first protocol (OSFP).

KEY POINTS

- Dijkstra's shortest path algorithm calculates the shortest distance between two vertices (nodes) on a graph data type.
- Graphs are made up of vertices (or nodes) and edges, which are the connections between them.
- Dijkstra's shortest path algorithm only works on weighted graphs with positive values on the edges.
- Dijkstra's shortest path algorithm can be implemented using the values of a two-dimensional array.

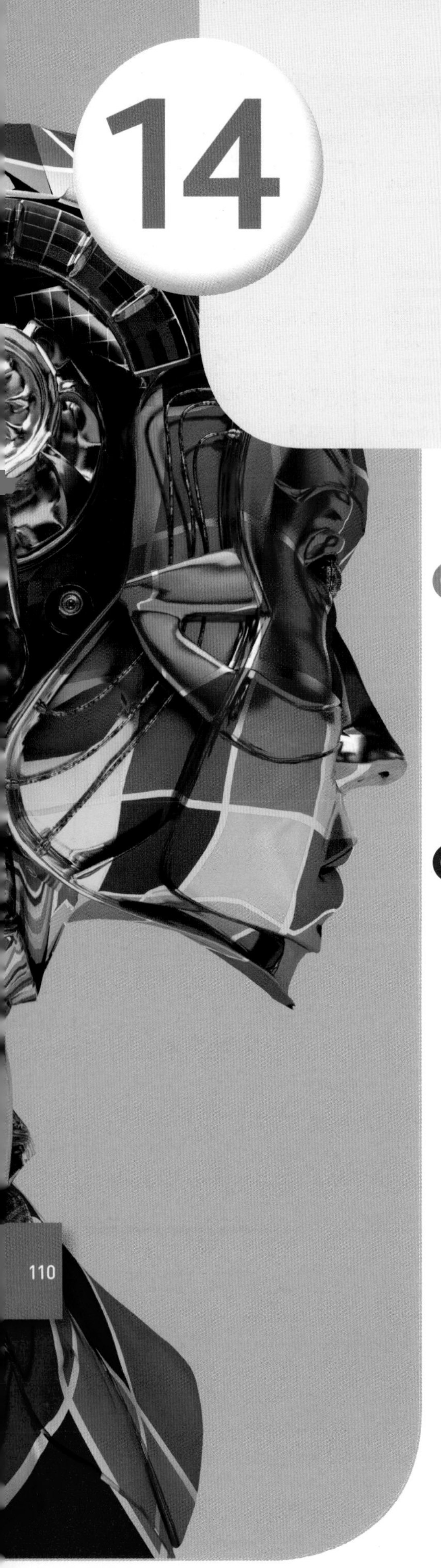

14 Search algorithms – binary, binary tree and linear search

A level only

SPECIFICATION COVERAGE

3.3.4 Searching algorithms

LEARNING OBJECTIVES

In this chapter you will learn:

- what a linear search is and how it could be implemented
- what a binary search is and how it could be implemented
- what a binary tree search is and how it could be implemented
- to compare the efficiency of the different search methods.

INTRODUCTION

One of the main benefits of using a computer is the ability to search. Consider how many everyday activities involve searching, for example:

- a cash machine searches for your bank account details to find how much (or little) money you have left in the account
- the computerised till at your local supermarket searches for the cost of the goods you are buying
- a search engine on the Internet looks for the cheapest holiday to the Algarve.

Most searches are carried out on data storage systems, but they are used in other applications as well, for example, in the find and replace process on a word processor. A simple search might just look for one keyword, but most search routines allow you to construct more complex queries using logic statements such as OR, AND and NOT.

There are a number of different searching algorithms that can be used. Which one you choose depends to a large extent on the data being searched in terms of its size and complexity. The efficiency of algorithms is usually represented using Big O notation and there is more on this in Chapter 22.

KEYWORD

Linear search: a simple search technique that looks through data one item at a time until the search term is found.

Linear search

A **linear search** works by looking at every item or set of data until the details that you are searching for are found or you fail to find it altogether. The efficiency of a search can be strongly influenced by the way that the data is organised. If there is no logical or rational method in the way the data has been stored then the only option is to use a linear search. This is the simplest but also the least efficient method.

You might use a linear search to find a book on a bookshelf – you know it is there somewhere but unless the books are organised in some way, say by title or author, then you will have to check every title until you find the one you want. A search might be coded something like this:

```
Repeat
  Look at the Title
Until Title is the one you want OR there are no more books
```

Figure 14.1 Unsorted books

The efficiency of the search also depends on the size of the dataset being searched and where the search item is within it. The best-case scenario is that it is near the beginning in which case it could find the result quickly. However, in the worst-case scenario the search item may be near the end of the dataset in which case it could take a long time. The speed of the algorithm therefore is proportionate to the size of the dataset.

Below is a section of commented code from Visual Basic showing a linear search. It is looking for a text string defined by `txtFind.Text`. Note that it also carries out a count to work out how many 'looks' it has to do to find the data:

```
Private Sub btnSearch_Click(ByVal sender As System.Object, ByVal e As System.EventArgs) Handles btnSearch.Click
    Dim CountLinear As Integer = 0
    txtTraceBinary.Text = ""
    txtTraceLinear.text = ""
    ' linear search
    Do
      CountLinear = CountLinear + 1
```

```
      txtTraceLinear.Text = txtTraceLinear.Text &
      CountLinear & "-" & grdTable.Rows(CountLinear).
      Cells(0).Value & vbCrLf
    Loop Until CountLinear = RowCount Or grdTable.
    Rows(CountLinear).Cells(0).Value = txtFind.Text
    txtLinearLooks.Text = CountLinear
    ' match found?
    If grdTable.Rows(CountLinear).Cells(0).Value =
    txtFind.Text Then
      lblResult.Text = "Match Found"
    Else
      lblResult.Text = "No Match Found"
    End If
  End Sub
End Class
```

Binary search

> **KEYWORD**
>
> **Binary search**: a technique for searching data that works by splitting datasets in half repeatedly until the search data is found.

If the data you want to look through is in some sort of logical order then you might be able to use a technique called a **binary search**. This method works in the same way as the children's game where someone thinks of a number between say 1 and 100 and you have to guess what it is by being told if your guesses are higher or lower than the number.

A logical person would start with 50, because they could then discount half of the numbers straight away. Guessing half way into the middle of the remaining numbers (either 25 or 75) will allow half of the remaining numbers to be discarded and so on. Each time you make a guess you halve the number of options that are left to you, and you alter the range within which the answer must be.

These 15 cells contain 15 numbers arranged in ascending order:

1	2	3	4	5	6	7	8	9	10	11	12	13	14	15

Figure 14.2

Use this method to find the number 51 which is contained in one of these cells. Start in the middle – block 8.

1	2	3	4	5	6	7	8	9	10	11	12	13	14	15
x	x	x	x	x	x	x	37							

Figure 14.3

Block 8 contains the number 37, so blocks 1 to 8 can now be discarded. Half way between 9 and 15 is 12 so look there next.

1	2	3	4	5	6	7	8	9	10	11	12	13	14	15
x	x	x	x	x	x	x	37				57	x	x	x

Figure 14.4

Block 12 contains the number 57 so blocks 12 to 15 can be discarded. Half way between blocks 9 and 11 is block 10 so look there.

1	2	3	4	5	6	7	8	9	10	11	12	13	14	15
x	x	x	x	x	x	x	37		51		57	x	x	x

Figure 14.5

Block 10 contains the number we are looking for. This has taken three 'looks' to find the missing number.

This pseudo-code shows how you might set out the process in a program. In this case the record number that needs to be found is stored in a variable called `FindMe`.

```
FindMe stores the record title that we are searching
for
LowestPointer ← 1
HighestPointer ← NumberofRecords
Do
  MiddlePointer ← (LowestPointer + HighestPointer) / 2
    If Record at MiddlePointer < FindMe Then
      LowestPointer ← MiddlePointer + 1
    End If
  If Record at MiddlePointer > FindMe Then
    HighestPointer ← MiddlePointer - 1
  End If
Until Record at MiddlePointer = FindMe OR
LowestPointer = HighestPointer
```

The pointers `LowestPointer` and `HighestPointer` point to the first and last locations in the file where the record you are looking for might be located. The pointer `MiddlePointer` stores the number roughly half way between the two extremes.

At first this seems like a very slow system, but in fact it is very efficient. If you want to search through just three records it will take a maximum of two 'looks' before you find a match and with seven records you will need three 'looks' and so on. If you have one million records you would need to take a maximum of just 20 'looks', and it would take a maximum of 33 looks to find one person in the world which currently has a population of over six billion.

Below is a section of commented code from Visual Basic showing a binary search. It is looking for a text string defined by **txtFind.Text**. Note that it also carries out a count to work out how many 'looks' it has to do to find the data.

```
Private Sub btnSearch_Click(ByVal sender As
System.Object, ByVal e As System.EventArgs)
Handles btnSearch.Click
  Dim MinNode As Integer = 0
  Dim MaxNode As Integer = RowCount
  Dim LookNode As Integer
  Dim LastNode As Integer
  Dim CountBinary As Integer = 0
  txtTraceBinary.Text = ""
  txtTraceLinear.text = ""
  ' binary search
  Do
    CountBinary = CountBinary + 1
    LastNode = LookNode
    ' calculate midpoint of remaining nodes
    LookNode = Int(MinNode + MaxNode) / 2
    ' determine which half of remaining nodes to
    discard
    If grdTable.Rows(LookNode).Cells(0).Value >
    txtFind.Text Then
      MaxNode = LookNode
    Else
      MinNode = LookNode
    End If
    txtTraceBinary.Text = txtTraceBinary.Text
    & LookNode & "-" & grdTable.Rows(LookNode).
    Cells(0).Value & vbCrLf
  Loop Until grdTable.Rows(LookNode).Cells(0).Value
  = txtFind.Text Or LastNode = LookNode
  txtBinaryLooks.Text = CountBinary
```

Binary tree search

Binary trees are often used in programs where data is very dynamic, which means that data is constantly entering and leaving the tree. Where a binary tree has been used the process of searching it is similar to the binary search method described above except that rather than looking through a list of data items, it must traverse the tree and look at the data item stored at each node. In this routine the variable **FindMe** contains the name we are looking for within the data stored in the tree and you will remember that the 'Root node' is the node the tree is built from.

```
CurrentNode ← RootNode
Repeat
  If Current_Node > Find_Me then
    Move left to child node
  Else
    Move right to child node
  End If
Until CurrentNode equals FindMe Or
CurrentNode has no children
```

KEYWORD

Binary tree search: a technique for searching a binary tree that traverses the tree until the search term is found.

The following section of commented code shows how the **binary tree search** is carried out. `txtSearchFor.Text` is the name of the variable that will hold the text strings being searched for.

```
Private Sub txtSearchFor_KeyDown(ByVal sender As
Object, ByVal e As System.Windows.Forms.KeyEventArgs)
Handles txtSearchFor.KeyDown
    ' check for enter key press
    If e.KeyCode = Keys.Enter Then
      Dim ThisNode As Integer = 1
      txtOutput.Text = "Root - " & NodeValue(1) &
      vbCrLf
      Do Until NodeValue(ThisNode) = txtSearchFor.
      Text Or ThisNode = 0
        ' move to node at left pointer
        If txtSearchFor.Text < NodeValue(ThisNode)
        Then
          ThisNode = PointerLeft(ThisNode)
          txtOutput.Text = txtOutput.Text & "L - " &
          ThisNode & " - " & NodeValue(ThisNode) &
          vbCrLf
        End If
        ' move to node at right pointer
        If txtSearchFor.Text > NodeValue(ThisNode) Then
          ThisNode = PointerRight(ThisNode)
```

```
            txtOutput.Text = txtOutput.Text & "R - " &
            ThisNode & " - " & NodeValue(ThisNode) &
            vbCrLf
          End If
        Loop
        If NodeValue(ThisNode) = txtSearchFor.Text Then
          txtOutput.Text = txtOutput.Text & "FOUND"
        Else
          txtOutput.Text = txtOutput.Text & "NOT FOUND"
        End If
      End If
    End Sub
End Class
```

A binary tree search is similar to the in-order tree traversal that we looked at in the previous chapter.

Practice questions can be found at the end of the section on page 132.

KEY POINTS

- There are three main search algorithms: binary, binary tree and linear.
- A linear search starts at the beginning of the data and goes through each item until it finds the search item.
- A binary search works by splitting the data in half each time until it finds the search item.
- A binary tree search traverses a binary tree until it finds the search item.
- The selection of an appropriate search method depends on the how much data is being searched and how it is organised.
- Different search algorithms have different time complexities, meaning that some will be more efficient than others.

TASKS

1 Explain how the three main search algorithms work: linear, binary and binary tree search.

2 Explain the circumstances where you might use a binary search compared to a linear search.

3 Why might a programmer use a binary tree structure?

4 Why is a binary search considered to be more efficient than a linear search on large datasets?

STUDY / RESEARCH TASKS

1 Write code to implement, populate and search a binary tree.

2 Write code to carry out a linear search on a text string input by the user.

3 Write code to carry out a binary search on a set of numeric data.

4 Research other search techniques and the circumstances under which they might be used.

5 Find out about the Google search algorithm and explain its advantages and limitations compared to other web search algorithms.

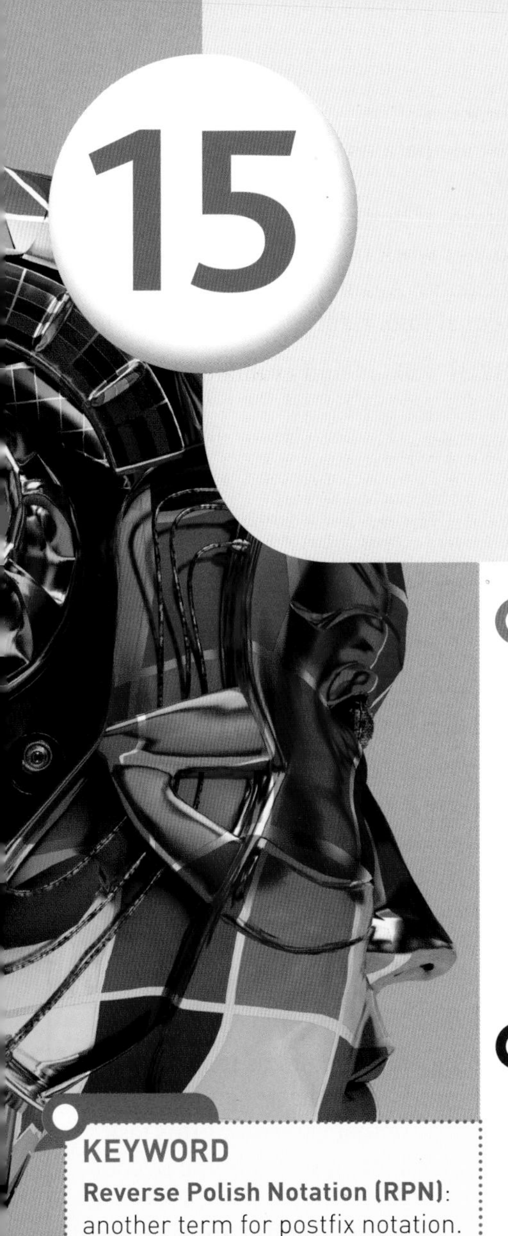

15 Reverse Polish Notation

A level only

SPECIFICATION COVERAGE

3.3.3 Reverse Polish

LEARNING OBJECTIVES

In this chapter you will learn:

- how to evaluate mathematical expressions
- the difference between infix, prefix and postfix expressions
- what Reverse Polish Notation (RPN) is and how it is used
- how to convert expressions from infix to postfix and vice versa
- how to trace an RPN algorithm
- how RPN can be implemented.

INTRODUCTION

Reverse Polish Notation (RPN) is a way of writing mathematical expressions in a format where the operators are written after the operands. For example, the expression: 5 + 3 becomes 5 3 +. The main advantages of this method are that it eliminates the need for brackets and it puts the expression in a sequence that is more convenient for an interpreter. To get a fuller understanding of RPN you need to know how mathematical expressions are constructed and the sequence in which they are evaluated.

KEYWORD

Reverse Polish Notation (RPN): another term for postfix notation.

Evaluating expressions

To start with a simple example, if we have the expression 5 + 3, we know to add the 3 to the 5 to create the result. This is known as an **infix** expression because the **operator** (+) is between the operands (5 and 3).

This gets slightly more complicated where the expression is longer. For example, 3 * 2 + 5 is another infix expression, which we would evaluate by multiplying 3 and 2 and then adding 5 to the result to get 11. We evaluate it in this way according to certain rules, which tell us which part of the expression to evaluate first.

Brackets (or parentheses) are often used in expressions to make these rules clearer. For example, (3 * 2) + 5 makes the sequence we must use much clearer. These rules are sometimes referred to as **BODMAS**, which means Brackets, Order, Division, Multiplication, Addition, Subtraction. This means:

KEYWORDS

Infix: expressions that are written with the operators within the operands, e.g. 2 + 3.

Operator: the mathematical process within an expression.

BODMAS: a methodology for evaluating mathematical expressions in a particular sequence.

- Evaluate the expression inside the brackets first.
- Then evaluate any orders, which are calculations like squares or square roots.
- Carry out any division or multiplication. If both appear in the expression then they have equal importance so you work from left to right.
- Then carry out any addition or subtraction. Again, if both operators appear in an expression, they have equal importance so work left to right.

If we had the infix expression: $3 + (18 / 3^2 * 3) - 4$ and evaluated it using these rules we would:

- Evaluate the expression in the brackets first:
  - Square the 3 to get 9.
  - Work out 18 / 9 to get 2.
  - Multiply 2 * 3 to get 6.
- Now we can carry out the addition 3 + 6 to get 9.
- Then subtract the 4 to get an answer of 5.

Polish and Reverse Polish Notation

KEYWORDS

Polish Notation: another way of describing prefix notation.

Interpreter: software that translates and executes programs line by line by converting programming statements either into machine code or by calling instructions to carry out the high-level language statements.

Operand: a value within an expression.

Polish Notation was invented by Polish mathematician Jan Lukasiewicz in 1924 and therefore pre-dates computers as we know them. It was developed as a way of simplifying mathematical expressions, eliminating the need for brackets completely, while still producing expressions without any ambiguity as to how they should be processed. In the 1950s, it was adapted to become Reverse Polish Notation (RPN) because it was evident that this way of writing expressions was a convenient format for an **interpreter** as it evaluates lines of programming code.

When code is written using a programming language, it has to be converted from that language into machine code (0s and 1s) so that it can be processed. The interpreter is a piece of software that carries out this task by parsing each line of code. This means that it analyses each line of code to check that it adheres to the rules of the language, known as the syntax. When parsing expressions, the interpreter analyses the **operands** first and then the operators. Therefore, it needs the operators to be on the right-hand side of the expression.

- Polish Notation (also known as prefix) is a method of rearranging an expression so that all of the operators are on the left and the operands are on the right. For example: 7 + 3 becomes + 7 3.
- Reverse Polish Notation rearranges an expression so that all the operators are on the right-hand side of the operands. So 7 + 3 becomes 7 3 +.

Converting expressions

KEYWORDS

Prefix: expressions that are written with the operators before the operands, e.g. + 2 3

Postfix: expressions that are written with the operators after the operands, e.g. 2 3 +

Notice that if you do change an infix notation to either **prefix** or **postfix**, you do not change the order of the operands within the expression. In the example above, the operands must appear in the order 7 followed by 3.

Where there are brackets in an expression, the same rule applies to RPN; you evaluate the expression in the brackets first. For example, the infix expression (5 + 4) / (4 – 1) would have an RPN of 5 4 + 4 1 – /.

- Notice how this is made up of two parts. The 5 + 4 is evaluated first and the RPN is created for this part of the expression: 5 4 +.
- The second part of the expression of / (4 – 1) is then evaluated and becomes 4 1 – /. Notice that the 4 – 1 is evaluated first as this is in brackets and the final operator is the divide, which will then divide the contents of the two bracketed expressions.
- Therefore, 5 + 4 = 9, 4 – 1 = 3 and 9 / 3 = 3.

Table 15.1 shows some more examples of expressions in infix format with their equivalent postfix notation.

Table 15.1 Table to show conversion from infix to postfix notation

Infix	Postfix	Result	Explanation
(6 * 3) – 1	6 3 * 1 –	17	Multiply 6 by 3 to get 18 and then minus 1
(6 * 2) / 3	6 2 * 3 /	4	Multiply 6 by 2 to get 12 and then divide by 3 to get 4
(4 * 3) / (6 * 2)	4 3 * 6 2 * /	1	Multiply 4 by 3 to get 12, then multiply 6 by 2 to get 12, divide the two answers to get 1.

It is possible to convert infix to postfix and vice versa. For example, the postfix notation 3 4 + would equate to an infix notation of 3 + 4. Similarly:

- The postfix expression 18 3 / 2 + would become the infix expression (18 / 3) + 2
- The postfix expression 20 5 / 6 2 + – would become the infix expression (20 / 5) – (6 + 2)

Evaluating RPN expressions

The most common method for evaluating postfix notation is to use a stack. Consider the infix expression (2 * 3) + 5. The postfix notation would be 2 3 * 5 +. To evaluate this using a stack:

1. Push 2 onto the stack.
2. Push 3 onto the stack.
3. Push * onto the stack.
4. As * is an operator (multiply) we need to pop this and all of the operands currently in the stack (2 and 3) and evaluate the expression 2 3 *
5. Push the answer (6) back onto the stack.
6. Push 5 onto the stack.
7. Push + onto the stack.
8. As + is an operator (plus) we need to pop 6 5 + and evaluate the expression.
9. Push the result (11) onto the stack.

The stack could be visualised during the process as shown in Figure 15.1.

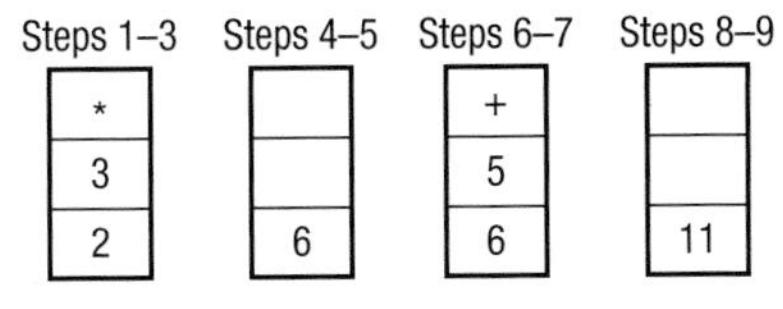

Figure 15.1 Representation of a stack implementing RPN

The following code shows how an expression in the format A operand B could be converted from infix to postfix notation. String-handling expressions are used to identify and extract each part of the expression that are stored temporarily before being rearranged into postfix notation.

```
Public Class frmConvert
    Public Source As String
    Public BracketL As String
    Public BracketR As String
    Public Expression As String
    Public Number1 As Integer
    Public Number2 As Integer
    Public Operand As String
    Public MainOperand As String
    Public Pointer As Integer
    Public AddToOutput As String
    Private Sub btnParse_Click(ByVal sender As System.
    Object, ByVal e As System.EventArgs) Handles
    btnParse.Click
        txtResult.Text = ""
        MainOperand = ""
        'store original expression in variable 'Source'
        Source = txtSource.Text
        Do
            ' remove trailing spaces
            Source = Trim(Source) + " "
            ' analyse next character in variable 'Source'
            Select Case Mid(Source, 1, 1)
                Case "("
                    ' extract part of main expression that is
                    in brackets
                    AddToOutput = Brackets(Source)
                Case "+", "/", "*", "-"
                    ' next character is an operand
                    ' store for now, add to end of expression
                    MainOperand = Mid(Source, 1, 1)
                    AddToOutput = ""
                    Source = Mid(Source, 3, 255)
                Case Else
```

```
            ' next character(s) in expression is
            numeric
            Pointer = InStr(Source, " ")
            Number1 = Mid(Source, 1, Pointer - 1)
            ' source contains remainder of the
            original expression
            Source = Mid(Source, Pointer + 1, 255)
            AddToOutput = Number1 & " "
        End Select
        txtResult.Text = txtResult.Text & AddToOutput
      Loop Until Len(Source) < 2
      txtResult.Text = txtResult.Text & MainOperand
   End Sub
   Private Function Brackets(ByVal SplitMe)
      ' extract contents of brackets
      BracketL = InStr(source, "(")
      BracketR = InStr(source, ")")
      Expression = Mid(source, BracketL + 1, BracketR
      - BracketL - 1)
      ' source holds the remainder of the original
      expression
      source = Mid(source, BracketR + 1, 255)
      'expression is presumed to be in form A operand B
      BracketL = InStr(Expression, " ")
      Number1 = Mid(Expression, 1, BracketL - 1)
      Operand = Mid(Expression, BracketL + 1, 1)
      Number2 = Mid(Expression, BracketL + 3, 255)
      'return RPN formatted expression
      Brackets = Number1 & " " & Number2 & " " &
      Operand & " "
   End Function
End Class
```

You may have noticed a similarity between the terminology used in this chapter and the terminology used to traverse a binary tree. In fact there is a direct relationship between the two:

- **In-order traversal** of a binary tree for an expression would produce an expression in infix format.
- **Post-order traversal** would produce an expression in postfix format or Reverse Polish Notation (RPN).
- **Pre-order traversal** would produce an expression in prefix format or Polish notation.

KEYWORDS

In-order traversal: a method of extracting data from a binary tree that will result in an infix expression.

Post-order traversal: a method of extracting data from a binary tree that will result in postfix expressions.

Pre-order traversal: a method of extracting data from a binary tree that will result in prefix expressions.

Working with the same example, a binary tree can be built using the postfix expression. In this case, 2 3 * 5 + could be represented in a binary tree as shown in Figure 15.2.

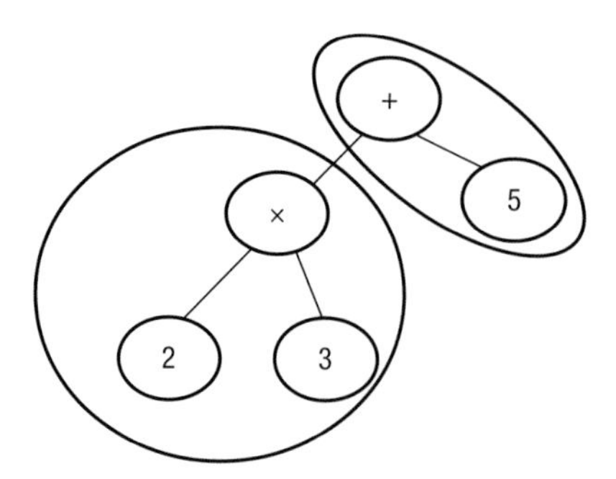

Figure 15.2 A binary tree showing the two parts of the mathematical expression

Note that the left subtree carries out the multiplication and the right subtree carries out the addition:

- A post-order traversal traverses the left subtree, traverses the right subtree and then visits the root, giving the result 2 3 * 5 +.
- An in-order traversal traverses the left subtree, visits the root and traverses the right subtree giving the result 2 * 3 + 5.
- A pre-order traversal visits the root, traverses the left subtree and then traverses the right subtree giving the result + * 2 3 5.

Applications of RPN

The code used in this chapter has been produced in Visual Basic, which is considered to be a general purpose imperative language. Some languages are specifically designed to be stack-oriented and would therefore be ideally suited to this application. In these cases, the interpreter or compiler checks all of the syntax with reference to the postfix (or RPN) notation of each expression. Perhaps the most common of these is PostScript, which is used to create **vector graphics**. This works by pushing operands onto the stack until an operator is pushed on. At that point it pops the operands off the stack with the operator, performs the calculation and pushes the answer back to the stack.

> **KEYWORD**
>
> **Vector graphics**: an image made up of objects and coordinates.

RPN is closer to the way in which computers actually carry out computations. You can look at infix as the way in which humans work, that is, we expect an operand followed by an operator. Postfix is the way in which processors work in that they are made up of a series of registers and units all of which carry out different functions. For example, one register will store values, while another (the arithmetic logic unit) carries out calculations. Therefore, it needs to know all the operands first so it can put them into the appropriate registers. At this point the processor needs to know which operators to use so it knows what to do with the operands.

As you know, there are many different types of programming languages. Some of these are high level, which means that the programmer can write in code that is similar to normal language. Others are low-level, which means that they are closer to the machine code (or 0s and 1s) that processors actually use. Some of these low-level languages such as bytecode use postfix notation .

Practice questions can be found at the end of the section on page 132.

TASKS

1 Convert the following expressions from infix to postfix (Reverse Polish) notation.
   a) 5 * 6 **b)** (5 * 4) – 3 **c)** (6 * 3) / (2 + 4)

2 Convert the following expressions from postfix to infix.
   a) 12 4 / 2 + **b)** 4 4 * 2 2 * + **c)** 24 6 / 3 2 + 2 /

3 Draw a binary tree for the expression (5 + 6) * 3.

4 What would be the result of the following traversals on the tree you made for question 3?
   a) in-order **b)** post-order **c)** pre-order

5 What is the purpose of Reverse Polish Notation?

6 Explain why infix notation is used by humans whereas postfix notation may be used by an interpreter or compiler.

KEY POINTS

- Reverse Polish Notation (RPN) is a way of writing mathematical expressions in a format where the operators are written after the operands.
- RPN is useful as it puts expressions in a format that can be used more straightforwardly by an interpreter.
- Infix refers to expressions that are in the order that humans work with, e.g. 5 + 3.
- Postfix refers to expressions that are in RPN, e.g. 5 3 +.
- Prefix refers to expression where the operators are first, e.g. + 5 3.
- RPN can be evaluated using a stack.

STUDY / RESEARCH TASKS

1 Research programming languages that use either prefix or postfix notation. Why do they use this particular form of notation?

2 Write code to convert:
   a) infix expressions to postfix
   b) postfix expressions to infix.

3 Find out how Java converts high-level code to bytecode using postfix notation.

4 Why was RPN used as a way of programming early calculators?

5 Explain how a stack can be used to convert an infix expression to a postfix expression.

16 Sorting algorithms–bubble and merge

A level only

SPECIFICATION COVERAGE

3.3.5 Sorting algorithms

LEARNING OBJECTIVES

In this chapter you will learn:

- what a bubble sort is and how to implement one
- what a merge sort is and how to implement one
- how to compare the efficiency of the two sorting methods.

INTRODUCTION

Sorting is one of the most common processes you would normally want to carry out on a set of data. Sorting simply means that the data are put into a particular order, typically alphabetical or numerical in either ascending or descending order.

There are lots of different ways of sorting data, and one of the skills that programmers need, is to decide which method suits their needs best. Some are particularly good when there are a lot of data to sort, others are particularly good when the data are almost, but not quite in the right order, and so on.

KEYWORD

Bubble sort: a technique for putting data in order by repeatedly stepping through an array, comparing adjacent elements and swapping them if necessary until the array is in order.

Bubble sort

If the data are held in an array you can sort the data by comparing each element in the array with the following element. If the first item is bigger than the second then you swap them. If you repeat this process enough times the data will eventually be sorted in ascending order.

In this example the data are stored in an array called **Storage** and the array holds **NumberOfRecords** records. The numbers at the start of each line are there to help with the explanation – they are not part of the algorithm.

```
1 For Loop1 = 1 To NumberOfRecords – 1
2   For Loop2 = 1 To NumberOfRecords – 1
3     If Storage(Loop2)> Storage(Loop2 + 1) Then
```

```
4         Temporary ← Storage(Loop2)
5         Storage(Loop2) ← Storage(Loop2 + 1)
6         Storage(Loop2 + 1 ) ← Temporary
7       End If
8     Next Loop2
9 Next Loop1
```

Suppose the array **Storage** had eight elements.

Element	1	2	3	4	5	6	7	8
Data	12	3	16	9	11	1	6	8

Figure 16.1

The algorithm would work like this:

- For now we will ignore lines 1 and 9 in the algorithm and start with lines 2 to 8.
- Lines 2 to 8 are a For/Next loop – a form of **iteration**. In this case the process is going to be repeated seven times. The instructions that are going to be repeated are lines 3 to 7.
- Line 3 compares each element in the array with its neighbour. So the first time the loop is processed **Loop2** contains the value 1 so the contents of **Storage(1)** is compared with the contents of **Storage(2)**. In this case these values would be 12 and 3, respectively.
- 12 is greater than 3 so lines 4, 5 and 6 are carried out and the values are swapped round to leave the array as shown in Figure 16.2.
- The value of **Loop2** is now incremented to 2 so the contents of **Storage(2)** is compared and swapped if necessary with the contents of **Storage(3)** and so on.
- This whole process of comparing and swapping carries on until all the elements in the array have been examined. At the end of the loop the array now looks like Figure 16.3.
- As you will have noticed this isn't very sorted yet. That is why lines 1 and 9 are there. They now repeat the process all over again until the array is finally sorted as shown in Figure 16.4.

KEYWORD

Iteration: repeating the same process several times in order to achieve a result.

1	2	3	4	5	6	7	8
3	12	16	9	11	1	6	8

Figure 16.2

1	2	3	4	5	6	7	8
3	12	9	11	1	6	8	16

Figure 16.3

1	2	3	4	5	6	7	8
1	3	6	8	9	11	12	16

Figure 16.4

This algorithm is called a bubble sort because each time the algorithm carries out one pass of the array the larger numbers are bubbling to one end of the array and the smaller ones to the opposite end.

This first example is actually very inefficient – it gets carried out regardless of whether it needs to be or not, and there is a lot of unnecessary work for the computer to do.

The second algorithm below carries out exactly the same process, but in a more sophisticated way. This time the algorithm uses a flag called **CompletedFlag** to record whether or not a swap has been made. If no swaps have been made then the data must be sorted so there is no point in carrying on.

```
Repeat
 CompletedFlag ← True
 For Counter = 1 To NumberOfRecords - 1
   If Storage(Counter) > Storage(Counter + 1) Then
     Temporary ← Storage(Counter)
     Storage(Counter) ← Storage(Counter + 1)
     Storage(Counter + 1) ← Temporary
     CompletedFlag ← False
   End If
 Next
Until CompletedFlag = True
```

The Visual Basic code below shows a bubble sort routine for text strings.

```
Private Sub btnSort_Click(ByVal sender As System.
Object, ByVal e As System.EventArgs) Handles
btnSort.Click
  Dim Loop1 As Integer
  Dim Loop2 As Integer
  Dim TempStore As String
  Dim RowsToSort As Integer
  RowsToSort = grdDataIn.RowCount - 2
  For Loop1 = 1 To RowsToSort - 1
    For Loop2 = 1 To RowsToSort - 1
    'compare each value in the table with the
    following value
    'changing the inequality will sort high to low
    If grdDataIn.Rows(Loop2).Cells(0).Value >
    grdDataIn.Rows(Loop2 + 1).Cells(0).Value Then
        'swap values to move larger values to later
        cells
        TempStore = grdDataIn.Rows(Loop2).Cells(0).
        Value
        grdDataIn.Rows(Loop2).Cells(0).Value =
        grdDataIn.Rows(Loop2 + 1).Cells(0).Value
        grdDataIn.Rows(Loop2 + 1).Cells(0).Value =
        TempStore
      End If
    Next
  Next
End Sub
```

KEYWORD

Merge sort: a technique for putting data in order by splitting lists into single elements and then merging them back together again.

Merge sort

A **merge sort** is classified as a 'divide and conquer' algorithm, which breaks a problem down into smaller and smaller units until it gets to a level where the problem can be solved. What this means in the case of the sort routine is that if you had a list with one element it is, by definition, sorted. Therefore, if you start with a large list of elements, all you need to do is break the list down into a series of smaller lists each containing one single element. You can then compare the lists and merge them back together to produce a sorted list.

The merge process works as follows. Assume you have two lists that are already sorted in order:

List 1	List 2
3	2
5	6
8	9
10	12

Figure 16.5

- Compare the first element of each list. In this case 3 would be compared to 2. Put the lowest number in the new merge list. In this case we move the 2. Our lists would now look like this:

List 1	List 2
3	~~2~~
5	6
8	9
10	12

Merge list = 2

Figure 16.6

- Repeat the process comparing the first element in each list and placing the lowest item in the merge list. We now have 3 compared to 6, so our lists will now look like this:

List 1	List 2
~~3~~	~~2~~
5	6
8	9
10	12

Merge list = 2, 3

Figure 16.7

- Repeat this process until there is only one element left and put this at the end of the list. You now have one list containing the sorted elements.

To sort an unordered list, you first need to break the list down. For example, if we have a list with eight elements as shown:

5	3	8	10	9	2	6	12

Figure 16.8

- Split the list into half.

5	3	8	10

9	2	6	12

Figure 16.9

- Keep splitting the list in half until each list only has one element:

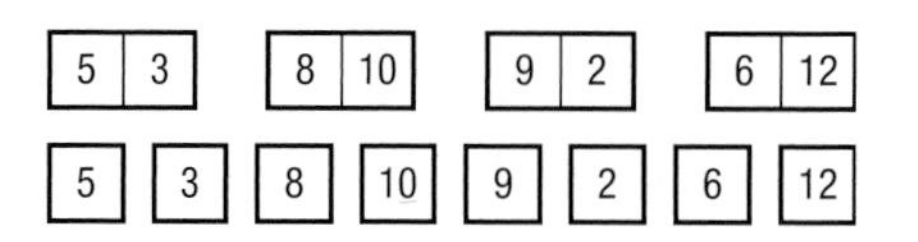

Figure 16.10

You now effectively have eight lists, all containing one element. We need to merge a pair of lists at a time until we have one complete list.

- Compare the first two lists, which are 5 and 3 and put the lowest number first. We get:

3	5

Figure 16.11

- Compare the next two lists, which are 8 and 10 and put the lowest number first. We get:

8	10

Figure 16.12

- Do this for the next two pairs of lists:

2	9

6	12

Figure 16.13

We now have four lists:

3	5

8	10

2	9

6	12

Figure 16.14

- Repeat the process merging these lists together. Start by comparing the first element in each list and putting the lowest first as shown earlier. For the first pair of lists:
  - Comparing 3 and 8, we would put 3 in the merge list.
  - Comparing 5 and 8 we would put 5 in the merge list.
- We then merge the 8 and the 10, which we know are already in the right order to get:

3	5	8	10

Figure 16.15

- Repeat this process for the other two lists and you get:

2	6	9	12

Figure 16.16

- Now merge these two lists together in the same way to get:

2	3	5	6	8	9	10	12

Figure 16.17

This is an efficient method of sorting where there are lots of elements in the original list. This is because the algorithm works by halving the lists each time. However, in terms of space, the merge sort will require more space than a bubble sort to create the intermediary lists and the final merge list. There is more on the efficiency of algorithms in Chapter 23.

As you have probably worked out, you can use a loop to split down the elements as many times as required to create single-element lists. Each pair of lists can then be compared and merged as many times as needed to reconstruct the list in the correct order. The following code shows the process in Visual Basic.

```
Public Sub MergeSort(ByVal ptrFirst As Integer,
ByVal ptrLast As Integer)
  Dim ptrMiddle As Integer
  If (ptrLast > ptrFirst) Then
   ' split list in half and carry out recursive call
   ptrMiddle = (ptrFirst + ptrLast) \ 2
   MergeSort(ptrFirst, ptrMiddle)
   MergeSort(ptrMiddle + 1, ptrLast)
   ' Merge the results.
   Merge(ptrFirst, ptrMiddle, ptrLast)
  End If
End Sub
' Merge two sorted sublists.
Public Sub Merge(ByVal beginning As Integer, ByVal
ptrMiddle As Integer, ByVal ending As Integer)
  ReDim TempStore(RowCount)
  Dim CountLeft As Integer
  Dim CountRight As Integer
  Dim counterMain As Integer
  ' Copy the array into a temporary array
  For LoopCount = 1 To RowCount
    TempStore(LoopCount) = DataStore(LoopCount)
  Next
  ' CountLeft and CountRight point to next item to
  save from left / right halves of the list
  CountLeft = beginning
  CountRight = ptrMiddle + 1
  ' counterMain is the index where we will put the
  next item in the merged list.
  counterMain = beginning
  Do While (CountLeft <= ptrMiddle) And (CountRight
  <= ending)
   ' Find the smaller of the two data at the top of
   the left and right lists
```

```
    If (TempStore(CountLeft) <= TempStore(CountRight))
    Then
        ' smaller value is in left half
        DataStore(counterMain) = TempStore(CountLeft)
        CountLeft = CountLeft + 1
    Else
        ' smaller value is in right half
        DataStore(counterMain) = TempStore(CountRight)
        CountRight = CountRight + 1
    End If
    counterMain = counterMain + 1
  Loop
  ' copy any data from the end of the list
  If CountLeft <= ptrMiddle Then
    ' copy from left half
    For LoopCount = 1 To ptrMiddle - CountLeft + 1
      DataStore(counterMain + LoopCount - 1) =
      TempStore(CountLeft + LoopCount - 1)
    Next
  Else
    ' copy from right half
    For LoopCount = 1 To ending - CountRight + 1
      DataStore(counterMain + LoopCount - 1) =
      TempStore(CountRight + LoopCount - 1)
    Next
  End If
End Sub
```

Notice that the code uses recursion, where a subroutine calls itself. In this example the MergeSort subroutine calls itself.

Practice questions can be found at the end of the section on page 132.

KEY POINTS

- Sorting means that the data is put into a particular order, typically alphabetical or numerical in either ascending or descending order.
- There are different algorithms that can be used to sort data.
- If the data is held in an array you can sort the data by comparing each element in the array with the data in the following element.
- A merge sort is classified as a 'divide and conquer' algorithm, which breaks a problem down into smaller and smaller units until it gets to a level where the problem can be solved.

TASKS

1 Explain how you could use a bubble sort and a merge sort to sort the following list of data:

12, 3, 4, 8, 2, 6, 10, 5

2 Explain why a bubble sort requires data to be stored in an array.

3 Would a bubble sort or merge sort be the quickest way of sorting this list? Explain your answer.

4 What if there were one million items in a list? Which would be the quickest then? Explain you answer.

5 Bubble sort uses 'iteration' and a merge sort uses 'recursion'. What do these terms mean and why are they needed?

STUDY / RESEARCH TASKS

1 Write code to carry out a bubble sort and a merge sort on a list of characters.

2 Research other algorithms that could be used to sort data. Explain the circumstances where you might use one rather than the other.

Section Three: Practice questions

1 Two methods for searching a dataset are a binary search and a linear search.

a) Write pseudo-code to show how a binary search works.

b) Explain how efficient this method is on a large ordered set of data, compared to a linear search.

2 The following graph shows the distance between five towns.

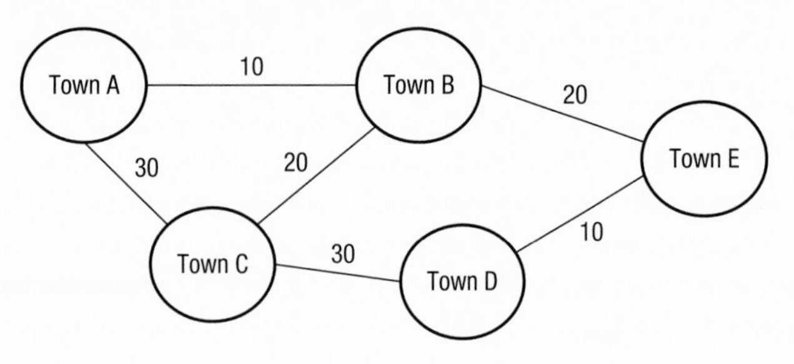

a) Show this as an adjacency matrix.

b) Show this as an adjacency list.

3 Convert the following infix expressions into Reverse Polish Notation (RPN).

a) 5 + 6

b) 4 * 7 + 12

4 Explain one advantage of using RPN over infix notation.

5 A binary search tree is used to store data arriving in this sequence: d, c, a, f, b, g.

a) Draw a binary search tree to show how this could be implemented.

b) Write pseudo-code to add data to this binary search tree.

c) Write pseudo-code to carry out a traversal that will put this data into alphabetical order.

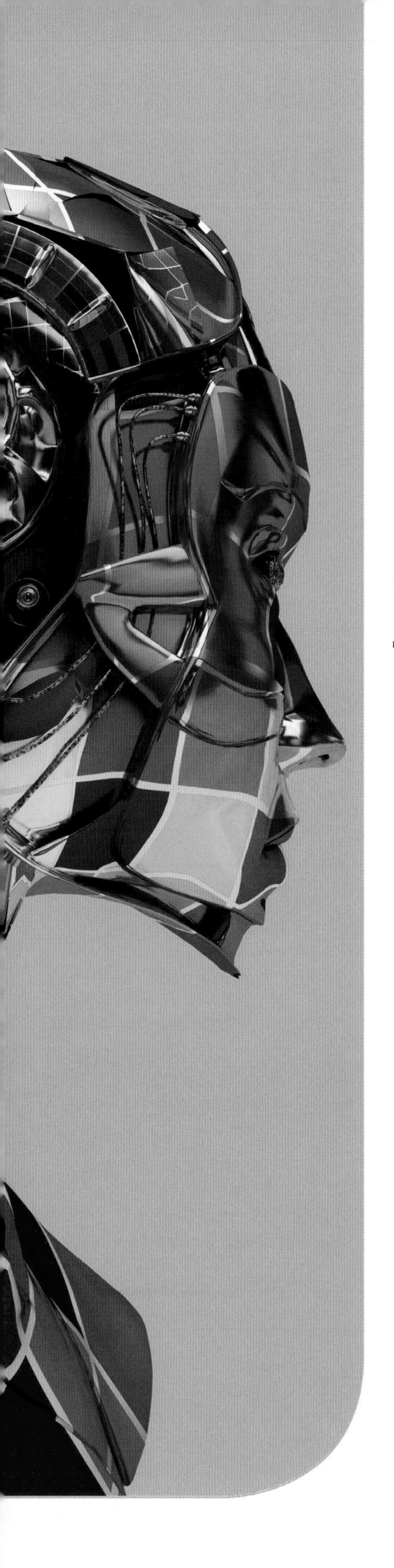

Section Four: Fundamentals of computational thinking

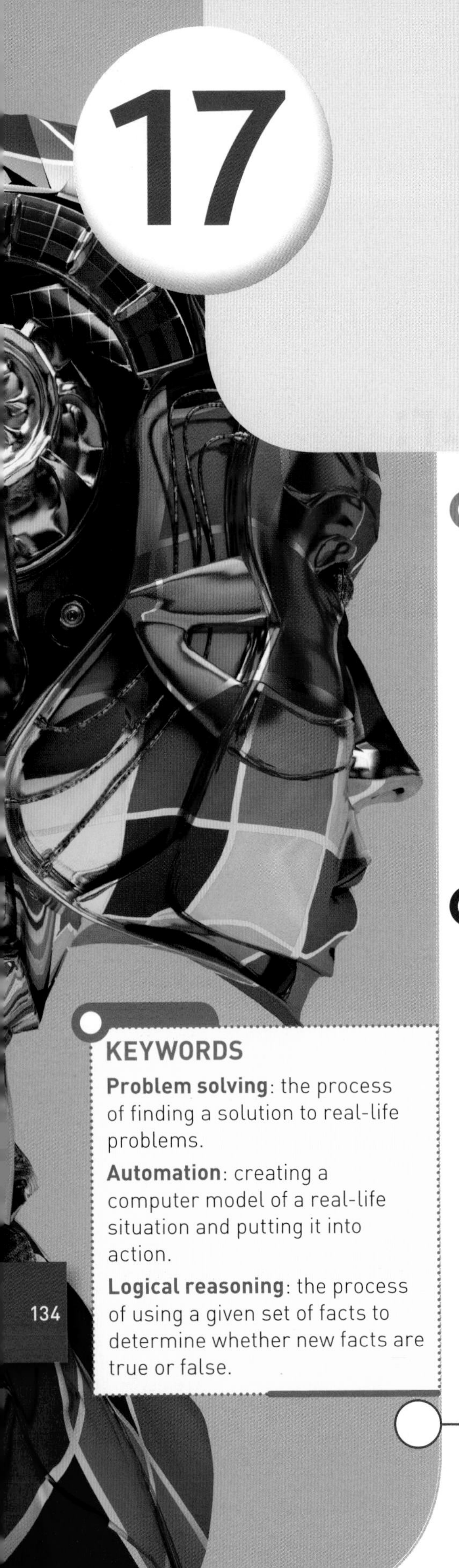

17 Abstraction and automation

SPECIFICATION COVERAGE

3.4.1 Abstraction and automation

LEARNING OBJECTIVES

In this chapter you will learn how:

- to use logical reasoning
- to take a systematic approach to problem solving
- to define problems using abstraction
- algorithms can be used to solve problems
- to hide the complexity of a problem from the user
- to decompose a problem and compose a solution
- to use computer models to recreate real-life problems and solutions.

INTRODUCTION

Computing is about processing data to solve a problem, for example doing calculations using data. The modern definition of the term implies the use of computer technology although much of the work needed to solve the problem is actually done away from the computer itself. There are techniques that programmers can use to help with **problem solving** and in this chapter we will be looking at:

- abstraction, which is the concept of picking out common concepts in order to reduce the problem to its essential defining features, ignoring less significant details
- **automation**, which is the process of creating a computer model and putting it into action.

The focus of this chapter is on problems that typically require mathematical calculations to solve them, as opposed to information processing systems.

KEYWORDS

Problem solving: the process of finding a solution to real-life problems.

Automation: creating a computer model of a real-life situation and putting it into action.

Logical reasoning: the process of using a given set of facts to determine whether new facts are true or false.

Logical reasoning

Logical reasoning is the process of using a given set of facts to determine whether new facts are true or false. More formally it is concerned with the concept of deductive reasoning which originates from the study of mathematics and philosophy which identifies rules or premises and then applies these to statements to come to a conclusion.

Using logical reasoning is an important skill for you as computer scientists as it is closely related to the issue of solving problems. Logical reasoning helps you to understand the nature of problems, to identify the facts that are relevant to the problem and to then be able to draw conclusions. It also enables you to identify new facts that you can deduce are true based on existing facts.

For example, we might have the following fact: 'Alex is a boy'. From this we might conclude that Alex is a boy's name. We might have another fact: 'Alex is a girl'. From this we might conclude that Alex is a girl's name. By combining the two facts we might reach a more accurate conclusion that Alex can be a boy's or a girl's name.

Consider the following example:

> Four friends sit at a concert together. Jane was sitting in seat A3. Kian was sitting to the right of Jane in seat A4. In the seat to the left of Jane was Ravi. Dev was sitting to the left of Ravi. Which seat is Dev sitting in?

To reason this out you will notice that the seats are sequentially ordered and there are just four people, so the answer is going to be somewhere between A1 and A6. Ravi is to the left of Jane so must be in seat A2 and Dev is to the left of Ravi so must be in seat A1.

The example below shows some facts that we might use when developing a satnav system. It shows how new facts can be determined from existing ones:

- Motorways have higher speed limits than single-lane roads.
- Single-lane roads have speed limits of between 30 mph and 60 mph.
- Dual carriageways have the same speed limits as motorways.
- Most roads in urban areas are single carriageway.

From these facts, we could determine further facts:

- A single-lane road could have a 40 mph speed limit.
- The speed limit on motorways must be more than 60 mph.
- Subject to traffic conditions, a journey would typically be quicker on a dual carriageway than on a single-lane road.
- Journeys through towns are likely to be slower that journeys along motorways.

Problem solving

People have been solving problems using computational thinking for thousands of years. In simple terms, it concerns identifying a problem and then working out the steps required to solve it. In doing this, you need to take account of any constraints that would impact on the solution. The objective is always to create the most efficient solution to any given problem and to be able to apply the solution to other, similar problems.

For example, humans have always travelled. The problem in this scenario is how to get from the start to the destination in the quickest and easiest way. Before humans were able to write things down, they would give each other verbal instructions on how to get from A to B. The next piece of technology to be used was the map, which provided a more efficient and reusable method for solving the problem. Maps have become increasingly sophisticated over the years to include more information for the user, although a certain amount of skill is required in order to read them.

The latest technology is the 'satnav', which combines large datasets, often being updated in real time with traffic information, being transmitted wirelessly to small portable devices using a very simple user interface that only requires the user to input the destination and then follow the verbal and visual instructions.

Figure 17.1 A very early map

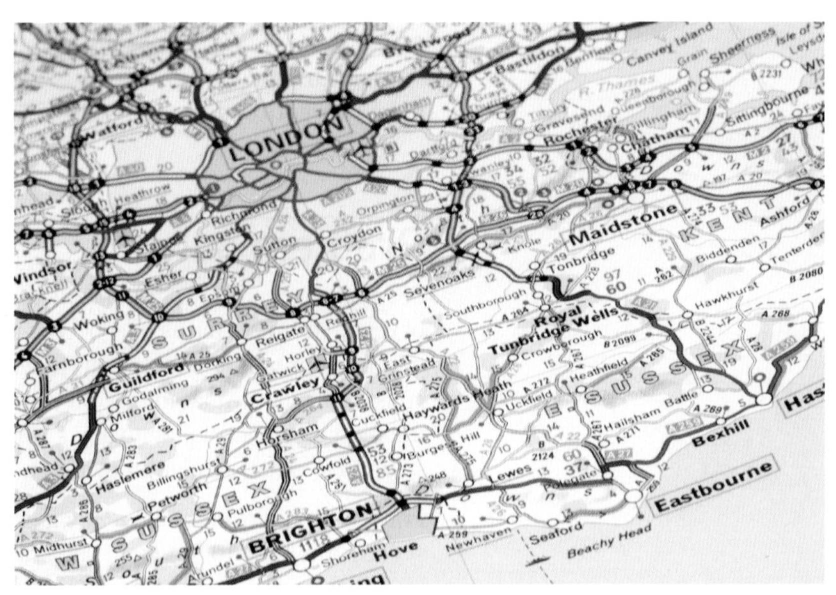

Figure 17.2 A modern map

Figure 17.3 A satnav

Defining and solving problems

This example demonstrates some of the key aspects of problem solving. In the first instance the problem needs to be clearly defined. In this case the problem is quite simple: How do I get from A to B? The solution however, is complex. As a user, all you want to do is type in the destination and then follow the instructions. However, as a computer scientist, you need to consider a large range of issues and constraints in order to come up with a solution. For example:

- How to define the start and end points of each journey, e.g. town name, grid reference.
- What form of transport will be used, e.g. car, cycle, walking.
- What routes will be used, e.g. roads, ferries.
- What data is available, e.g. road networks, traffic information.
- How data is kept up to date.
- How to calculate the quickest or shortest route to the destination.
- How to recalculate the route in case of traffic jams or road closures.
- What communication channels will be available to transmit the relevant data.
- How to present the information to the user in the most user-friendly way.

Having identified the problem you would then develop the most efficient solution, which may require several iterations. One of the key aspects of computing is that solutions must be checked to ensure that they do solve the problem. With our satnav example, extensive testing will be undertaken in-house by the manufacturers of the devices. They will also beta-test by getting users to use the systems in real-life situations before releasing the technology to the general public. The manufacturers will then constantly review feedback from customers in order to refine the technology.

Algorithms

You have already come across algorithms in Section One. Basically an algorithm is a step-by-step procedure for carrying out a particular task.

> **KEYWORD**
> **Algorithm**: a sequence of instructions.

Algorithms are the building blocks of computer programs and ultimately all problems are solved by writing algorithms. To take a very simple example, to calculate how long it takes to do a journey you could use the following algorithm:

```
TimeDeparted = 15:00
TimeArrived = 16:00
Drivetime = TimeArrived - TimeDeparted
```

This is an example of pseudo-code, which is a way of expressing the algorithm without having to use any specific programming language. The start point for programmers is often to work out what algorithms are needed to solve a problem and then to write these in pseudo-code during the planning stage. This can be a time-consuming process depending on the complexity of the solution.

Programmers use a technique called hand-tracing or dry running to work through their code. This means that they follow the code line by line to work out what is happening. This can help them to identify any problems with the code before it is implemented. There is an example of dry running in Chapter 5. Most programs are made up of multiple related procedures so it is important to identify how these link together to create the program as a whole.

When all of the procedures have been identified, the pseudo-code can be converted into proper programming code using whatever language is considered the most appropriate for solving the problem.

Abstraction

The concept of abstraction is to reduce problems to their essential features. Another way of explaining abstraction is that it is the process of finding similarities or common aspects about the problem, while ignoring differences. This is a useful concept for programmers as they can view the problem from a high level, concentrating on the key aspects of designing a solution whilst ignoring the detail, particularly during the initial design stages.

Once a solution has been identified for the current problem, a feature of abstraction is that the abstraction from one problem can be applied to another similar problem, which shares the same common features.

Broadly speaking there are two main type of abstraction.

Representational abstraction

> **KEYWORD**
> **Representational abstraction**: the process of removing unnecessary details so that only information that is required to solve the problem remains.

This is the process of removing unnecessary details until it is possible to represent the problem in a way that can be solved. This level of abstraction could be described as viewing the 'big picture' – working out what is relevant to solving the problem and what is unnecessary detail that can be ignored.

With the satnav problem, at a basic level, the problem can be reduced to finding the shortest distance between point A and point B. An abstraction of that would be to:

- identify point A and point B in some way
- identify the connecting paths between A and B
- calculate the shortest path between A and B.

Now that the abstraction is complete a solution can be created to solve the problem. For example, a variation of Djikstra's shortest path algorithm could be developed. In addition, related problems can be solved using the same abstraction.

Some information that would be found on a map would not be required to find a shortest route. For example, the location of rivers, railway lines and landmarks could be ignored, so the map stored by the satnav would be an abstraction of the real location.

Abstraction by generalisation/categorisation

This is the process of placing aspects of the problem into broader categories to arrive at a hierarchical representation. This involves recognising common characteristics of representations so that more general representations can be developed. We have already seen this concept applied with object-oriented programming in Chapter 6, where subclasses are defined from the characteristics of a base class.

> **KEYWORD**
>
> **Abstraction by generalisation/categorisation**: the concept of reducing problems by putting similar aspects of a problem into hierarchical categories.

For example, to represent information about cars and buses, we recognise that they have a lot in common so generalise/categorise them both as vehicles. When programming using an object-oriented language we can represent this generalisation using inheritance.

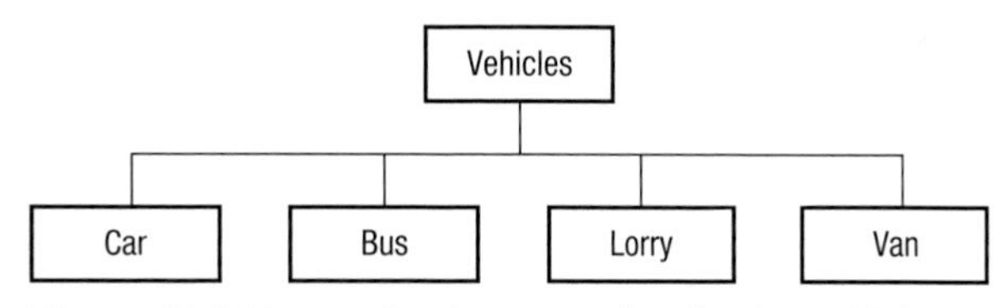

Figure 17.4 Abstraction by generalisation for vehicles

The principle of abstraction can also be applied to various elements of computing including:

- Procedural abstraction: This is the concept that all solutions can be broken down into a series of procedures or subroutines. This in fact is the way that all procedural languages work, enabling the programmer to identify the main processes needed to complete the task and to contain these within procedures. At the design stage it is sufficient for the programmer to work out what each procedure will do without defining how it will do it. The procedure may in turn call other procedures although it does not need to know how these work in order to call them. This is the basis for **top-down design** that we looked at in Chapter 5. Other considerations include what event will trigger the procedure, how procedures will link together, including any possible side effects, and how errors will be handled.
- Functional abstraction: Similar to procedural abstraction, **functional abstraction** focuses on common functions that can be used to solve problems. Functions are a feature of procedural languages and the cornerstone of functional programming, where all the main processes are defined in terms of functions. Functions can be created for any common procedure and functions can be built on top of other functions producing higher levels of abstraction. All the program needs is for the parameters to be input into the function in order to generate a result. Using functions reduces complexity as the function only needs to be written once.

> **KEYWORDS**
>
> **Top-down design**: related to the modular approach, this starts with the main system at the top and breaks it down into smaller and smaller units a bit like a family tree.
>
> **Functional abstraction**: breaking down a complex problem into a series of reusable functions.

- Data abstraction: This is the process of organising and structuring data in a way that produces a particular view of the data that is useful for the programmer. Almost all data is abstracted, hence the term abstract data types that we looked at in Chapter 7. For example, a queue is an abstract data type, which may be made up of an array. By abstracting the data into a queue, all the programmer has to do is push and pop to the queue without having to worry about the structure of the underlying dataset. Another feature of **data abstraction** is that the data can be implemented in different ways. For example, once data is abstracted into an array, it could be used to create other abstract data types such as stack or a binary tree. This is known as data composition where data objects are combined in order to create a compound structure.
  Data abstraction involves separating the actual implementation from the interface. In the satnav example, the algorithm needs to find the shortest journey between two points. The interface provides this information but how it is implemented is hidden. It doesn't matter whether the data is being stored in an array, as a vector or in a relational database, providing the relevant answer is provided by the program.
- **Problem abstraction**: This is the process of reducing a problem down to its simplest components until the underlying processing requirements that solve the problem are identified. By doing this, these underlying processes can be applied to solve analogous problems. For example, satnavs use vectors (see Chapter 11) and a variation of Dijkstra's algorithm (see Chapter 13). Neither of these concepts were developed specifically for the satnav, but both have been adopted to create the required solution. Therefore, the underlying principles used to solve one problem been applied to different problem with similar characteristics.
  Another example is the use of graphs in general. In Chapter 9 we looked at the use of graphs to explain relationships between nodes. Many problems have been solved using graph theory, as the underlying requirements of the problems are the same even though they may not appear to be. For example, graphs are used to optimise the transmission of data on computer networks, to model atomic and chemical structures, to predict the spread of disease and to analyse social networks.

> **KEYWORD**
> **Data abstraction**: hiding how data is represented so that it is easier to build a new kind of data object, e.g. building a stack from an array.

> **KEYWORD**
> **Problem abstraction**: removing unnecessary details in a problem until the underlying problem is identified to see if this is the same as a problem that has already been solved.

Information hiding

In broad terms, **information hiding** is the process of providing a definition or interface of a system or object, whilst keeping the inner workings hidden. A common example of the principle is the car. All cars have a common interface in that they have a steering wheel, gearbox, pedals etc. By operating this common interface it is possible to operate the car. The actual mechanics of how the car works is hidden. In fact, the mechanics of how the car works may change without it impacting on the interface. For example, changing from a petrol to a hybrid engine does not change the basic principles of how to drive a car.

An example in computing is where a common interface such as a GUI is used. With our satnav example, the interface prompts the user to input an end point. The complexity of calculating the route is hidden. If the way in which the route was calculated changed, it would not necessarily affect the interface. In this way, information hiding separates the user interface from the actual implementation of the program.

> **KEYWORD**
> **Information hiding**: the process of hiding all details of an object that do not contribute to its essential characteristics.

More specifically when programming, information hiding can be used to define a set of behaviours on a dataset, where the data can only be accessed through those behaviours. It is not possible for other parts of the program to access the dataset directly. This prevents unintended damage to the dataset and also means that how the dataset is stored can be changed without affecting any programs that use it, as they do not access it directly.

Information hiding is closely related to the concept of encapsulation where data and behaviours are stored together within a class or object. Encapsulation can be seen as a method of implementing the information-hiding principle.

Decomposition/Composition

> **KEYWORDS**
>
> **Decomposition**: breaking down a large task into a series of subtasks.
>
> **Composition**: building up a whole system from smaller units. The opposite of decomposition.

A broad definition of **decomposition** is breaking large complex tasks or processes down into smaller, more manageable tasks. Abstraction techniques will be used in order to decompose the system requirements.

Procedural decomposition is the process of looking at a system as a whole and then breaking that down into procedures or subroutines needed to complete the task. This process is very similar to the idea of the top-down approach we looked at in Chapter 5 where each main task is identified, then the subtasks that make up each task. Depending on the complexity of the system, subtasks may be further subdivided until the designer reaches a level of detail that is sufficient to start building the system.

Procedural **composition** is then the process of creating a working system from the abstraction. This involves:

- writing all the procedures and linking them together to create compound procedures
- creating data structures and combining them to form compound structures.

A satnav system could be decomposed as follows:

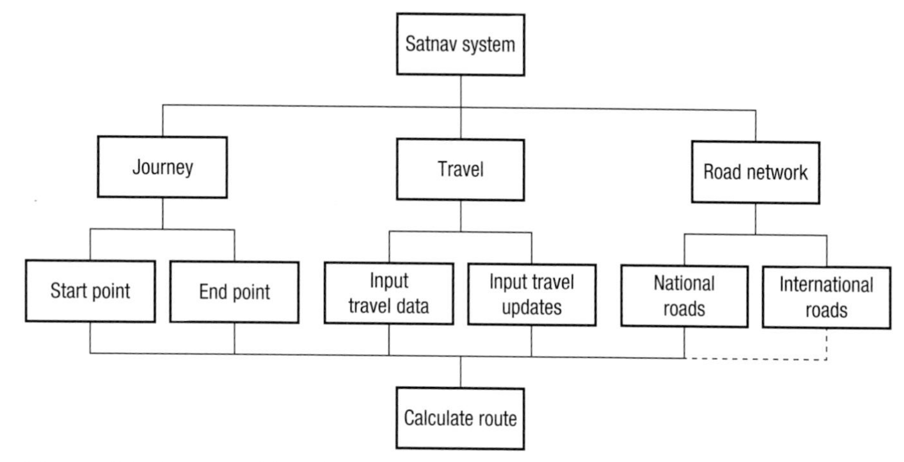

Figure 17.5 Decomposition of a satnav system

Automation

Automation in this context is the process of creating computer models of real-life situations and putting them into action. Most computer programs are created to solve real problems. One of the objectives of creating computer systems is to create elegant solutions to difficult problems. The key to this is:

- understanding the problem
- being able to create suitable algorithms
- building the algorithms up into program code
- using appropriate data in order to solve the problem.

For example, computer models are widely used to analyse traffic flows and to control traffic lights across road networks. Major cities and towns often have severe traffic congestion and by controlling the traffic lights it is possible to keep traffic moving more freely.

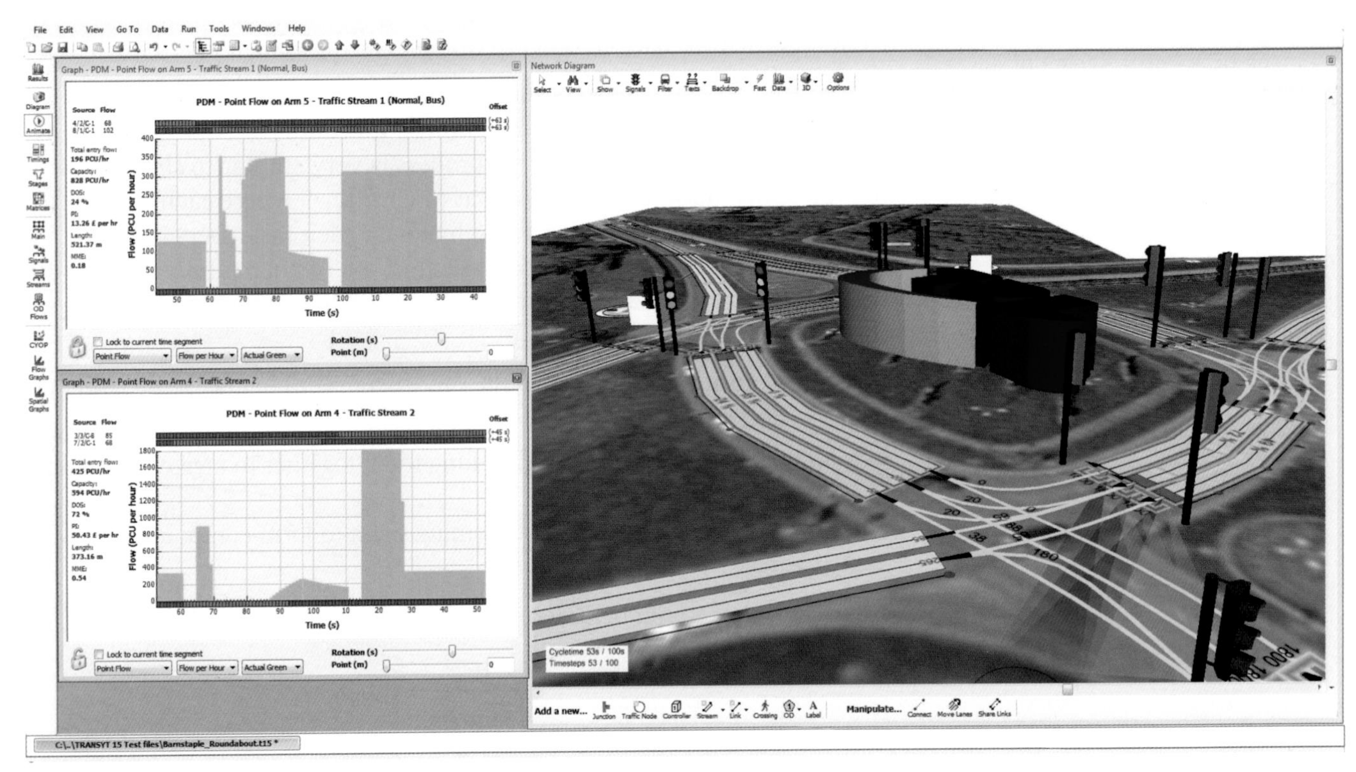

Figure 17.6 Computer model for traffic flow system

The screenshot in Figure 17.6 is from a software model called TRANSYT and demonstrates the problem. Where one set of lights is on green you may assume the traffic is flowing freely. However, by definition it means that there is likely to be a queue of stationary traffic (or pedestrians) waiting at a red light. Where there are several sets of traffic lights close together, the problem becomes more difficult to solve.

Therefore, the outcome is simply to keep traffic moving as freely as possible around the network. The solution is far more complex. The designer needs to consider:

- the location of all traffic lights
- the number of roads that meet at each set of traffic lights
- how many lanes of traffic there are at each set of lights
- how much traffic there is on each of the lanes

- what time of day it is, i.e. is it rush-hour and whether people are generally heading into or out of the city
- whether the lights control a pedestrian crossing as well as a road.

There are probably many other considerations, but the challenge for the designer is to identify the key factors that will make the model accurate. In addition they need to consider what data to use and where to get it from. As a minimum they will need data for:

- the roads in the network
- the physical location of the lights
- the volume of traffic on the road, which will either be historical or real-time data.

Having collected all of this data, code must be written to optimise traffic flows, which involves switching the signals and leaving them on green or red for the correct amount of time. For example, if there is a busy main road with heavy traffic, more time on green must be allowed at the expense of traffic on the side roads.

Using automated models in this way requires constant calibration of the model. This means that the designers need to see how well their modelled system works in real life. If traffic is not flowing as expected they need to make changes either to their algorithm or to their data in order to make the model more accurate.

Practice questions can be found at the end of the section on pages 179 and 180.

TASKS

1 From the following facts, use logical reasoning to determine further facts that you know to be true:
   a) every cat eats mice
   b) some animals that eat mice are fat
   c) all mice carry diseases
   d) mice can run fast.

2 You are asked to work out the timetable for all the students in the sixth form.
   a) What factors do you need to take into account in order to solve this problem?
   b) Give an example of representational abstraction and abstraction by category / generalisation in this scenario.
   c) Explain how you might decompose the problem.

3 Define the following terms:
   a) procedural abstraction
   b) functional abstraction
   c) data abstraction
   d) problem abstraction.

4 Define information hiding and give an example of where it might be used.

5 Create a specification for a model to simulate any or all of the following scenarios:
   a) the likelihood of a particular team winning a competition
   b) the speed at which a concert hall could be evacuated in the event of a fire alarm
   c) how many people and households there will be in the UK in 2050.

STUDY / RESEARCH TASKS

1 Explore the concept of computational thinking and consider examples that predate the invention of the computer.

2 Models have been built to predict how many gold medals Britain will win during Olympic Games. Find out what variables go into these models and how accurate the predictions have been.

3 Research what data and algorithms are used in order to predict:

  a) weather

  b) tornados

  c) tsunamis.

4 Find examples, other than those in this chapter, where an algorithm developed to solve one problem has been used to solve a different problem.

KEY POINTS

- Logical reasoning is the process of using a known set of facts to determine whether new facts are true or false.
- Problem solving is identifying a problem and then working out the steps required to solve it.
- Simple problems may have complex solutions.
- An algorithm is a step-by-step procedure for carrying out a particular task.
- Abstraction reduces unnecessary detail instead focusing on the essential features that will solve the problem.
- Information hiding is the process of hiding all details of an object that do not contribute to its essential characteristics.
- Decomposition is breaking large complex tasks or processes down into smaller, more manageable tasks.
- Composition is then the process of creating a working system from the abstraction.
- Automation in this context is the process of creating computer models of real-life situations and putting them into action.

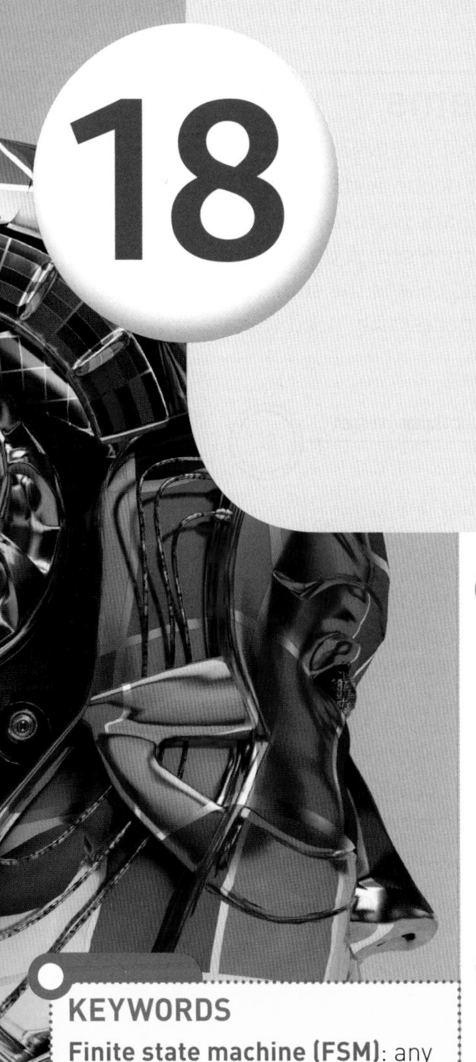

18 Finite state machines

SPECIFICATION COVERAGE

3.4.2 Finite state machines

LEARNING OBJECTIVES

In this chapter you will learn:

- what a finite state machine is and how it can be used
- how to use state transition diagrams
- how to use state transition tables.

A-level students will learn:

- how to use finite state machines with outputs.

KEYWORDS

Finite state machine (FSM): any device that stores its current status and whose status can change as the result of an input. Mainly used as a conceptual model for designing and describing systems.

Finite: countable.

INTRODUCTION

In general terms a **finite state machine (FSM)** is any device that can store its current status (or state) and can change state based on an input. The FSM may receive further inputs, which in turn change the state again. There are a **finite** (countable) number of transitions that may take place. Some FSMs also have outputs, one type of which is called a Mealy machine. Knowledge of these is only required for A level and is covered at the end of the chapter.

As a simple example, an automated door is an FSM:

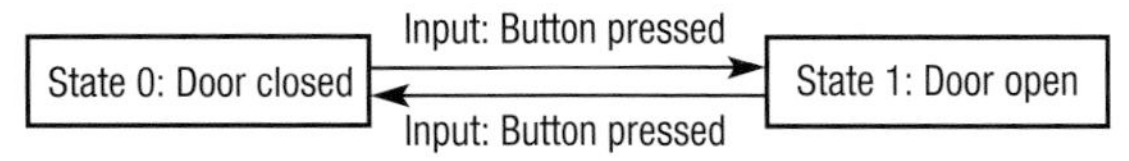

Figure 18.1 A simple example of an FSM

Finite state machines are common in everyday life and include any devices where there are a predefined set of steps and outcomes involved in the operation of the machine.

In practice, finite state machines are used as a conceptual model to design and describe systems. They are particularly useful at the design stage as they force the designer to think about every possible input and how that changes the state of the machine. As a result they are commonly used to develop computer systems or design logic circuits and can also be used to check the syntax of programming languages.

There are two main ways of representing an FSM: a state transition diagram or a state transition table.

State transition diagrams

KEYWORDS

State transition diagram: a visual representation of an FSM using circles and arrows.

Accepting state: the state that identifies whether an input has been accepted.

State transition diagrams use circles to represent each state and arrows to represent the transitions that occur as the result of an input. For example, a ticket machine in a car park requires two inputs: money to be put in and the green button to be pressed. A double circle represents the accepting or goal state, which in this case is the state that is required in order to issue a ticket. FSMs do not necessarily need to have an **accepting state**.

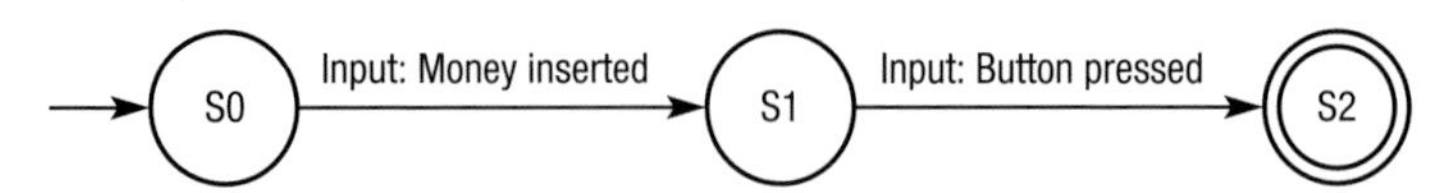

Figure 18.2 A simple state transition diagram

In this case:

- S0 is the machine in its idle state, waiting for an input.
- S1 is its state after the money has been put in.
- S2 is its state after the button has been pressed. This is the accepting state.

The FSM has sequence and memory in that each transition is based on the one before. For example, the button can only be pressed after the money has been inserted. Whole systems or individual procedures can be modelled using state transition diagrams. For example, the procedure for logging onto a computer network might look like this:

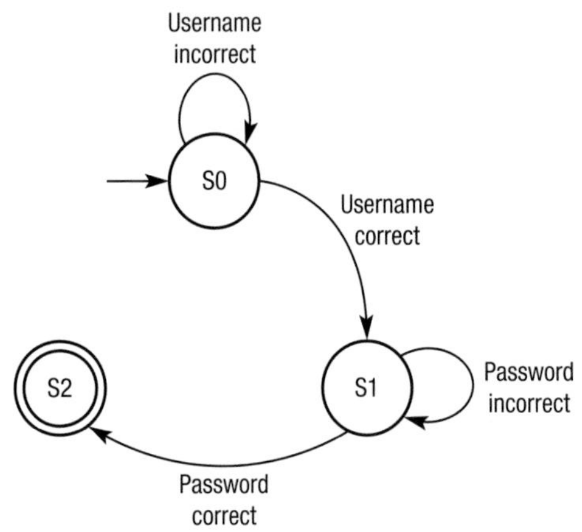

Figure 18.3 State transition diagram to show the process of logging onto a computer network

This shows the dependency of one state on the next. If the username is correct it can change to the next state. If the password is correct it can move on to the accepting state. Without the correct inputs the state will not change.

The example in Figure 18.4 shows an FSM that is used to check that the rules of a programming language are being followed. It is a simplified example using just the letters a, b and c, though in real life the FSM could be set up to represent all of the acceptable words and combinations of words usable in any particular programming language. Notice the addition of a start arrow.

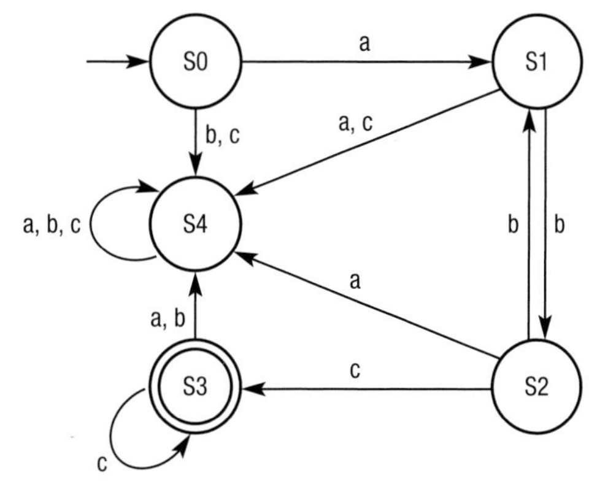

Figure 18.4 State transition diagram to show syntax rules

Looking at the diagram you can see whether certain combinations of letters are acceptable or not. For example:

- abc is an acceptable combination.
- abcc is an acceptable combination.
- acb is not acceptable. This would end in S4.
- abca is not acceptable as S3 is the accepting state so the final letter must be a c. This would end in S4.

State transition tables

KEYWORD

State transition table: a tabular representation of an FSM showing inputs, current state and next state.

The same information can also be represented as a table. These show the input and the current state, which is the state before the input. It then shows the state after the input. For example, a **state transition table** for the automated door in Figure 18.1 would be:

Input	Current state	Next state
Button pressed	Door closed	Door open
Button pressed	Door open	Door closed

The table for the ticket machine in Figure 18.2 would be:

Input	Current state	Next state
Money inserted	S0	S1
Button pressed	S1	S2

The table for processing the letters in Figure 18.4 would be:

Input	Current state	Next state
a	S0	S1
b	S1	S2
c	S1	S4
a	S1	S4
b	S2	S1
c	S2	S3
a	S2	S4
b	S0	S4
c	S0	S4
a	S3	S4
b	S3	S4
c	S3	S3
a	S4	S4
b	S4	S4
c	S4	S4

KEYWORDS

Mealy machine: a type of finite state machine with outputs.

Cipher: an algorithm that encrypts and decrypts data, also known as code.

Shift cipher: a simple substitution cipher where the letters are coded by moving a certain amount forwards or backwards in the alphabet.

A/D
S0
B/E
C/F

Figure 18.5 State transition diagram with outputs

Finite state machines with outputs

A level only

Some finite state machines will produce output values based on the input values. An example of this is a **Mealy machine**, named after the man who invented it. An example of an application of a Mealy machine could be a simple **cipher** where the letter input becomes transformed into another letter. Figure 18.5 shows a simple three-letter **shift cipher**, where A becomes D, B becomes E and so on.

This is a simple example that shows the concept of the Mealy machine where the current state is transformed to a new state with the output being shown. Mealy machines were originally devised to define electronic circuits and are commonly used to express bitwise operations. For example, Figure 18.6 shows a right arithmetic shift on a binary value, which will have the effect of halving the value.

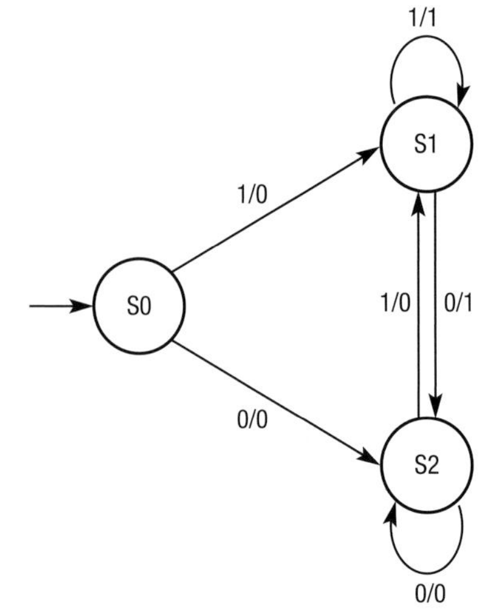

Figure 18.6 A Mealy machine performing a right shift

The state transition table for this would be:

Input	Current state	Output	Next state
0	S0	0	S2
1	S0	0	S1
0	S1	1	S2
1	S1	1	S1
0	S2	0	S2
1	S2	0	S1

Practice questions can be found at the end of the section on pages 179 and 180.

KEY POINTS

- A finite state machine (FSM) is a concept that shows a device that stores the current state and how the state will change as the result of an input.
- FSMs are used as a conceptual model for designing and describing computer systems.
- FSMs can be used to check the syntax of language including programming languages.
- State transition diagrams are a visual way of showing how states change as the result of an input.
- State transition tables are an alternative way of showing how states change as the result of an input.
- A special type of FSM called a Mealy machine can also produce outputs.

TASKS

1 Looking at Figure 18.4, which of these are acceptable?

a) caabb

b) bac

c) aaabbccc

d) abc

e) aabbca

2 Draw a state transition diagram and table that evaluates whether a sequence of bits has an even number of zeros.

3 What is a Mealy machine?

4 Draw a state transition diagram from the following table:

Input	Current state	Output	Next state
1	S0	A	S1
2	S2	B	S1
3	S1	C	S3
4	S3	D	S2

5 Draw a state transition diagram that will carry out a bitwise XOR operation.

STUDY / RESEARCH TASKS

1 Draw a state transition diagram and table that flips 0s to 1s within a binary string.

2 Draw a state transition diagram and table to control the pointer in a stack.

3 Research other practical applications of the FSM.

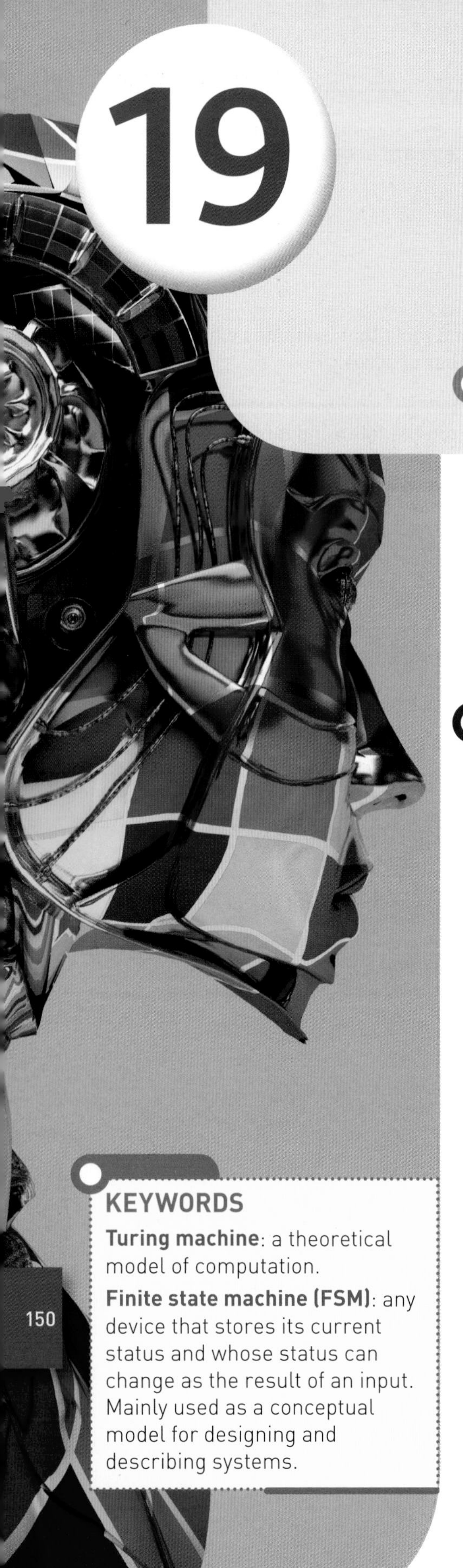

19 The Turing machine

A level only

SPECIFICATION COVERAGE

3.4.5 A model of computation

LEARNING OBJECTIVES

In this chapter you will learn:

- that a Turing machine is a theoretical model for identifying whether a problem is computable
- how the Turing machine works
- how state diagrams can be used to represent the workings of the Turing machine
- how a universal machine can be constructed.

INTRODUCTION

The **Turing machine** is a theoretical model developed by Alan Turing in 1936 as a way of trying to solve what was called 'the decision problem'. In simple terms, the problem was whether it was theoretically possible to solve any mathematical problem within a finite number of steps given particular inputs. Turing developed a theoretical machine that was able to carry out any algorithm and in doing so essentially produced a model of what is computable.

It is worth noting that the Turing machine was devised as a concept rather than as an actual machine and its invention predates microprocessors and computing as we know it today. Scientists have since created physical machines according to Turing's model and software simulations have also been made.

KEYWORDS

Turing machine: a theoretical model of computation.

Finite state machine (FSM): any device that stores its current status and whose status can change as the result of an input. Mainly used as a conceptual model for designing and describing systems.

A Turing machine is a **finite state machine (FSM)** with the ability to read and write data to an unlimited tape. It can be visualised as shown in Figure 19.1.

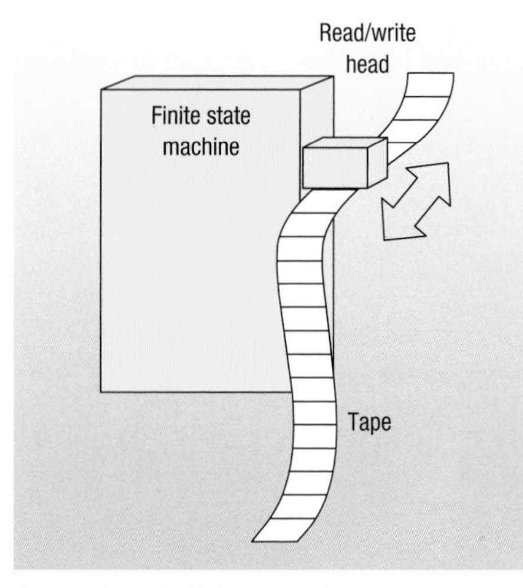

Figure 19.1 A Turing machine

The basics of its operation are:

- The tape is divided into an infinite number of cells. The tape is used as memory.
- Each cell will contain a symbol. This could be a character or number. Commonly the contents of the tape are binary digits, i.e. 0s and 1s, and the blank symbol,

sometimes shown as a □. The acceptable symbols are known as the **alphabet**.

- The **read/write head** can either read what is in the cell or write into the cell. It can also erase the current contents of the cell, effectively overwriting the contents.
- The tape can move left or right one cell at a time so that every cell is accessible by the read/write head.
- The machine can halt at any point if it enters what is known as the **halting state**, or if the entire input has been processed.

Despite being invented before computers as we know them now, you can see that this model sounds very much like a modern PC, with the tape representing memory and the cells representing memory addresses. Moving the tape through the read/write head will produce sequences of characters, which is analogous to a computer executing instructions in a program.

To represent programs in Turing machines you need:

- a **start state** – the state of the machine at the start of the program
- a halting state – the state that will stop the program running
- an alphabet – this is a list of the acceptable symbols that can go into each cell
- movement – the ability to move the head so that you can read/write to every cell
- a **transition function** – indicating what should be written at each cell and whether to move left or right based on the input read.

Controlling the machine is represented through state diagrams very similar to the **state transition diagrams** that we looked at for FSMs. You can visualise the tape as follows:

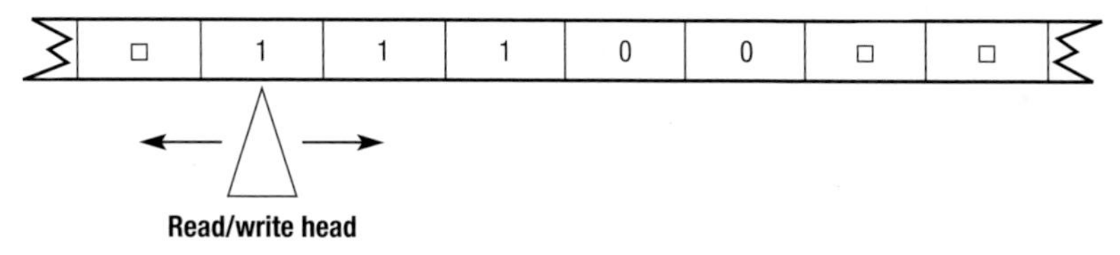

Figure 19.2 The tape and read/write head

At the moment the read/write head is either reading or writing a 1 in the current cell. In the examples used in this chapter the read/write head starts with the left-most non-blank location. Note that the symbols in the other cells are 0s, 1s or blanks. The arrows indicate that the head can move left or right.

Figure 19.3 shows a state transition diagram for the transition function of a Turing machine.

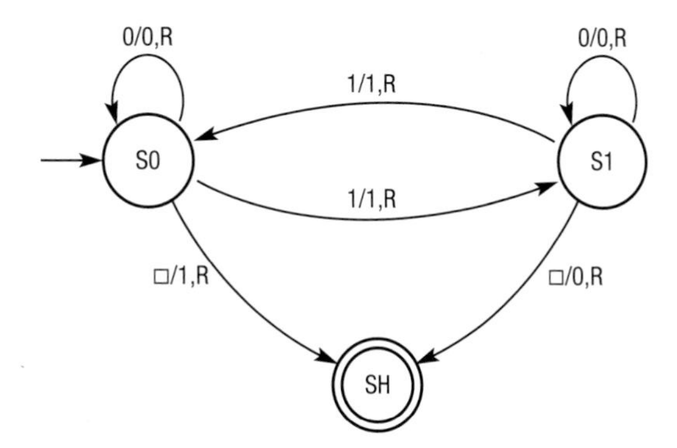

Figure 19.3 State diagram to show the operation of a Turing machine

KEYWORDS

Alphabet: the acceptable symbols (characters, numbers) for a given Turing machine.

Read/write head: the theoretical device that writes or reads from the current call of a tape in a Turing machine.

Halting state: stops the Turing machine.

Start state: the initial state of a Turing machine.

Transition function/rule: a method of notating how a Turing machine moves from one state to another and how the data on the tape changes.

State transition diagram: a visual representation of the transition function of a Turing machine.

- S0 represents the starting state.
- SH represents the halting state.
- This diagram has one other state, S1, although it would be possible to have as many states as required to represent an algorithm.
- Moving from one state to the next requires a transition function or transition rule. These are determined by the arrows in the transition diagrams. The rules are shown on the diagram in the format 1/1,R which means in this case that if the input symbol is a 1, keep the symbol on the tape as a 1 and then move the head right. Another example is □/0,R which means if the input symbol is blank, change it to a 0 and move the head right.
- Writing the transition rules effectively creates an algorithm.

It is useful to be able to trace the steps that the machine will go through, looking at the current state after each movement.

The Turing machine is in state S0 and the input symbol is 1. Therefore the rule 1/1,R is applied which writes a 1 to the tape, moves the head right and changes to state S1. The machine will now look like this:

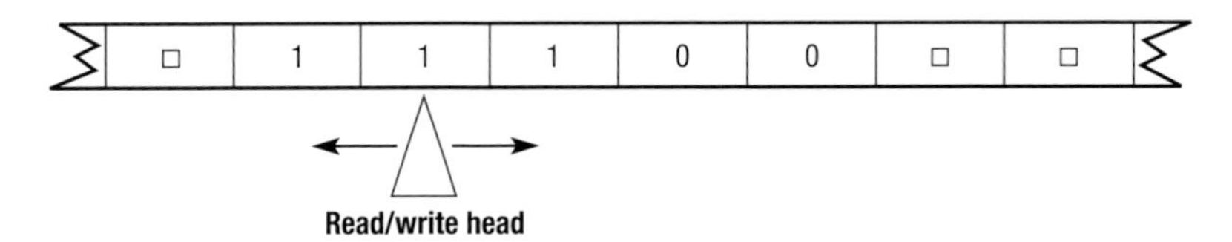

We are now on S1 in the state diagram. The rule 1/1,R is applied which writes a 1 to the tape, moves the head to the right and changes back to state S0:

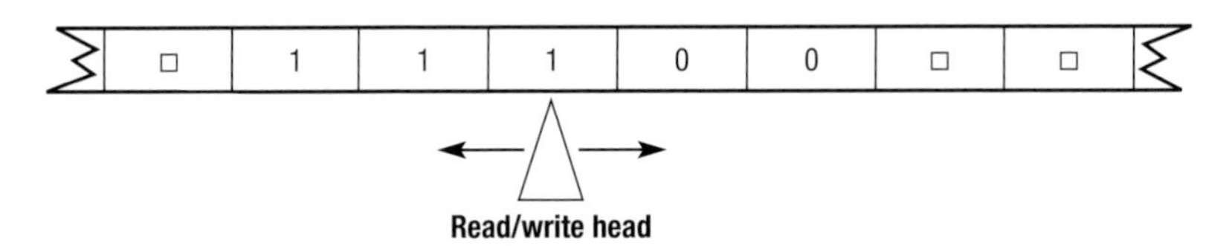

The next step reads a 1, writes a 1, changes to state S1 and moves the head right. Then the next two steps read a 0, so write a 0, move right and stay in state S1 so the tape now looks like this:

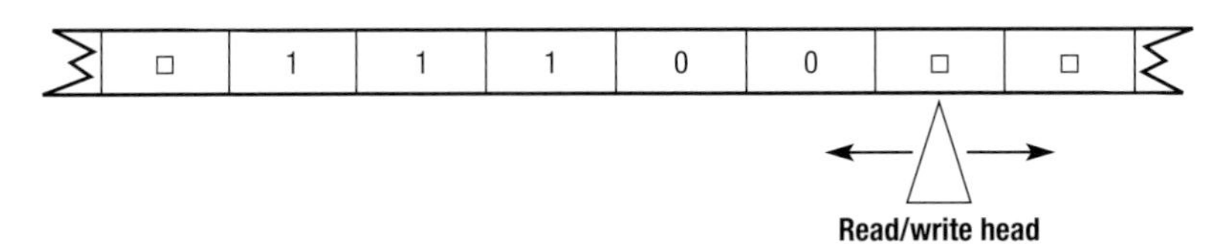

With the machine in this position, we are now in state S1 reading a □. The rule now is to write a 0 to the tape, move right and move to the halting state. The program now stops as the algorithm is complete and the final contents of the tape are:

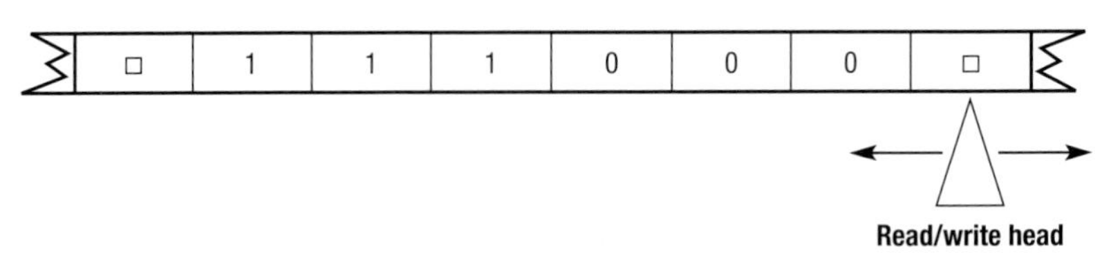

This algorithm is an odd parity generator that ensures the number of 1s in a binary string is odd:

- The start point is the left-most non-blank symbol in the binary string.
- Scanning the digits, if there are an even number of 1s then a 1 needs to be added to ensure that it becomes odd.
- When a blank is reached, if there are already an odd number of 1s, a 0 is added to maintain this number as odd.
- The program then halts.

This could be represented by the following **instruction table**:

KEYWORD

Instruction table: a method of describing a Turing machine in tabular form.

State	Read	Write	Move	Next state
S0	□	1	R	SH
S0	0	0	R	S0
S0	1	1	R	S1
S1	□	0	R	SH
S1	0	0	R	S1
S1	1	1	R	S0

Another way of notating the Turing machine is to show the transition rules in the following format:

δ (Current State, Input Symbol) = (Next State, Output Symbol, Movement)

The transition rules for the odd parity generator therefore could be written with the following functions:

δ (S0, □) = (SH, 1, R)

δ (S0, 0) = (S0, 0, R)

δ (S0, 1) = (S1, 1, R)

δ (S1, □) = (SH, 0, R)

δ (S1, 0) = (S1, 0, R)

δ (S1, 1) = (S0, 1, R)

Note the use of □ for a blank cell and the use of L and R for left and right. Other notation may be used, for example, it is common to use a B to represent a blank and left and right arrows (← →) to show movement.

Universal machine

At a theoretical level you can use the Turing machine on any problem that is computable. The limitation of the Turing machine is that every process we need to carry out requires its own Turing machine to do it. For a large program, this could quickly become problematic. You could view this as a black box model where you define the inputs, the process carried out, and the output:

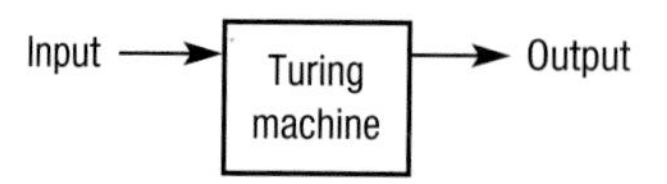

If we want to add A and B:

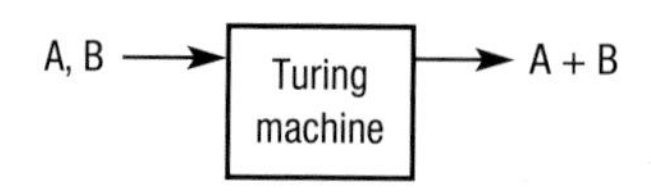

To multiply A and B:

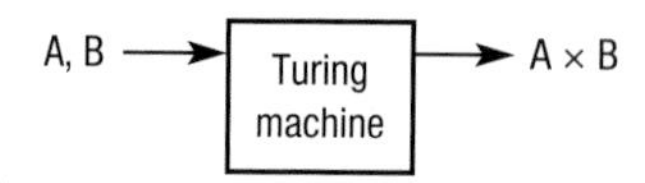

KEYWORD

Universal machine: a machine that can simulate a Turing machine by reading a description of the machine along with the input of its own tape.

Every computation requires its own Turing machine and its own tape on which to work. However, it is inevitable that you will want to combine the results of several computations and this is why Turing developed the concept of a **universal machine**.

Rather than defining each individual process within a single machine, the universal machines takes two inputs:

- a description of all the individual Turing machines required to perform the calculations
- all the inputs required for the calculations.

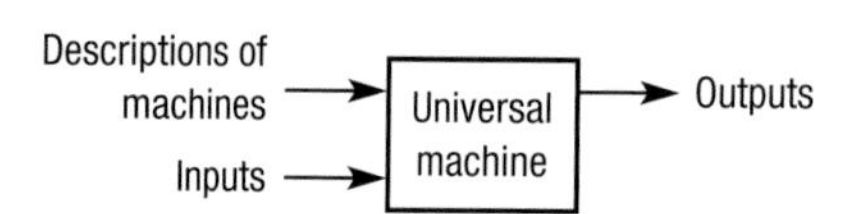

Figure 19.4 The universal machine

Perhaps the easiest way to think of it is as a series of individual Turing machines all linked together that can take any input and perform any calculation defined by any of the component machines.

This is stored on one tape (rather than lots of individual tapes) with one block of cells containing the instructions and one block of cells containing the inputs. The result of this is a machine that can simulate any number of Turing machines with their corresponding inputs and produce a range of outputs.

This is often seen as the earliest form of the stored program concept (see Chapter 33) where instructions and data are stored in the same place in memory. This is one of the key principles of modern computing even though it was devised as long ago as 1936. As Turing machines break down processes to small steps, this equates to many of the other techniques we have looked at in this section, such as decomposition, where large problems are broken down until a programmable solution can be found.

The concept of the Turing machine and the universal machine is closely linked with the concept of computable problems (see Chapter 22) in that they define what is computable. If it is possible to describe a Turing machine to solve a problem, then it will be possible to write an algorithm to solve it.

Practice questions can be found at the end of the section on pages 179 and 180.

KEY POINTS

- The Turing machine is a theoretical machine that is able to carry out any algorithm and in doing so essentially produces a model of what is computable. It works with a tape of an infinite length split into cells.
- Each cell has a value in it, typically a 0, 1 or a blank, but could have any symbols.
- The read/write head can move in any direction along the entire length of the tape.
- The read/write head reads and writes values to the cells.
- A universal machine is a machine that can simulate any other Turing machine by processing a description of how the other Turing machine works, that is, its transition function, that is stored on the tape alongside the data that is to be processed.

TASKS

1 Define the components of a Turing machine.

2 Why did Alan Turing develop the Turing machine?

3 Why did he develop the universal machine?

4 Identify two states that a Turing machine should have.

5 Draw a state transition diagram and instruction table and write the transition rules (functions) for the following:

a) Create a counter that starts at 1 and adds 1 each time.

b) Perform a bit universion, i.e. change 0s to 1s and 1s to 0s.

c) Carry out a unary addition in the format: 1 + 1 = 11, or 11 + 1 = 111, or 1 + 11 + 1 = 1111.
The alphabet will consist of blank, 1 and +.

6 Choose one of the above and draw a series of diagrams to show the current state of the tape after each step.

7 Why are the Turing machine and universal machines still relevant to modern computing?

STUDY / RESEARCH TASKS

1 Research the 'Busy Beaver' and how it can be solved using a Turing machine.

2 Find a Turing machine simulator on the Internet and use it on the algorithms described in question 5 above.

20 Regular and context-free languages

A level only

SPECIFICATION COVERAGE

3.4.2.3 Regular expressions

3.4.2.4 Regular languages

3.4.2 Context-free languages

LEARNING OBJECTIVES

In this chapter you will learn:

- what regular expressions are and how they can be used to define and search sets
- how regular expressions can be used on text strings
- how to search using strings of regular expressions
- how context-free languages can be used to describe the syntax of a programming language
- how Backus–Naur Form (BNF) is used
- how syntax diagrams are used.

INTRODUCTION

A **regular language** is one that can be represented using regular expressions. Regular expressions contain strings of characters that can be matched to the contents of a set allowing you to find patterns in data. They are a powerful tool for searching and handling strings. They also provide a shorthand definition of the contents of the set.

KEYWORD

Regular language: any language that can be described using regular expressions.

Regular expressions

To start with an example, the expression a|b|c is a **regular expression**, which means that the set will contain either an 'a' or a 'b' or a 'c'. A set is a collection of data that is unordered and contains each item at most once. It is written as follows showing the name of the set and the contents within the brackets:

- alphabet = {a, b, c, d, e, f, g, ...}
- integers = {0, 1, 2, 3, 4, 5, 6, 7, 8, 9, ...}

KEYWORD

Regular expression: notation that contains strings of characters that can be matched to the contents of a set.

The contents of a set are typically characters and numbers and a regular expression can be used to define and search the set. There are regular expressions for handling text strings (covered in this chapter) and for handling numbers (covered in the next chapter). There is also a relationship between regular expressions and finite state machines in that all regular expressions can be expressed as state transition diagrams for an FSM and vice versa, and there is more on this in this chapter.

Common regular expressions are shown in Table 20.1.

Table 20.1 Regular expressions with examples of outputs

Regular expression	Meaning	Strings produced
a|b|c	a or b or c	a b c
abc	a and b and c	abc
a*bc	Zero or more a followed by b and c	bc abc aabc aaabc
(a|b)c	a or b and c	ac bc
a+bc	One or more a and b and c	abc aabc aaabc
ab?c	a and either zero or one b and c	ac abc

You can see that there may be many permutations of strings that might be produced by a regular expression. For example, a*bc could produce a text string with an infinite number of the letter 'a' at the beginning, but it would have to end in bc.

Consider the following examples:

- (a|b|c)d*e would produce the following possible strings: ade, ae, be, bde, ce and cde. There could in fact be an infinite number of the letter 'd' but it would have to end in an e.
- (a+(b|c))d would produce abd or aabd or aaabd (and so on with the preceding 'a') or the 'b' could be replaced by a 'c' in each string.

Perhaps the easiest way to understand all the permutations is to produce a state transition diagram.

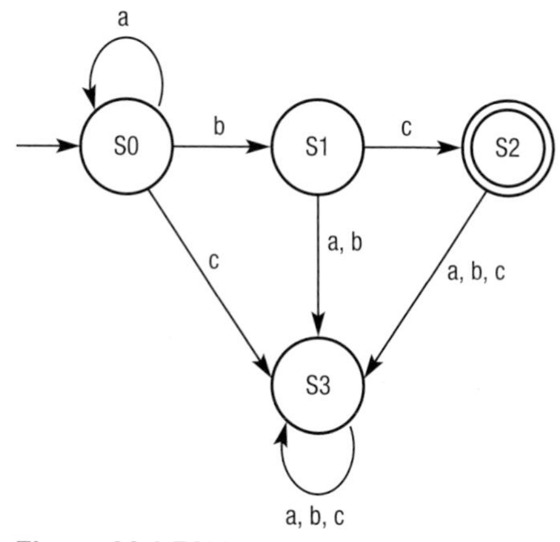

Figure 20.1 FSM to represent the regular expression a*bc

Figure 20.1 shows that 'a' can repeat any number of times and it must then be followed by 'b' and a 'c'. S2 represents the accepting state, so the last letter produced by this expression has to be a 'c'.

If you consider the expression a(bc)* it means that 'a' will be the first letter followed by any number of 'bc'. So 'a' would be one outcome as would abc or abcbc or abcbcbc. This could be represented as a state transition diagram as shown in Figure 20.2.

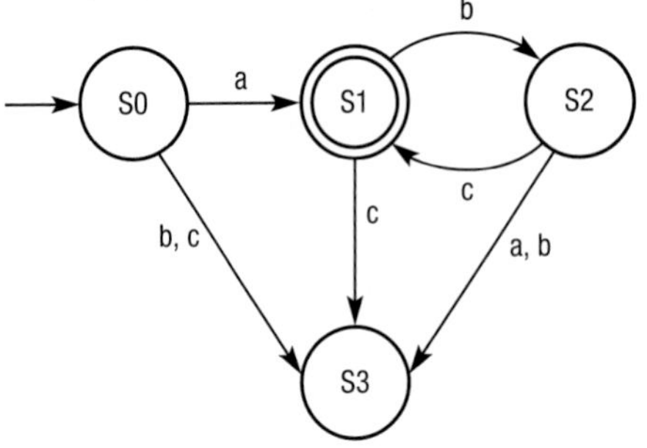

Figure 20.2 FSM to represent the regular expression a(bc)*

Here the string must start with an 'a' and end with a 'bc'. It is not possible for the outcome to be a 'c' on its own.

It is also possible to write the regular expression from the state transition diagram.

Consider the FSM in Figure 20.3; this diagram can only produce an 'a' or 'b' followed by a 'c', which would be written as (a|b)c.

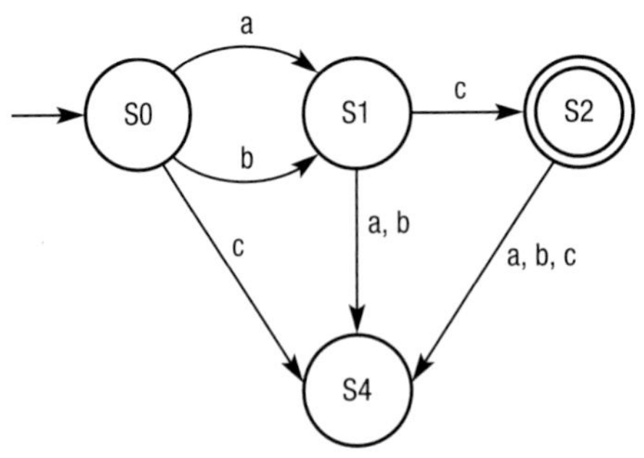

Figure 20.3 FSM to represent the regular expression (a|b)c

Searching strings

The power of regular expressions in this context is in using them to identify patterns in strings. Common uses include data validation, find and replace or searching for files with a particular file name. Most programming languages support the use of regular expressions and although the syntax and delimiters used may vary, the underlying principle remains the same. There is a standard set of expressions defined by POSIX.

Table 20.2 Standard expressions

Expression	Definition	Example
.	Effectively a wildcard and matches any character	.ole would match to mole, hole, vole etc.
[]	Matches to a single character within the brackets	[mh]ole would match to mole and hole but not vole.
[^]	Matches to any character except those in the brackets	[^ m]ole would match to hole and vole but not mole.
*	Matches the preceding characters zero or more times	m*ole would match ole, mole, mmole, mmmole, etc.
{m,n}	Matches the preceding character at least *m* but no more than *n* times	a{2,5} would match to aa, aaa, aaaa and aaaaa.

The following is an extract of Visual Basic-based pseudo-code showing how a search string could be implemented using a regular expression to identify numbers 0 to 9:

```
'Set a variable that contains the search string
Dim myString As String = "Software Version 3"
'Define the alphabet
Dim regex = New Regex("[0-9]")
'Set a variable to store the matching characters
Dim match = regex.Match(myString)
'Where a match is found between the alphabet and
the search string, write the matching characters to
the screen
If match.Success Then
   Console.WriteLine(match.Value)
End If
```

This code would produce the output "3". To use this code you will need to import `System.Text.RegularExpressions` into your program.

Context-free languages

> **KEYWORD**
> **Context-free language**: an unambiguous way of describing the syntax of a language useful where the language is complex.

A **context-free language** is a method of describing the syntax of a language used where the syntax is complex. As we saw in the previous chapter, one of the applications of state transition diagrams is to check that the syntax rules of a language are being followed. The technique can be used to check that different components of the code are in the correct place.

Regular expressions map directly to state transition diagrams. However, there are situations where the grammar used within a language is too complex to be defined by regular expressions. The key problem with regular expressions is that they only work when matching or counting other symbols in a string where there is a finite limit.

For example, consider a binary palindrome. This is a binary number that is the same backwards as it is forwards, e.g. 01110. If you think about a palindrome in normal language, for example, 'anna' or 'level', it would not be possible to create a regular expression that describes the syntax as there is no regular expression that can describe how each letter is used. Similarly with binary palindromes, there is no regular expression for the patterns of zeros and ones.

Where the counting and matching is infinite, a context-free language is needed. Context-free languages can also support notation for recursion and are sometimes a clearer way of defining syntax even where regular expressions can be used.

> **KEYWORD**
> **Backus-Naur Form (BNF)**: a form of notation for describing the syntax used by a programming language.

Backus–Naur Form (BNF)

The concept of context-free languages is that rules can be written that define the syntax of the language which are completely unambiguous and that can work beyond the current state. One method for doing this is to use Backus–Naur notation, known as **BNF (Backus–Naur Form)**.

KEYWORDS

Set: a collection of symbols in any order that do no repeat.

Terminal: in BNF, it is the final element that requires no further rules.

In common with regular expressions, BNF produces a **set** of acceptable strings, which effectively describe the rules of the language. It uses a set of rules that define the language in the format:

```
<S> ::= <alternative1> | <alternative2> |
<alternative3>

<alternative1> ::= <alternative2> | <alternative4>

<alternative4> ::= terminal
```

BNF works by replacing the symbol on the left with the symbols on the right until the string is defined. The idea is to keep going until you reach a **terminal**, which is a rule that cannot be broken down any further. In the example above:

- each symbol or element is enclosed within angle brackets <>
- the ::= means 'is replaced with' and defines the rule for the symbol
- each symbol needs to be split down further until you reach a terminal.

To define integers, a BNF expression may look like this:

```
<integer> ::= <digit> | <digit> <integer>

<digit> ::= 0|1|2|3|4|5|6|7|8|9
```

This shows that an integer is defined as either a digit or a digit followed by another integer. A digit is defined as the numbers 0 to 9 and this is a terminal as there is no further rule needed to define digits. This expression would be recursive as integer is defined in terms of itself.

Consider a more complex example, for customer details held in a database:

```
<customerdetails> ::= <name> <address>

<name> ::= <title> <firstname> <lastname>

<address> ::= <housenumber> <streetname> <town>
<county> <postcode>

...

<housenumber> ::= <integer>

<integer> ::= <digit> | <digit> <integer>

<digit> ::= 0|1|2|3|4|5|6|7|8|9

...
```

This example shows how BNF can produce simple rules, which can be written as:

- customer details must be made up of name and address.
- name must be made up of title, first name and last name.
- address must be made up of house number, street name, town, county and postcode.
- house number must be made up of an integer.
- integer must be made up of a digit or another integer.

This is only a partial BNF definition and could be continued to further split down the name into a series of acceptable characters, or the postcode into a series of letters and numbers and so on.

KEYWORD

Syntax diagram: a method of visualising rules written in BNF or any other context-free language.

Syntax diagrams

Another way of representing BNF expressions or any kind of context-free language is a **syntax diagram**. These map directly to BNF and use the symbols:

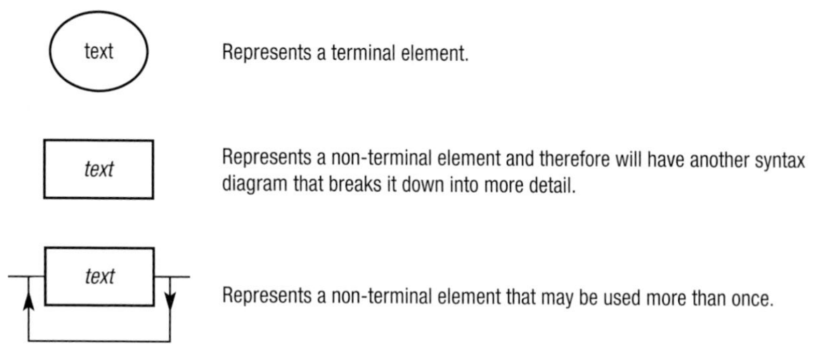

Figure 20.4 Syntax diagram symbols

Syntax diagrams are modular so there are likely to be many syntax diagrams required to represent a whole language. Each has an entry and exit point to identify the start and end of each particular part. For example, an integer would be represented as shown in Figure 20.5.

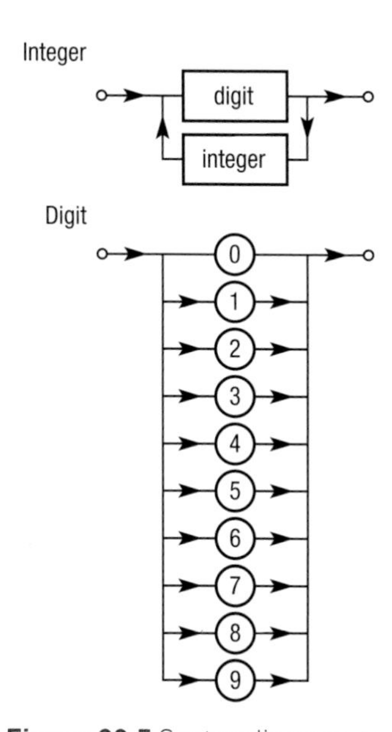

Figure 20.5 Syntax diagram to represent an integer

Figure 20.6 shows how you could break the customer details example into a series of syntax diagrams. This is just a partial diagram to demonstrate how to get to a terminal. In this case there are a series of non-terminal stages before you arrive at the terminal element, which are the actual characters that comprise a person's name.

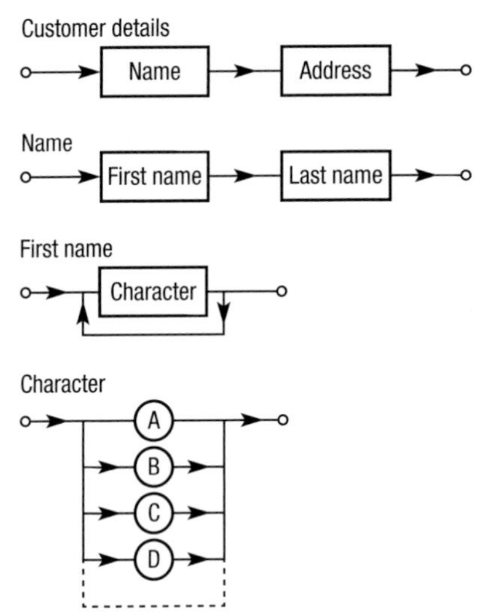

Figure 20.6 A syntax diagram for customer details

A worked example will help to explain this.

Suppose you wanted to create rules for a password that stated that for each character within the password the user must select either a capital letter, a lower case letter, a special character or a digit in any order. The BNF may look like this:

```
<character> ::= <uppercase>|<lowercase>|<number>|<spe
cialcharacter>

<uppercase> ::= "A" – "Z"

<lowercase> ::= "a" – "z"

<number> ::= 0|1|2|3|4|5|6|7|8|9

<specialcharacter> ::= *| - | _ | % | ^
```

Note that we have reached a terminal on all our elements.

Suppose we wanted to insist that the first character was a capital letter followed by any combination of other characters. The syntax diagram would look like Figure 20.7.

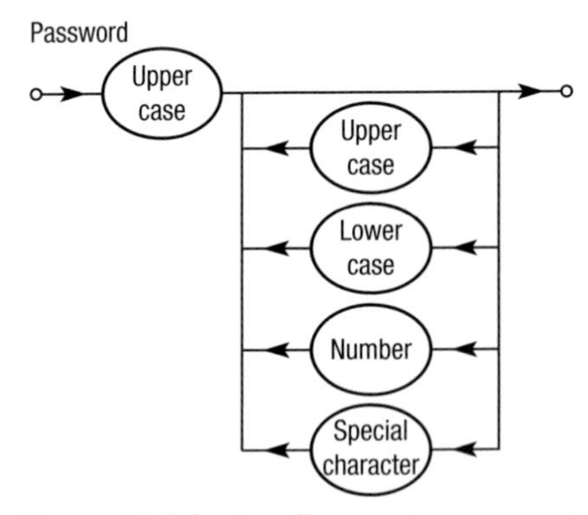

Figure 20.7 Syntax diagram to represent the creation of a valid password

Practice questions can be found at the end of the section on pages 179 and 180.

TASKS

1 Explain the difference between regular expressions and context-free languages.

2 Identify two text strings that would be acceptable for the following regular expressions:

a) a|b+c

b) (a|b)c*

c) a*b*c

3 Draw state transition diagrams for the three regular expressions in question 2.

4 Write a regular expression for each of the following descriptions that uses an alphabet of 0 and 1:

a) must start with a 1

b) any number of 0s followed by any number of 1s

c) any combination of 0s and 1s.

5 Use BNF to define the syntax of a car registration number in the format LLNN LLL.

6 Draw a syntax diagram for question 5.

KEY POINTS

- Regular expressions are a method of defining and searching sets of data.
- Regular expressions can be used to handle text and numeric strings.
- A regular language is one that uses regular expressions.
- Context-free languages are methods that can be used to describe the syntax of programming languages.
- Backus–Naur Form (BNF) is a notation for describing the syntax of a programming language in an unambiguous way.
- Syntax diagrams are a method of visualising the syntax of a particular programming language.

STUDY / RESEARCH TASKS

1 Use BNF to try and explain some of the rules of constructing a sentence using the English language.

2 Research Extended Backus–Naur Form (EBNF).

3 Look into the work of Panini and the Sanskrit language. How does this relate to BNF?

21 Maths for regular expressions

A level only

SPECIFICATION COVERAGE

3.4.2.2 Maths for regular expressions

LEARNING OBJECTIVES

In this chapter you will learn:

- how sets can be created using set comprehension
- how sets can be represented in a programming language
- what the empty set is and how it is used
- what operations can be carried out on sets
- how sets can be finite or infinite
- what a subset is and how they are created.

INTRODUCTION

In the previous chapter we looked at how to use regular expressions to define and search strings within a set. In this chapter we will be looking at how regular expressions can be used on sets of numbers.

Sets

As a reminder, a set is a collection of unordered values where each value appears only once in the set. The values in the set are sometimes referred to as elements, objects or members. The common format for representing a set is as follows:

A = {1, 2, 3, 4, 5}

where A is the name of the set and 1 to 5 are the values or elements within it. Any value can be represented in a set. For example:

- $\mathbb{N}$ = {0, 1, 2, 3, 4, 5, 6, 7, 8, 9, ...} where $\mathbb{N}$ represents **natural numbers**
- $\mathbb{Z}$ = {..., –3, –2, –1, 0, 1, 2, 3, ...} where $\mathbb{Z}$ represents integers.

There is more on different sets of numbers in Chapter 27.

Set comprehension

In the example for A above, the set is defined by listing the actual numbers within the set. It is also possible to define the contents or members of a set using **set comprehension**. This means that the set is defined by the properties that the members of the set must have. This is sometimes called **set building**.

KEYWORDS

Natural number: a positive whole number including zero.

Set comprehension: see Set building.

Set building: the process of creating sets by describing them using notation rather than listing the elements.

KEYWORD

Member: describes a value or element that belongs to a set.

For example:

$A = \{x \mid x \in \mathbb{N} \wedge x \geq 1\}$

where:

- A is the name of the set
- the curly brackets { } represent the contents of the set
- x represents the actual values of the set that will be defined after the pipe |
- the pipe | means 'such that', meaning that the equation after the x defines the values of x
- $\in$ means 'is a **member** of'
- $\mathbb{N}$ is all of the natural numbers, e.g. 0, 1, 2, 3 etc
- $\wedge$ means 'and'
- ≥ 1 means greater than or equal to one.

In this case we have used set comprehension to create a set of values that are the natural numbers from 1 upwards, which we could show as {1, 2, 3, 4, ...}. Note the use of the ellipsis (...) to indicate that the sequence continues.

Let's look at another example:

$A = \{x \in \mathbb{R} \mid x = x^2\}$

In this case x is a real number where the value of x is the same as the value of x^2. There are only two values that meet this criterion, which are 0 or 1. Therefore A = {0, 1}.

Set comprehension is a powerful tool as it can be used to define complex sets of values, without having to define every value. It means that the set can be represented in an efficient and compact form. For example:

$A = \{2x \mid x \in \mathbb{N}\}$ would produce the elements {0, 2, 4, 6, 8, ...}

$A = \{x^2 \mid x \in \mathbb{N}\}$ would produce the elements {0, 1, 4, 9, 16, ...}

Let's look at an example using binary values. Standard notation for binary values would be:

$\Sigma = \{0, 1\}$

This shows that the elements of the set of binary values are 0 or 1. To build a set of all binary strings containing two bits, you could use Σ^2 where the ^{2} indicates two bits. Therefore:

$\Sigma^2 = \{00, 01, 10, 11\}$

$\Sigma^3 = \{000, 001, 010, 011, 100, 101, 110, 111\}$

Again this shows how simple notation can be used to define a large number of elements. In some cases the number of elements may be infinite. For example:

$A = \{0^n1^n \mid n \geq 1\}$ would produce the set {01, 0011, 000111, 00001111, ...} representing all binary stings with an equal number of 0s and 1s with the 0s preceding the 1s.

Representing sets in programming languages

Most programming languages have set-building routines enabling you to create sets either by entering values or using set comprehension techniques. For example:

- In Python you can write the code `a = set ([0, 1, 2, 3])` where the contents of the square brackets form the set.. Alternatively, you could write the code `a = set ([x**2 for x in [1, 2, 3]])`, which would produce the set {1, 4, 9}.
- In Haskell you can write the code `[1..100]`, which makes a list containing the values 1 to 100. This can be combined with other functions to define a more complex list, for example `[x*2 | x <- [1..100]]`, will produce a list of the doubles of all the numbers between 1 and 100.
- In C# you could write the code `IEnumerable<int> numbers = Enumerable.Range (0,9)` to produce the integers 0 to 9. The code `var evens = from num in numbers where num % 2 == 0 select num;` would then extract the even numbers.

The empty set

KEYWORD

Empty set: the set that contains no values.

There is a special set known as the **empty set** which is represented either as {} or as Ø. The empty set has no elements. However, it is not to be confused with zero. The easiest way to think about it is as a container that could contain something, but it is empty.

Consider the following question: How many countries begin with the letter X? If you tried to put this answer in the container there would be nothing to put in so you might put a zero in the container. The problem with this is that the container is no longer empty as we now have one element in it – a zero.

Therefore in scenarios where an operation results in no answer we can use the empty set. Consider the following operation:

$A = \{1, 3, 5, 7, 9, ...\}$

$B = \{2, 4, 6, 8, 10, ...\}$

$A \cap B = \emptyset$

The $\cap$ represents intersection so this equation is looking for elements that are in the first set that can also be found in the second set. As there aren't any, the answer is represented as the empty set by the Ø symbol.

Finite and infinite sets

KEYWORDS

Finite set: a set where the elements can be counted using natural numbers up to a particular number.

Infinite set: a set that is not finite.

Cardinality: the number of elements in a set.

Countable set: a finite set where the elements can be counted using natural numbers.

We have already seen that some sets may contain a finite number of elements, or that they may contain an infinite number of elements. For example:

- $A = \{1, 2, 3, 4, 5\}$ is a **finite set** with five elements
- $A = \{1, 2, 3, 4, 5, ...\}$ is an **infinite set** made up of natural numbers
- $A = \{x \mid x \in \mathbb{N} \wedge x \geq 1\}$ is an infinite set of natural numbers greater than zero.

Where a set is finite it has **cardinality**, which means that it can be counted using natural numbers. It is also referred to as a **countable set**. The cardinality of a set is simply the number of elements in the set. We may also refer to this as the size of the set. For example, the first set above has a cardinality of five as it has five elements. The empty set has a cardinality of zero.

KEYWORDS

Countably infinite sets: sets where the elements can be put into a one-to-one correspondence with the set of natural numbers.

Cartesian product: combining the elements of two or more sets to create a set of ordered pairs.

Union: where two sets are joined and all of the elements of both sets are included in the joined set.

Intersection: describes which elements are common to both sets when two sets are joined.

Difference: describes which elements differ when two sets are joined together.

Infinite sets do not have cardinality as we do not know the total size of the set. However, for some infinite sets it is possible to go through the process of counting the elements, even though you would never reach the end. These are described as **countably infinite sets** as they can be counted off against the natural (countable) numbers.

Set operations

It is possible to join two or more sets together to create a new set. This is known as the **Cartesian product**. For example:

A = {a, b, c}

B = {1, 2, 3}

A × B would produce a set of all possible ordered pairs where:

- the first member of A is paired with the first member of B
- then the first member of A is paired with the second member of B and so on for every member of B
- then the second member of A is paired with the first member of B
- the process is repeated until every member of A has been paired with every member of B.

The resulting set or Cartesian product of the two sets would be: {(a,1), (a,2), (a,3), (b,1), (b, 2), (b,3), (c,1), (c,2), (c,3)}. This could also be written as:

$A \times B = \{x , y \mid x \in A \wedge y \in B\}$

Notice that the cardinality of the output set is always going to be the same as the product of the two input sets. In this case, A and B both had a cardinality of three, which means that our output set will have a cardinality of nine (3 × 3).

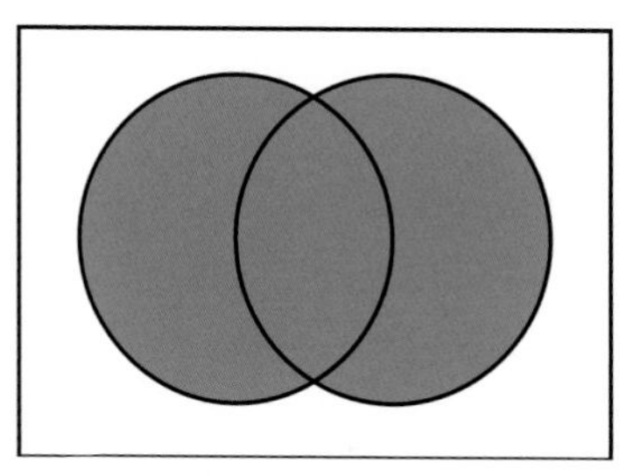

Figure 21.1 Venn diagram to represent the union of two sets

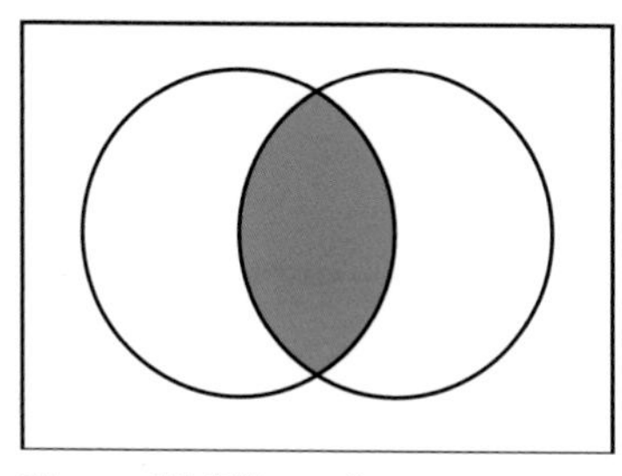

Figure 21.2 Venn diagram to represent the intersection of two sets

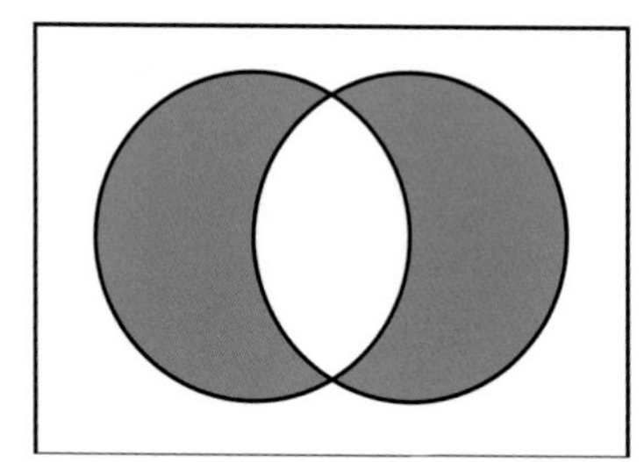

Figure 21.3 Venn diagram to represent the difference between two sets

When working with two or more sets there are different ways of defining the relationship between the members of the two sets. There are three main operations:

- **Union**: This means joining together two or more sets so that the new set is a combination of both sets. This can be represented as $A \cup B$. For example, if A = {0, 1, 3, 5, 7, 9} and B = {0, 2, 4, 6, 8} then $A \cup B$ will be {0, 1, 2, 3, 4, 5, 6, 7, 8, 9}. Note that the 0 is the only value common to both sets and will only appear once in the combined set. This could be represented visually as shown in Figure 21.1.
- **Intersection**: This means when two sets are joined together, the resulting set contains those elements that are common to both. It can be represented as $A \cap B$. For example, if A = {1, 3, 5, 7, 9} and B = {1, 3, 4, 6, 8} then $A \cap B$ would be {1, 3}. This could be represented visually as shown in Figure 21.2.
- **Difference**: This means that when two sets are joined together the resulting set contains elements that are in either set, but not in their intersection. This can be represented as $A \ominus B$ or $A \Delta B$. For example, if A = {1, 2, 3, 4, 5, 6, 7, 8} and B = {2, 4, 6, 8, 10} then the difference would be {1, 3, 5, 7, 10}. This could be represented visually as shown in Figure 21.3.

Subsets

Where all of the elements of one set are also contained within another set, it is said to be a **subset**. For example, A = {1, 3, 5, 7, 9, ...} is a subset of B = {1, 2, 3, 4, 5, ...} as all of the elements of A are contained within B. This can be shown as $A \subset B$ where the symbol means 'is a **proper subset** of'. The definition of a proper subset is one that has fewer elements than the set. In this case, the subset A contains just the odd numbers from the other set. Similarly we could say that $A \subset B$ where A = {1, 2, 3, 4, 5] and B = {1, 2, 3, 4, 5, 6} as there is at least one number in B that is not in A.

The notation of a subset (as opposed to a proper subset) is $A \subseteq B$ and the distinction is that two sets that are the same can be said to be subsets of one another. For example, where A = {1, 2, 3, 4, 5} and B = {1, 2, 3, 4, 5} we can say that A is a subset of B because it contains everything within B.

KEYWORDS

Subset: a set where the elements of one are entirely contained within the other; can include two sets that are exactly the same.

Proper subset: where one set is wholly contained within another and the other set has additional elements.

Practice questions can be found at the end of the section on pages 179 and 180.

TASKS

1 What is cardinality?
2 Use set comprehension to represent all real numbers greater than zero.
3 Use set comprehension to define all negative integers.
4 Use set comprehension to define the cube of all natural numbers.
5 Explain why the empty set is not the same as zero.
6 Use examples to explain the difference between union, intersection and difference when joining two or more sets together.
7 Define the Cartesian product that would result from A = {x, y, z} and B = {1, 2}.
8 What is the cardinality of a set resulting from the Cartesian product of A with 8 elements and B with 9 elements?
9 Use an example to explain a proper subset.

KEY POINTS

- A set is a collection of unordered, non-repeated data items of the same type.
- Set comprehension is the process of building a set using an expression.
- Programming languages support set-building techniques.
- The empty set is a set with no values in it.
- Sets can have a finite or infinite number of elements in them.
- Sets can be combined in different ways to create new sets.

STUDY / RESEARCH TASKS

1 Write code to define sets for:
   a) all rational numbers greater than zero
   b) all negative integers
   c) the cube of all natural numbers.

   Research other ways in which sets can be joined together including supersets, power sets and complements.
2 Find out about 'Hilbert's paradox of the Grand Hotel' to help you understand the concept of countably infinite sets.

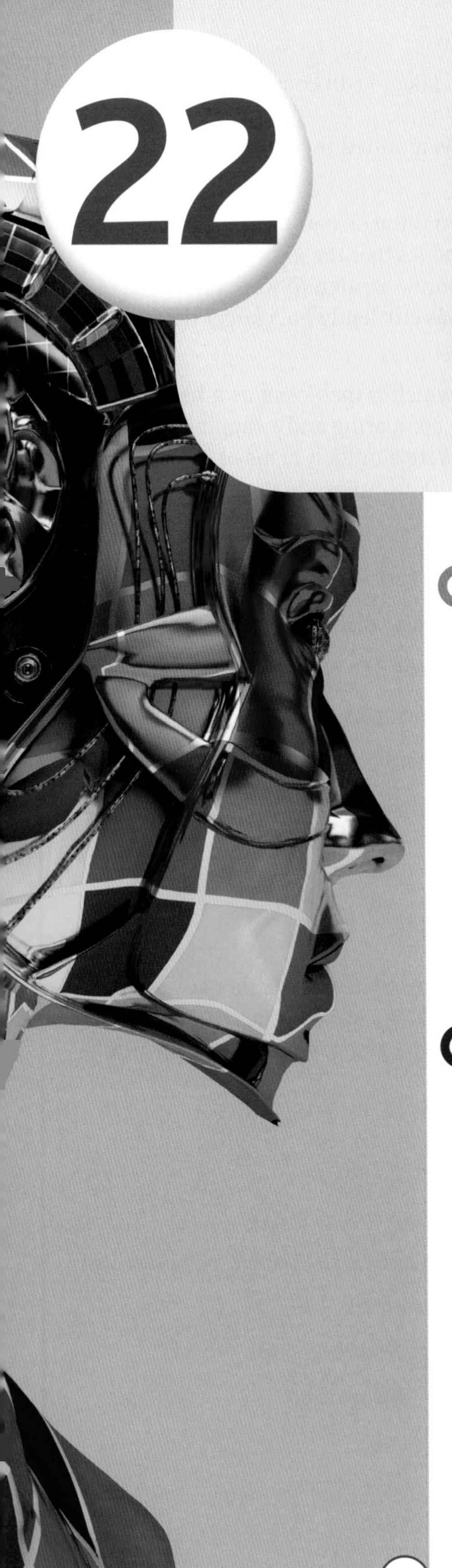

22 Big O notation and classification of algorithms

A level only

SPECIFICATION COVERAGE

3.4.4 Classification of algorithms

LEARNING OBJECTIVES

In this chapter you will learn:

- some algorithms are more efficient than others
- how to classify algorithms by their time and space complexity
- how the basic mathematical functions describe time and space complexity work
- how Big O notation classifies algorithms in terms of their time and space complexity work
- that some problems are tractable (solvable within a reasonable amount of time on a computer) and some are intractable (not solvable within a reasonable amount of time on a computer).

INTRODUCTION

An algorithm is a sequence of steps designed to perform a particular task. As we have seen, programs are constructed from algorithms, which may comprise a few lines of code or whole blocks of code, depending on the complexity of the problem. For example, an A-level project might have a few hundred lines of code. A typical phone app has around 100 000 lines of code and according to some estimates, the latest version of MS Office uses around 45 million lines of code. This illustrates that there is usually a relationship between the size and scope of the problem and the size and the scope of the code needed to provide a solution.

This chapter looks at how you can classify algorithms by their complexity in terms of how much time code takes to achieve a result and how much memory it requires. The method of describing the complexity of algorithms is called Big O notation.

KEYWORD

Algorithm: a set of instructions required to complete a particular task.

Classifying algorithms

Faced with a problem, a programmer may come up with different **algorithms** that provide a solution. One of the objectives for writing good code is to produce an efficient solution. Efficiency is usually measured in terms of time and space:

- Time: how long does the algorithm take to run compared to other algorithms.
- Space: how much space (memory) is required by the algorithm compared to other algorithms.

The key consideration is what is called input size or problem size. Typically this is the number of parameters or values that the algorithm will be working on. For example, a search routine written to work on a dataset with only a few values may not work as efficiently on a larger dataset with hundreds of values.

The code below shows a bubble sort which is inefficient as it has to loop through the whole dataset every time comparing and swapping two adjacent items on each pass. If there were a million items of data it may have to go round approximately $1\,000\,000^2$ times.

```
For Loop1 = 1 To NumberOfItems - 1
  For Loop2 = 1 To NumberOfItems - 1
    'if the following name is smaller then swap
    If NameStore(Loop2) > NameStore(Loop2 + 1) Then
      SwapData = NameStore(Loop2)
      NameStore(Loop2) = NameStore(Loop2 + 1)
      NameStore(Loop2 + 1) = SwapData
    End If
  Next
Next
```

The code could be made more efficient by checking whether any swaps have taken place in each pass of the data. If a swap has taken place between two data items than they do not need to be compared again on the next iteration of the loop:

```
Do
SwapData = ""
  For Loop2 = 1 To CountTo
    If NameStore(Loop2) > NameStore(Loop2 + 1) Then
      SwapData = NameStore(Loop2)
      NameStore(Loop2) = NameStore(Loop2 + 1)
      NameStore(Loop2 + 1) = SwapData
    End If
  Next
  CountTo = CountTo - 1
Loop Until SwapData = ""
```

In this chapter we will look at how to compare algorithms by analysing the time and space requirements in response to changes in input size.

Functions

Comparing algorithms uses a technique called Big O notation. This uses standard mathematical functions. Before we look at Big O, we need to understand the functions it uses.

A **function** simply relates an input to an output. For example $f(x) = x^2$ is an example of a function. It means that you take the input value for the function, x, and produce an output, which in this case is the squared value of x. The set of values that can go into a function is called the **domain** and the set of values that could possibly come out of it is called the **codomain**. The set of values that are actually produced by the function is called the range. It is always the case that the range will be a subset of the codomain.

Another important mathematical concept for understanding Big O is that of permutations. If you consider an item of data such as a text string, or a number, there are different ways in which the characters or digits can be put together. For example:

- A word with two letters has two permutations: 'to' can be 'to' or 'ot'.
- A word with three letters has six permutations: 'dog' can be 'dog', 'dgo', 'odg', 'ogd', 'gdo' or 'god'.
- A word with four letters can have 24 permutations. Rather than work through them the basic formula is that there are four ways to pick the first letter, followed by three ways to pick the second letter followed by two ways to pick the third letter and finally one way to pick the last letter. This could be shown as $4 \times 3 \times 2 \times 1$, which gives 24.
- A word with five letters therefore has $5 \times 4 \times 3 \times 2 \times 1 = 120$ permutations.

This is called the **factorial** function and can be denoted by $n!$ Where n is an integer. For our five-letter word, we could show this as 5! which means $5 \times 4 \times 3 \times 2 \times 1 = 120$.

KEYWORDS

Function: relates each element of a set with the element of another set.

Domain: all the values that may be input to a mathematical function.

Codomain: all the values that may be output from a mathematical function.

Factorial: the product of all positive integers less than or equal to n, e.g. 3! is $3 \times 2 \times 1$.

Big O notation

Big O notation is a method of describing the time and **space complexity** of an algorithm. It looks at the worst-case scenario by essentially asking the question: how much slower will this code run if we give it 1000 things to work on instead of 1? For example, if you had a bubble sort routine that compared the first two items of data and swapped them if necessary, then compared the next two items of data and so on, this might work quite quickly on a list of 10 items. But what if you asked it to sort a list of 1 million items?

Big O notation provides a measure of how much the running time requirements of the code will grow as the magnitude of the inputs changes. Big O calculates the upper bound, which is the maximum amount of time it would take an algorithm to complete. The notation refers to the order of growth, also known as the order of complexity, which is a measure of how much more time or space it takes to execute the code as the **input size** increases. The format is a capital letter O followed by a function. All of the explanations below relate specifically to **time complexity**, rather than space complexity.

KEYWORDS

Space complexity: the concept of how much space an algorithm requires.

Input size: in Big O notation the size of whatever you are asking an algorithm to work with, e.g. data, parameters.

Time complexity: the concept of how much time an algorithm requires.

KEYWORD

Constant time: in Big O notation where the time taken to run an algorithm does not vary with the input size.

Big O notation uses five main classifications:

- O(1), known as **constant time**, means that the algorithm will always execute in exactly the same amount of time regardless of the input size. Accessing an array would be an example of this as each element of the array is accessed directly by referring to its position. Therefore, it would not take any longer to access a single element if there were one or ten million items in the array. Note that O(1) does not necessarily mean that the code will run quickly, it just means that it will take the same amount of time (it is constant) regardless of the input.
  If you were to represent this as a graph, it might look like Figure 22.1.

  This shows that however much the input size increases, the time taken to run the algorithm remains the same.

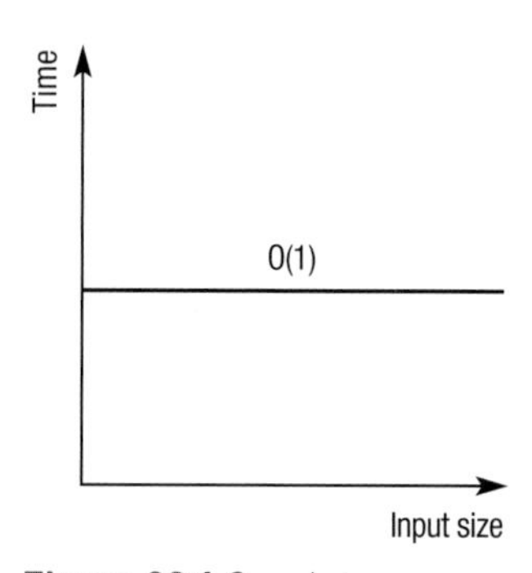

Figure 22.1 Graph to represent constant function

KEYWORD

Linear time: in Big O notation where the time taken to run an algorithm increases in direct proportion with the input size.

- O(*N*) represents a **linear** function and it means that the runtime of the algorithm will increase in direct proportion with the input size. For example, there could be a relationship where if you input twice as much data, the algorithm will take twice as long.
  This could be represented as $y = x$ where every change in x produces a corresponding change in y. The linear relationship may be more complex. Another example of O(N) is $y = 2x$ which means that every change in x would produce double this change in y. This could be represented graphically as shown in Figure 22.2.

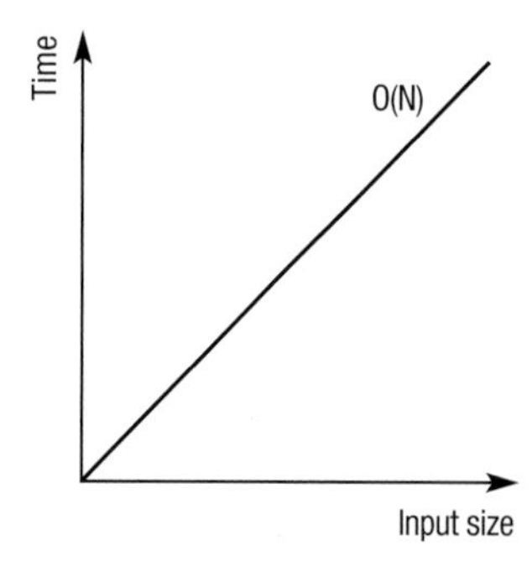

Figure 22.2 Graph to represent linear function

Looping around a list would be an example of this because the code needs to access every element of the list. If you increase the size of the list the amount of time taken to carry out the loop will increase by a linear amount.

- O(N^2) is an example of a polynomial function and it means that the runtime of the algorithm will increase proportionate to the square of the input size. To take a simplified example, let's say that one item takes 1 second (1^2). Ten items therefore would take 100 seconds (10^2) and 100 items would take 10 000 seconds (100^2). This could be represented as $y = x^2$. This could be represented graphically as shown in Figure 22.3.

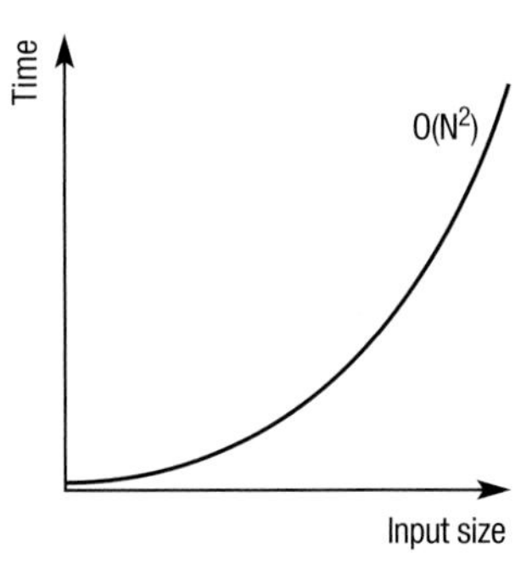

Figure 22.3 Graph to represent polynomial function

Iterative or nested statements such as bubble sorts and insertion sorts are examples of these as the program has to go back through itself again with each iteration. The following code shows a bubble sort using a loop:

```
Private Sub btnSort_Click(ByVal sender As System.
Object, ByVal e As System.EventArgs) Handles
btnSort.Click

    Dim Loop1 As Integer

    Dim Loop2 As Integer

    Dim TempStore As String

    Dim RowsToSort As Integer

    RowsToSort = grdDataIn.RowCount - 2

    For Loop1 = 1 To RowsToSort - 1

      For Loop2 = 1 To RowsToSort - 1

        'compare each value in the table with the
         following value

        ' changing the > operator will sort high
          to low

        If grdDataIn.Rows(Loop2).Cells(0).Value >
        grdDataIn.Rows(Loop2 + 1).Cells(0).Value Then

          'swap values to move larger values to
           later cells

          TempStore = grdDataIn.Rows(Loop2).Cells(0).
          Value

          grdDataIn.Rows(Loop2).Cells(0).Value =
          grdDataIn.Rows(Loop2 + 1).Cells(0).Value

          grdDataIn.Rows(Loop2 + 1).Cells(0).Value =
          TempStore

        End If

      Next

    Next

  End Sub
```

KEYWORDS

Exponential time: in Big O notation where the time taken to run an algorithm increases as an exponential function of the number of inputs. For example, for each additional input the time taken might double.

Logarithmic time: in Big O notation where the time taken to run an algorithm increased or decreases in line with a logarithm.

- $O(2^N)$ is an example of an **exponential** function where the runtime will double with every additional unit increase in the input size. To take a simplified example, one item might take 1 second, two items would take 2 seconds, three items 4 seconds and so on. Ten items of data would take 512 seconds. This can be expressed as $y = 2^x$ where x is each individual item being input and y represents the time taken. Obviously the amount of time taken to process each input will start to become unworkable. Problems with an exponential order of growth are often referred to as intractable problems, which means that they can't be solved with a computer in a reasonable time. There is more on this later in the chapter.

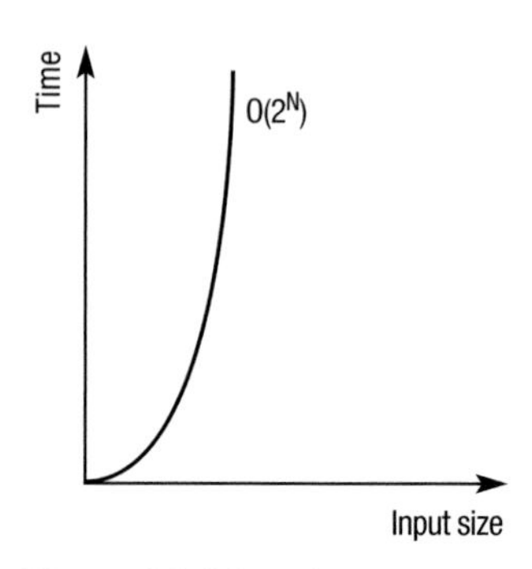

Figure 22.4 Graph to represent exponential function

- $O(\log N)$ represents a **logarithmic** function which uses an exponent to raise the value of a base number in order to produce the desired number. For example, with $y = \log_2 x$, this means we are using base 2. If $x = 8$ then y must equal 3 as you have to multiply 2 by 2 by 2 or 2^3 in order to get 8. In base ten, $y = \log_{10} x$, if $x = 10000$ then y must equal 4 as $10 \times 10 \times 10 \times 10$ or 10^4 equals 10000.

The usefulness of this can be shown by looking at a binary search. These work by continually splitting the data in half until the item being searched for is found (see Chapter 9). It is called a binary search as it splits the data in two each time. Therefore it has a time complexity of $O(\log_2 N)$ as each subsequent split of the data takes less time as only half the data is being searched. The number of data items is halving each time so even large datasets can be searched with a relatively small number of comparisons:

Table 22.1

Number of data items	Number of comparisons needed in a binary search
2	1
4	2
8	3
16	4
128	8
65356	16
1048576	20
4294967296	32

Table 21.1 shows the way in which the size of the input data that can be searched increases exponentially with each comparison. Therefore to carry out a binary search of over 4 billion items would only actually take up to 32 comparisons within the dataset.

This could be represented graphically as shown in Figure 22.5.

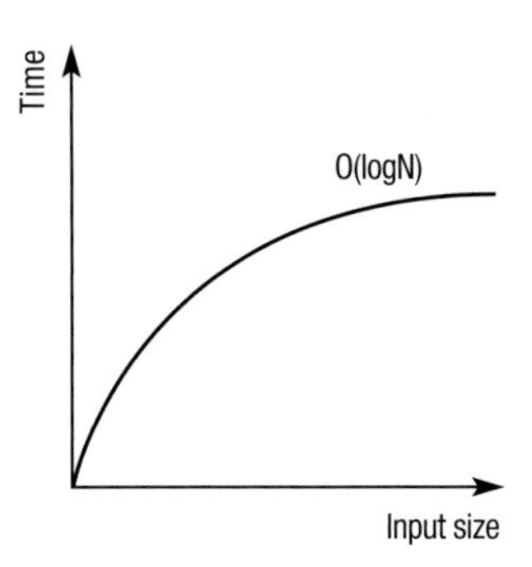

Figure 22.5 Graph to represent logarithmic function

The value of using Big O notation is that you can find the most efficient solution for your problem. Remember that Big O uses the upper bound so is a good measure of how scalable your solution is, that is, how efficient it will be as the input size increases. As a broad rule of thumb:

- An O(1) algorithm scales the best as it never takes any longer to run.
- An O(log*N*) algorithm is the next most efficient.
- An O(*N*) algorithm is the next most efficient.
- An O(N^2) algorithm is a polynomial and is considered to be the point beyond which algorithms start to become intractable. Note that the superscript number could be any value.
- An O(2^N) algorithm is the least efficient and considered intractable.

Deriving the complexity of an algorithm

It is possible to derive the time complexity of an algorithm by looking at the contents of the code. For example:

- An algorithm that requires no data and contains no loops or recursion, such as a simple assignment statement or comparison statement, will have a time complexity of O(1).
- An algorithm that loops through an array accessing each data item once will have a time complexity of O(*N*).
- An algorithm with inner and outer loops will be polynomial with runtime increasing depending on the depth of the nesting and the number of loops. In this case it will be O(N^2).
- The addition of a loop within the inner loop would alter the **polynomial time** to O(N^3).
- An algorithm that uses recursion to call itself could have a time complexity of O(a^N).

It is not always easy to work out the precise time complexity of an algorithm as it may depend on how many times parts of the algorithm may run, based on the outcome of conditional statements. For example, with an `If` statement, part of the algorithm will only need to run if a condition is true. Therefore the actual number of times that some of the program code executes will depend upon the data that is input to the algorithm when it is carried out.

KEYWORD

Polynomial time: in Big O notation where the time taken to run the algorithm is a polynomial function of the input size, e.g. the square of the input size.

With Big O notation, to describe the complexity of a problem, the usual practice is to quote the worst-case scenario for the most efficient algorithm. However, in choosing a suitable algorithm it is possible to make a comparison between the worst, best and average cases based on time complexity.

Some common algorithms are known to have certain time complexities as shown in Table 22.2.

Table 22.2 Common algorithms with time complexity in Big O notation

Complexity	Algorithms
$O(1)$	Indexing an array
$O(\log N)$	Binary search
$O(N)$	Linear search of an array
$O(N^2)$	Bubble sort Selection sort Insertion sort
$O(2^N)$	Intractable problems

Tractable and intractable problems

KEYWORDS

Tractable problem: a problem that can be solved in an acceptable amount of time.

Intractable problem: a problem that cannot be solved within an acceptable time frame.

Heuristic: with algorithms it is a method for producing a 'rule of thumb' to produce an acceptable solution to intractable problems.

A **tractable problem** is one that is said to be solvable in polynomial time. In simple terms this means that the algorithm that solves the problem runs quickly enough for it to be practical to solve the problem on a computer.

Intractable problems are those which are theoretically possible to solve, but cannot be solved within polynomial time. The problem may be solvable if the input size is small, but as soon as the input size increases it is considered impractical to try and solve it on a computer. A classic example is 'the travelling salesman' problem, which is primarily a conceptual problem that has been tackled by mathematicians and computer scientists for over 100 years:

- A salesman has to travel between a number of cities.
- The distance between each pair of cities is known.
- He must visit each city just once and then return to his start point.
- He must calculate the shortest route.

On the face of it, there may be a simple solution to this problem, particularly if there are only a small number of cities to visit. However, as the number of cities increases, the permutations of routes grow at a much faster rate.

There are many similar problems that involve calculating distances between pairs of points. For example, some analysis looks at points on a circuit board in order to optimise data transmission. As we have seen, the time complexity of a problem is typically concerned with looking at the worst-case scenario, so must consider thousands of points. To date, algorithms have been created that can calculate the actual shortest distance between around 85 000 pairs of points. These algorithms have a very large time complexity that go well beyond polynomial time and are therefore considered intractable.

Faced with intractable problems, programmers often produce **heuristic** algorithms. Having accepted that the perfect solution is not possible, a solution that provides an incomplete or approximate solution is seen as being preferable to no solution at all. Often this will involve ignoring certain complex elements of the problem, or accepting a solution that is not optimal.

Heuristics often uses 'rules of thumb' and therefore cannot guarantee accurate results for every possible set of inputs. Instead they produce results that may be accurate for common uses of the program, but less accurate where the program is being used with less likely inputs. The objective with a heuristic algorithm is to produce an acceptable solution in an acceptable time frame, where the optimum solution would simply take too long.

For example, several heuristic solutions have been developed for the travelling salesman problem that theoretically enable millions of cities to be considered. None of these compare every possible pair of cities so they do not create an actual solution. Instead they produce an approximate solution, which may well be very accurate. Estimates of some of these methods produce results that may be within 5% degree of accuracy of the optimum solution.

Unsolvable problems

Unsolvable problems are those which will never be solved regardless of how much computing power is available either now or in the future and regardless of how much time is given to solve it. The '**halting problem**' is an example of a problem that is proven to be unsolvable. In simple terms, the halting problem is whether it is possible to make a program to determine if a program will finish running for a particular set of inputs.

The question devised by Alan Turing in the 1930s asks whether it would be possible to write a computer program to solve the halting problem. The conclusion is that the problem is unsolvable, as it is proven that it cannot be done.

The halting problem is considered to be one of the first unsolvable problem ever identified and has led to the discovery of many other unsolvable problems. This area of computing has led to a general acceptance that there are some problems which:

- simply cannot be solved by computers (unsolvable problems)
- can theoretically be solved by computers but it is not possible within a reasonable time frame (intractable problems)

KEYWORDS

Unsolvable problem: a problem that it has been proved cannot be solved on a computer.

Halting problem: an example of an unsolvable problem where it is impossible to write a program that can work out whether another problem will halt given a particular input.

Practice questions can be found at the end of the section on pages 179 and 180.

TASKS

1. Why is it important to develop algorithms that take account of time and space requirements?
2. What is input size?
3. Explain why a problem might be more time consuming to solve as the input size increases.
4. Use an example to explain the domain, codomain and range within a function.
5. What is a factorial?
6. Describe the five main classifications used in Big O notation and give an example of an algorithm for each of the time complexities.
7. What is the most and least efficient time complexity according to Big O notation?
8. Why are some problems considered intractable?
9. What is the halting problem?

KEY POINTS

- Some algorithms are more efficient than others in terms of their time and space complexity.
- Time complexity is the amount of time an algorithm takes to solve a problem and is the main focus of study for A level.
- Space complexity is the amount of memory an algorithm takes to solve a problem.
- Big O notation is a method of comparing algorithms in terms of their time and space complexity.
- Big O notation uses standard mathematical functions to classify algorithms.

STUDY / RESEARCH TASKS

1. What is P and why is it important when designing algorithms?
2. Research other problems that are considered intractable such as:
   a) P = NP problem
   b) the k-server problem.
3. Write code for a binary search and a linear search. Explain the time complexity of each with reference to your code.
4. Research the space complexity of different data structures, e.g. array, list, linked list, binary tree.

Section Four: Practice questions

1 A holiday tour business wants to create a new computer system. They offer 30 different tours every year, each of which can have up to 50 people on it. The tour company organises the travel and hotel arrangements and lays on a number of excursions. They need to organise tour guides to accompany their customers and manage all of the payments from customers and to suppliers.

a) Explain how you could use abstraction by generalisation/categorisation to break this problem down.

b) Produce a hierarchy chart to show how you could decompose this problem.

c) List at least six items of data that the tour business will need to collect.

d) Give two examples of how information hiding could be used in this scenario.

2 The finite state machine (FSM) shown processes a language with an alphabet of a, b and c.

a) Which of these input strings would be accepted?

i) aaabc

ii) baabc

iii) aaaab

iv) abc

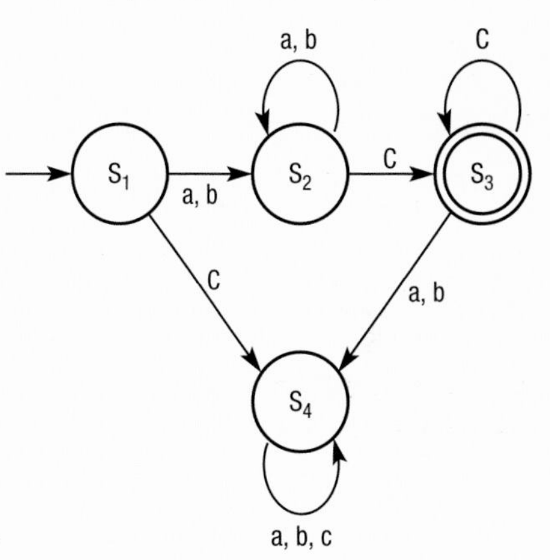

b) Which of the states is the accepting state?

c) Draw a transition table for this FSM.

d) Write a regular expression that would recognise the same language as this FSM.

3 Backus–Naur Form (BNF) can be used to define the rules of a language. In the example below BNF is used to define parts of the addresses.

```
<fulladdress> ::= <housenumber>_<street>_<town>_
<county>_<postcode>
...
<street> ::= <character> | <character> <street>
<character> ::= <A|B|D|E...>
...
<housenumber> ::= <digit> | <digit> <housenumber>
<digit> ::= <0|1|2|3|4...>
```

a) Give an example of where regular expressions could be used in the example above.

b) The BNF above is incomplete. Write all the rules needed for `<fulladdress>`, including a rule to define a postcode, which is made up of up of two characters, two integers and a space, an integer and two characters, e.g. LE11 1AA.

4 The common orders of time complexity are shown in the table.

Time complexity
$O(1)$
$O(n^2)$
$O(\log_2 n)$
$O(k^n)$
$O(n)$

a) Describe in words what O(1) means.

b) Which is the time complexity of an intractable problem?

c) What is meant by an intractable problem?

d) Which is the time complexity for a binary search?

e) Which is the time complexity for a linear search?

f) On average, would a binary search or a linear search be quickest on a list of just five items? Explain your answer.

5 A Turing machine is represented by the following transition table.

a) What is a Turing machine?

b) What is a Universal Turing machine?

State	Read	Write	Move	Next state
S0	1	0	R	S1
S0	0	1	R	S1
S0	B	0	R	SH
S1	1	1	R	S1
S1	0	0	R	S0
S1	B	1	R	SH

c) Draw a state transition diagram for the instructions in the table.

d) Write out the instructions in the format:

δ (Current State, Input Symbol) = (Next State, Output Symbol, Movement)

e) The Turing machine is carrying out a computation. Its starting state is S0 and the contents of the tape and location of the tape head are shown below. State SH is the halting state. Trace the computation, showing the contents of the tape, the current position of the read/write head and the current state as the input symbols are processed.

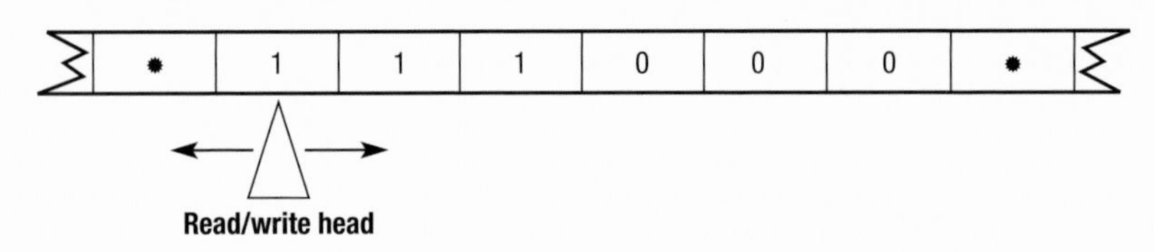

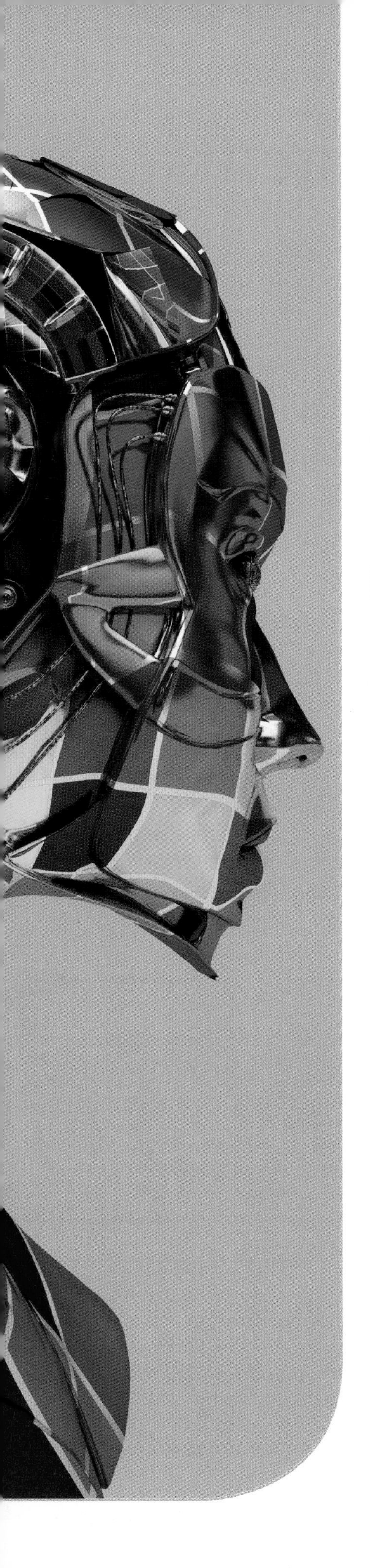

Section Five: Fundamentals of data representation

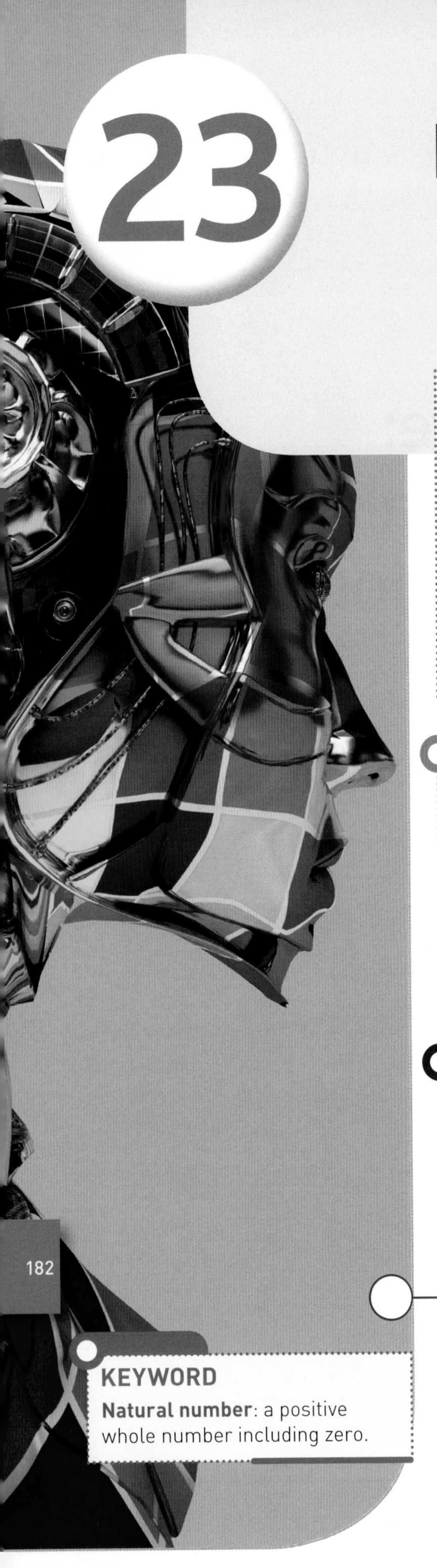

23 Number systems

SPECIFICATION COVERAGE

3.5.1.1 Natural numbers

3.5.1.2 Integer numbers

3.5.1.3 Rational numbers

3.5.1.4 Irrational numbers

3.5.1.5 Real numbers

3.5.1.6 Ordinal numbers

3.5.1.7 Counting and measuring

LEARNING OBJECTIVES

In this chapter you will learn:

- there are different types of numbers such as natural numbers, integers, rational and irrational numbers, real numbers and ordinals
- that numbers are used in different ways to produce different outcomes
- how an array is used to store numbers
- the importance of numbers for counting and measuring.

INTRODUCTION

A number is a unit of mathematical data used to count, quantify, label and measure. We are used to using standard number systems such as the decimal system and over the next few chapters you will learn about binary and hexadecimal too. In this chapter we will look at how different types or sets of numbers can be used in different ways in computing.

KEYWORD

Natural number: a positive whole number including zero.

Natural numbers

These are the most recognisable type of number as they are the numbers that we use every day for counting and ordering. Typically these are numbers made up of the decimal digits 0, 1, 2, 3, 4, 5, 6, 7, 8, 9. This is known as decimal or base 10 as there are ten different digits that we use. Using a single-digit number we can represent a maximum value of 9. We simply add a new digit, to create the numbers 10–99. We then continue adding digits to create hundreds, thousands and so on.

Each extra digit we add is worth ten times as much as the previous digit, as it is base 10. This is an important concept and will help you to understand the binary system in the next chapter. We can use this system to represent an infinite range of numbers, the basic principle being that there are only actually ten different digits on which all the numbers are based. It is a common mistake to view natural numbers as 1 to 10, whereas in fact they are 0 to infinity. The inclusion of 0 as a natural number has implications in computing. For example, if you set a counter when programming, the first instance of the counter will be 0 rather than 1.

The mathematical symbol for natural numbers is **N** or $\mathbb{N}$ so you might represent them as follows:

$\mathbb{N} = \{0, 1, 2, 3, 4, 5, 6, 7, 8, 9, \ldots\}$

Integer numbers

KEYWORD

Integer: any whole positive or negative number including zero.

An **integer** is a whole number whose value can be positive or negative. Zero is also classed as an integer. A whole number is one that does not contain a fractional part, which means there can be no fractions after the number and no decimal place values.

Negative values are indicated by the minus sign (–). The relationship between the negative and positive integers is that if you add the two you get back to zero. For example $-3 + 3 = 0$ or $3 + (-3) = 0$.

Integers are one of the standard data types that can be used when defining variables and constants in programming languages. In theory, there are an infinite number of integers that can be represented. In practice, different languages have variations of integer types depending on the size of the number that needs to be stored. In Visual Basic for example, the `Integer` type uses four bytes allowing integers in the range –2 147 483 648 through 2 147 483 647 to be created. It has another data type called `Long` to handle integers that will be outside this range.

The mathematical symbol for integer numbers is **Z** or $\mathbb{Z}$ so you might represent them as follows:

$\mathbb{Z} = \{\ldots, -3, -2, -1, 0, 1, 2, 3, \ldots\}$

Rational numbers

KEYWORD

Rational number: any number that can be expressed as a fraction or ratio of integers.

A **rational number** is one that can be expressed as a fraction, also known as a quotient. Fractions are made up of two integers with the value being the ratio between the two. The ratio can be expressed as a fraction or as its decimal equivalent. For example we know ½ as half, and it can also be represented as a decimal as 0.5.

The top integer is called the numerator and the bottom integer is called the denominator. The numerator can be any integer, which means that it can be any positive or negative whole number. The value of the fraction therefore can be negative or positive and it can be greater or less than one, or zero. Fractions can also result in the value of one. For example $\frac{5}{5} = 1$ or $\frac{8}{8} = 1$. This means that by definition all integers are rational numbers.

The only rule is that the denominator can be any integer apart from zero, as to divide by zero creates an undefined result. For example if you divided

3 by 0, the answer is not 0 as it is not possible to define how many times 0 goes into 3. This creates specific problems in computer programming as a program may not be able to continue when it comes across an undefined result. This is called a 'divide by zero' error and programmers need to be aware of it and trap it by writing appropriate code.

The mathematical symbol for rational numbers is **Q** or $\mathbb{Q}$.

Irrational numbers

An **irrational number** is any number that cannot be represented as a ratio of integers because the decimal equivalent would go on forever without repeating. A classic example of an irrational number is π. As a fraction, a widely used approximation is $\frac{22}{7}$. In decimal form, the number is infinite so it has to be truncated (cut off) to a set number of decimal places, for example to 3.14 or if more accuracy is required, to 3.1415926535.

> **KEYWORD**
>
> **Irrational number**: a number that cannot be represented as a fraction or ratio as the decimal form will contain infinite repeating values.

Some square roots and cube roots are irrational numbers. A square root can be represented as $x = n^2$. If you start with a simple example, the square root of 16 is 4. If you try to calculate the square root of 2, 3 or 99, you will find that it is impossible as there is no number that you can square that will exactly make 2, 3 or 99.

A feature of an irrational number is that the values after the decimal place do not repeat in a pattern. For example, there is no pattern to the decimal places in π. Recurring decimals do have a pattern and therefore are classed as rational numbers. For example, one-third = 0.333... recurring.

The implication in computing is that where irrational numbers need to be handled, the programmer needs to decide on the level of precision that is required and therefore how much memory to allocate to storing the value. In simple terms, this might mean defining the number of places to use after the decimal point. Computers handle this in a specific way using fixed and floating point numbers and there is more on this in Chapter 25.

Real numbers

A **real number** is any positive or negative value and can include a fractional part. Integers, rational numbers and irrational numbers are all real numbers. The defining feature of a real number is that the fractional part can be any length, allowing the number to represent a measurement to any level of precision and accuracy required.

> **KEYWORD**
>
> **Real number**: any positive or negative number with or without a fractional part.

For example on a scale between 1 and 10, real numbers can be used to break that down into units of 0.1. On a scale between 0 and 0.1, real numbers can be used to break that down into units of 0.01 and so on. There is an infinite range of numbers that can be generated.

Real numbers are used to measure continuous or infinitely changing values. For programmers, the issue is in defining how accurate and precise a numbers needs to be for the application. Consider the following examples:

1. A surveyor working on a road may need to take measurements ranging from several kilometres to several metres, e.g. 10.15 km.
2. An Olympic race timer for the 100 m may need to deal with a range from 8 to 15 seconds accurate to a thousandths of a second, e.g. 9.934 seconds.

3 A sensor taking readings of the core temperature in a nuclear power station may need to cope with numbers from 0 to 600 degrees accurate to five decimal places, e.g. 325.65744.

As with irrational numbers, computers deal with real numbers using fixed or floating point techniques that are described in detail in Chapter 25.

Ordinal numbers

KEYWORDS

Ordinal number: a number used to identify position relative to other numbers.

Cardinal number: a number that identifies the size of something.

Well-ordered set: a group of related numbers with a defined order.

Array: a data structure where data items are grouped together under a single identifier and are then accessed based on their position.

Ordinal numbers are those that identify the position of something within a list. For example: first, second and third. They are often used with **cardinal numbers**, which identify the size of the list. For example, you might say you were third out of 20.

In computing, ordinal numbers are used to identify the position (location) of data within an ordered set. Consider the following set:

S = {'Anne', 'Asif', 'John', 'Mary', 'Wanda', ...}

This set is said to be **well-ordered** as it has an internal structure that defines the relationship between the data items. In this case, the data is made up of names in ascending alphabetical order. The ordinal number is one that shows the order of the data so in this case the first item S(1) = Anne, the second item S(2) = Asif, the third item S(3) is John and so on.

This is a useful technique to locate and manipulate data within certain data structures such as lists, queues and arrays. Some programming languages support an ordinal data type, which will contain a value that can be counted and ordered.

A one-dimensional **array** called `Register` shows a list of names in a register:

Allen	Brown	Christie	Davali	Ennis

Ordinal numbers are used to assign the data as follows:

```
Register(1) = "Allen"
Register(2) = "Brown"
```

A table called `Exam` shows the results for each student:

Allen	Brown	Christie	Davali	Ennis
50	75	82	90	45

Ordinal numbers are used to extract the data as follows:

```
Exam(1,2) = 50
Exam(2,2) = 75
```

Counting and measurement

As we have seen, the basic use of numbers is to count and measure. We use different types of numbers in different ways depending on the task. For example, we:

- count using natural numbers as we only need to use positive whole numbers
- measure using real numbers as the range of numbers may be positive or negative and may require a fractional part.

The use of natural numbers to count is common in programming. For example:

- a counter may be used to keep track of how many times a loop statement is repeated
- the program counter in the processor keeps track of which instruction needs to be processed next
- a natural number is used to identify the location of data within a data structure
- a variable may be set up to keep count, for example, of the number of items in a stock control system, or the score in a computer game.

The use of real numbers to measure is common in programming. For example:

- CNC machines handle measurements that vary from millimetres to metres and must work to a high degree of accuracy
- microwave cookers control and measure both time and temperature
- power stations use data control systems to optimise the production of electricity
- robotics engineers use real-time measurements of the environment in which the robot is working.

Practice questions can be found at the end of the section on page 228.

TASKS

1 What are the defining features of:
   a) natural numbers
   b) integers
   c) rational numbers
   d) irrational numbers
   e) real numbers?

2 Why is it important for programmers to distinguish between the different types of numbers?

3 Use an example to explain the difference between an ordinal number and a cardinal number.

4 Explain why all integers are rational numbers.

5 Why does dividing a number by zero not result in a value of zero?

6 What type of numbers would be needed to record the following data?
   a) the number of runners in a race
   b) the position of runners as they finish a race
   c) the temperature on race day
   d) the time it took to run the race

7 Identify three computer applications that make use of:
   a) real numbers
   b) natural numbers.

KEY POINTS

- It is important to understand that there are different types of numbers.
- We use the word decimal to refer to numbers that are base 10, that is, made up of the numbers 0 to 9.
- Numbers are used to count and to measure (or quantify).
- Numbers used to count are called ordinals and are an important concept in computing, for example, when identifying the location of items within a list.
- As a programmer, you need to choose the right number type when working with data, e.g. whether to work with a number as an integer or a real number.

STUDY / RESEARCH TASKS

1 Find out about the irrational numbers called 'Euler's number' and 'the Golden Ratio'.

2 Apart from 2, 3 and 99, identify some other square roots that are irrational numbers.

3 Find out which number types are supported as data types in a programming language of your choice.

4 Why do programming languages often have several different data types for an integer?

5 The Ancient Greeks did not recognise the number zero. How did their number systems work without it?

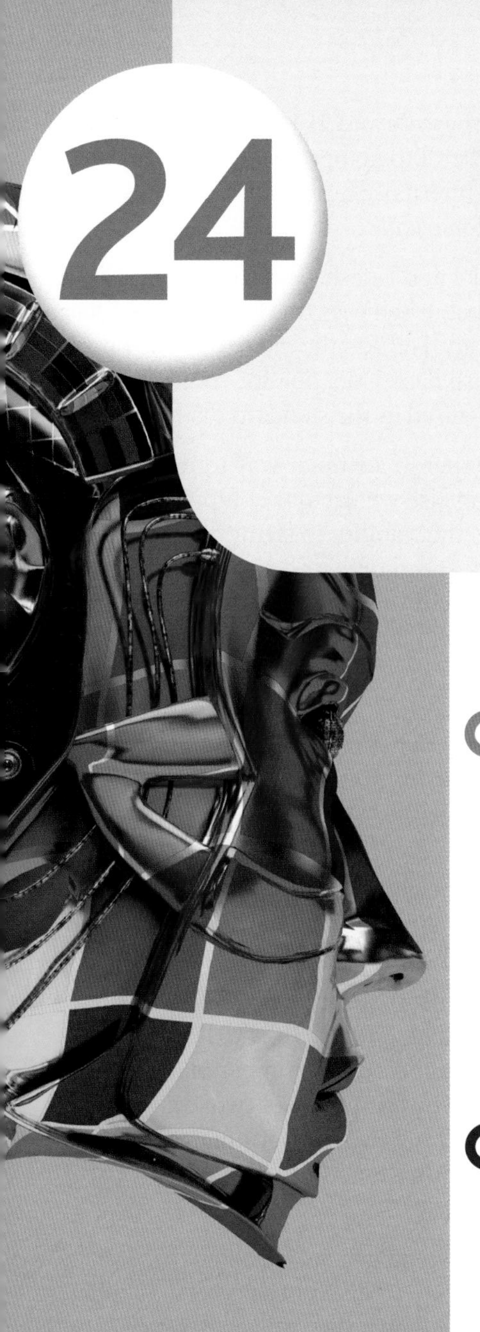

24 Number bases

SPECIFICATION COVERAGE

3.5.2.1 Number base

3.5.3.1 Bits and bytes

3.5.3.2 Units

LEARNING OBJECTIVES

In this chapter you will learn:

- the function of bits and bytes and how they are combined to form larger units
- how number bases work including binary, decimal and hexadecimal
- how to convert binary to decimal and vice versa
- how to convert binary to hexadecimal and vice versa
- how to convert decimal to hexadecimal and vice versa.

INTRODUCTION

Computers process data in digital form. This means that the data is represented as discrete values, in the form of zeros and ones. This is known as binary data and in this chapter you will discover how binary is used and how it relates to other **number bases** such as decimal and hexadecimal.

The bit

Computers process data in digital form. Essentially this means that they use microprocessors, also referred to as chips or silicon chips, to control them. A chip is a small piece of silicon implanted with millions of electronic circuits. The chip receives pulses of electricity that are passed around these microscopic circuits in a way that allows computers to represent text, numbers, sounds and graphics. But how?

It all comes down to the **bit**. A bit is a **bi**nary digi**t**. The processor can only handle electricity in a relatively simple way – either electricity is flowing, or it is not. This is often referred to as two states. The processor can recognise whether it is receiving an off signal or an on signal. This is handled as a zero (0) for off and a one (1) for on. Each binary digit therefore is either a 0 (no signal) or a 1 (a signal).

KEYWORDS

Number base: the number of digits available within a particular number system, e.g. base 10 for decimal, base 2 for binary.

Bit: a single binary digit from a binary number – either a zero or a one.

The processor now needs to convert these 0s and 1s into something useful for the user. Although it might be difficult to comprehend, everything you use your computer for is represented internally by a series of 0s and 1s. To help you understand this, think of Morse code.

Morse code only uses two signals – a dot and a dash. These two states can be used to create every letter in the alphabet. It achieves this by stringing dots and dashes together in different combinations. Perhaps the most well-known piece of Morse code is 'dot dot dot – dash dash dash – dot dot dot'. 'dot dot dot' is S and 'dash dash dash' is O; SOS is recognised as the standard distress call.

Computers string zeros and ones together in a similar way to represent text, numbers, sound, video and everything else we use our computers for. The really clever thing about computers is their ability to string zeros and ones together at very high speed. The clock speed of your computer indicates the speed at which the signals are sent around the processor. In simple terms, a clock speed of 2 GHz means that it will receive 2 billion of these on/off pulses per second.

The byte

KEYWORD

Byte: a group of bits, typically 8, used to represent a single character.

The first hint most students get of the nature of the **byte** is when they begin to measure the size of memory or disk space in terms of megabytes, gigabytes and terabytes. A single byte is a string of eight bits. Eight is a useful number of bits as it creates enough permutations (or combinations) of zeros and ones to represent every character on your keyboard. Follow this through:

- With one bit we have two permutations: 0 and 1.
- With two bits we have four permutations: 00, 01, 10 and 11. This could be represented as 2^2 or 2×2. As we increase the number of bits, we increase the number of permutations by the power of two.
- Three bits would give us 2^3 which is $2 \times 2 \times 2 = 8$ permutations.
- Four bits would give us 2^4 permutations which is $2 \times 2 \times 2 \times 2 = 16$ permutations.

If we stop at four you can see that 4 bits would give us enough permutations to represent 16 different letters of the alphabet, 16 different numbers, 16 different colours or 16 different sounds. If we move onto 8 bits, we get 2^8 which is 256 permutations. Therefore, 8 bits is enough to represent every letter in the alphabet and every keyboard character with a few to spare. 8 bits is referred to as a byte, which typically represents one character.

The basic equation here is that the more bits you use, the greater the range of numbers, characters, sounds or colours that can be created. Taking numbers as an example, as we have seen, 8 bits would be enough to represent 256 different numbers (0–255). As the number of bits increases, the range of numbers increases rapidly. For example 2^{16} would give 65 536 permutations, 2^{24} would give approximately 1.6 million and 2^{32} would give over 4 billion permutations.

Units

KEYWORD

Unit: the grouping together of bits or bytes to form larger blocks of measurement, e.g. GB, MB.

Larger combinations of bytes are used to measure the capacity of memory and storage devices. The size of the **units** can be referred to either using binary or decimal prefixes. For example, in decimal, the term kilo is commonly used to indicate a unit that is 1000 times larger than a single unit. So the correct term would be kilobyte (KB). In binary, the correct term is actually kibibyte (Ki) with 1024 bytes being the nearest binary equivalent to 1000.

Common units are shown below using both binary and decimal prefixes:

Table 24.1 Common binary and decimal units

Binary			Decimal		
kibibyte	Ki	2^{10}	kilobyte	KB	10^3
mebibyte	Mi	2^{20}	megabyte	MB	10^6
gibibyte	Gi	2^{30}	gigabyte	GB	10^9
tebibyte	Ti	2^{40}	terabyte	TB	10^{12}

Number bases

A number base indicates how many different digits are available when using a particular number system. For example, decimal is number base 10 which means that it uses ten digits: 0, 1, 2, 3, 4, 5, 6, 7, 8 and 9 and binary is number base 2 which means that it uses two digits: 0 and 1. Different number bases are needed for different purposes. Humans use number base 10 whereas computers use binary which is a form of digital data.

The number base determines how many digits are needed to represent a number. For example, the number 98 in decimal (base 10) requires two digits. The binary (base 2) equivalent is 1100010 which requires seven digits. As a consequence of this there are many occasions in computing when very long binary codes are needed. To solve this problem, other number bases can be used, which require fewer digits to represent numbers. For example, some aspects of computing involve number base 16 which is referred to as hexadecimal.

The accepted method for representing different number bases (in textbooks and exam questions) is to show the number with the base in subscript. For example:

- 43_{10} is decimal
- 1011_2 is binary
- $2A7_{16}$ is hexadecimal.

Hexadecimal

Hexadecimal or hex is particularly useful for representing large numbers as fewer digits are required. Hex is used in a number of ways. Memory addresses are shown in hex format as are colour codes. The main advantage of hex is that two hex digits represent one byte.

Consider the number 11010011_2. This is an 8-bit code which when converted to decimal equals 211_{10}. The same number in hex is $D3_{16}$. This basic example shows that an 8-bit code in binary can be represented as a two-digit code in hex. Consequently hex is often referred to as 'shorthand' for binary as it requires fewer digits.

As it is number base 16, hex uses 16 different digits: 0 to 9 and A to F. Table 24.2 shows decimal numbers up to 31 with their hex equivalents.

Table 24.2 Hex look-up table

Decimal	Hex	Decimal	Hex
0	0	16	10
1	1	17	11
2	2	18	12
3	3	19	13
4	4	20	14
5	5	21	15
6	6	22	16
7	7	23	17
8	8	24	18
9	9	25	19
10	A	26	1A
11	B	27	1B
12	C	28	1C
13	D	29	1D
14	E	30	1E
15	F	31	1F

There is scope for confusion here as humans rarely use letters as numbers. Also, the numbers in hex may convert to different numbers in decimal. For example, the number 16 in decimal is the equivalent of the number 10 (one zero) in hex.

Working with number bases

When performing any calculations, humans use number base 10 probably because we have ten digits on our hands. Commonly this system is known as decimal and uses 10 different digits: 0, 1, 2, 3, 4, 5, 6, 7, 8 and 9. When we get to 9 we add an extra digit to the left and start again. When we get to 99, we add a further digit to the left and so on. Each digit we add is worth ten times the previous digit. This is easier to understand if you think back to how you were taught maths at primary school.

The number 2098 is easy to understand in decimal terms. To state the obvious, it is made up of $(2 \times 1000) + (0 \times 100) + (9 \times 10) + (8 \times 1)$. When creating a number, we start with the units and add further digits as needed to create the number we want.

Binary is number base 2 and works on exactly the same principle. This time we only have two digits, 0 and 1. It has to be binary because computers only work by receiving a zero or one (off and on). So, 1 is the biggest number we can have with one bit. To increase the size of the number, we add more bits. Each bit is worth two times the previous bit because we are using number base 2. The table below shows an 8-bit binary number 10000111. Notice the value of each new bit is doubling each time, as binary is base 2.

128	64	32	16	8	4	2	1
1	0	0	0	0	1	1	1

Again, using the same principle as with decimal to work out the number we have:

(1 × 128) + (1 × 4) + (1 × 2) + (1 × 1). This adds up to 135.

Therefore 10000111 in binary = 135 in decimal.

Binary to decimal conversions

Binary numbers are converted to decimal integers as follows:

- Write down a binary number (e.g. 10000111).
- Above the number, starting from the least significant bit (LSB) write the number 1.
- As you move left from the LSB to the most significant bit (MSB) double the value of the previous number:

MSB							LSB
128	64	32	16	8	4	2	1
1	0	0	0	0	1	1	1

- Wherever there is a 1, add the decimal value: the above example represents one 128, one 4, one 2 and a 1 giving a total value of 135 (128 + 4 + 2 + 1 = 135). Therefore 10000111 in binary equals 135 as a decimal integer.

Decimal to binary conversions

To convert a decimal integer to a binary number, use the same method as above, but working the other way. For example, to convert the number 98:

- Write down the power of 2 sequence. (Eight bits are used here but you will notice that you only need seven for this example.)

MSB							LSB
128	64	32	16	8	4	2	1

  - Starting from the MSB put a 1 or 0 in each column as necessary to ensure that it adds up to 98 as follows:
  - 0 under 128
  - 1 under 64
  - 1 under 32
  - 0 under 16
  - 0 under 8
  - 0 under 4
  - 1 under 2
  - 0 under 1

Therefore 01100010 = 98.

Another way of carrying out this calculation is to carry out repeated divisions on the decimal number as follows:

- 98 divide by 2 = 49 with a remainder of 0
- 49 divide by 2 = 24 with a remainder of 1
- 24 divide by 2 = 12 with a remainder of 0
- 12 divide by 2 = 6 with a remainder of 0
- 6 divide by 2 = 3 with a remainder of 0
- 3 divide by 2 = 1 with a remainder of 1
- 1 divide by 2 = 0 with a remainder of 1

Notice that you keep dividing by 2 until there is nothing left to divide. Reading from the bottom this gives us 1100010 which equals 98. (Note that the leading zero is omitted.)

Check your answer by working it back the other way:

MSB							LSB
128	64	32	16	8	4	2	1
0	1	1	0	0	0	1	0

$64 + 32 + 2 = 98$

Decimal to hex conversions

A common approach to convert decimal integers to hex is to first convert the decimal to binary and then convert the binary to hex. Taking the decimal number 211 as an example:

- Work out the binary equivalent.

128	64	32	16	8	4	2	1
1	1	0	1	0	0	1	1

- Split the binary number into two groups of four bits and convert each into the hex equivalent.

8	4	2	1	8	4	2	1
1	1	0	1	0	0	1	1

Therefore $11010001_2 = 211_{10}$

$8 + 4 + 1 = D$ (the hex equivalent of 13) and $2 + 1 = 3$

Therefore $211_{10} = 11010011_2 = D3_{16}$

Hex to decimal conversions

The process here is to convert the hex to binary, and then the binary into decimal. Hex to binary conversions are the reverse of the above process. Take the hex number, and then convert each digit in turn into its binary equivalent using groups of four bits. Take $2A3_{16}$ as an example:

		2				A				3	
8	4	2	1	8	4	2	1	8	4	2	1
0	0	1	0	1	0	1	0	0	0	1	1

2 = 0010, A = 1010, 3 = 0011

Therefore 1010100011_2 is the binary equivalent of $2A3_{16}$

This binary code can then be converted into decimal in the usual way:

512	256	128	64	32	16	8	4	2	1
1	0	1	0	1	0	0	0	1	1

$512 + 128 + 32 + 2 + 1 = 675_{10}$

When carrying out a conversion, it is useful to remember the binary equivalent of the 16 digits used in hex as shown in Table 24.2.

Practice questions can be found at the end of the section on page 228.

TASKS

1 Explain why computers can only process data in binary form.

2 What is the biggest decimal integer you can represent with:
 a) 4 bits? **b)** 8 bits? **c)** 16 bits?

3 How many different permutations of numbers can you represent with:
 a) 4 bits? **d)** 20 bits?
 b) 8 bits? **e)** 24 bits?
 c) 16 bits?

4 Convert the following decimal numbers into binary:
 a) 10 **d)** 65
 b) 12 **e)** 165
 c) 15

5 Some programming languages use hexadecimal. Explain what hexadecimal is and what the benefits are of using this system compared to binary or decimal.

6 Convert the following hexadecimal numbers into binary:
 a) 10 **b)** 12 **c)** 1F **d)** F1

7 Convert the following hexadecimal numbers into decimal:
 a) E **b)** 21 **c)** 17 **d)** AB

8 Identify a situation where it would be appropriate to use the following units of measurement:
 a) kilobyte **b)** megabyte **c)** terabyte.

STUDY / RESEARCH TASKS

1 Write a program that converts binary to decimal and vice versa.

2 Write a program that converts hex to decimal and vice versa.

3 In computing we commonly use binary, decimal and hexadecimal. In the past, computing also used octal. Find out how it works, what it was used for and why it is not widely used in computing these days.

4 Ancient number systems did not use zero. Explain how a number system can work without a zero.

5 Apart from the ones you have already looked at, what other number bases are used, or have been used throughout history?

6 Why do we use base 12 and base 60 for telling the time rather than base 10?

7 Find a simulation of a binary watch online. See if you can learn to tell the time as quickly in binary as you can using decimals.

8 Identify a situation where it would be appropriate to use the following units of measurement:
 - exabyte
 - zettabyte
 - yottabyte.

KEY POINTS

- Computers process data in digital form, that is, as series of discrete values.
- 0s and 1s are called binary digits or bits.
- Bits are grouped together to create bytes.
- Bytes are grouped together to create kilobytes, megabytes, gigabytes and terabytes.
- Computing uses three main number bases: binary (base 2), decimal (base 10) and hexadecimal (base 16).
- You need to be able to convert between the three number bases.

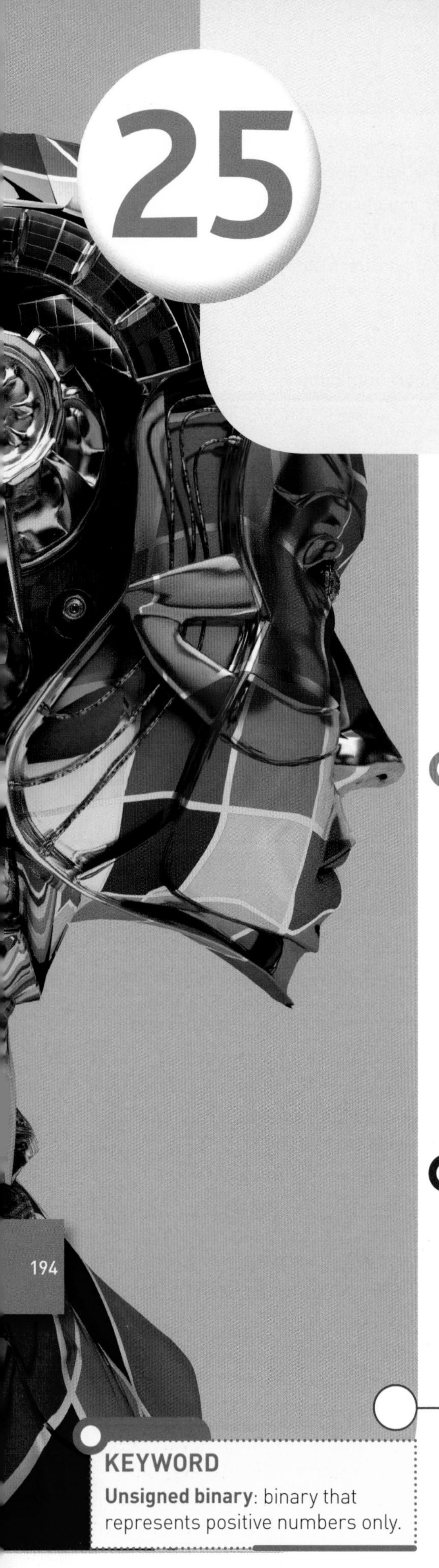

25 The binary number system

SPECIFICATION COVERAGE

3.5.4.1 Unsigned binary

3.5.4.2 Unsigned binary arithmetic

3.5.4.3 Signed binary using two's complement

3.5.4.4 Numbers with a fractional part

3.5.4.5 Rounding errors (A level only)

3.5.4.6 Absolute and relative errors (A level only)

3.5.4.7 Range and precision (A level only)

3.5.4.8 Normalisation and floating point form (A level only)

3.5.4.9 Underflow and overflow (A level only)

LEARNING OBJECTIVES

In this chapter you will learn:

- how to add and multiply positive binary integers
- how to work with signed binary numbers using two's complement
- how fixed numbers are used to represent fractions.

A-level students will learn:

- how floating point numbers are used to represent fractions
- what overflow and underflow are
- how to round binary numbers
- what precision and normalisation are with reference to binary numbers.

INTRODUCTION

In the previous chapter we looked at the common number systems and bases. We use different number bases, as humans tend to work with decimals and computers can only process data in binary. As computer science students, we need to know how binary works and how the computer carries out calculations in binary.

Adding unsigned binary integers

In the previous chapter we looked at converting between number bases. As computers can only process binary data, it is also important to understand how they carry out basic calculations using binary.

KEYWORD

Unsigned binary: binary that represents positive numbers only.

To add two numbers together in binary, first line up the numbers in the same way as you would do column addition in decimal:

```
 00110010+
 10110101
 --------
 11100111
 --------
  11
```

Now add the columns starting from the right-hand side, remembering that you can only use 0s and 1s:

- 0 + 0 will equal 0 so put 0 on the answer line
- 0 + 1 or 1 + 0 will both equal 1 so put 1 in the answer line
- 1 + 1 will equal 10 (one, zero) so you will put 0 in the answer line and carry the 1
- 1 + 1 + 1 will equal 11 (one, one) so you will put 1 in the answer line and carry the 1

You can check your answer back by converting all the numbers to decimal, carrying out the addition and then converting the answer back to binary. In this case, the first number is 50, the second number is 181, so the answer should be 231.

Multiplying unsigned binary integers

To multiply in binary, you multiply the first number by each of the digits of the second number in turn starting from the right-hand side (in the same way that you would do multiplication in decimal). This means you are either multiplying each digit by 0 or by 1, which will give you either a 0 or 1 as the answer. You then do the same for the next digit, shifting your answers to the left as you would in decimal multiplication.

You then carry out a binary addition to find the final answer. For example, to multiply 11011 by 11:

```
   11011×
      11
  -------
   11011
  110110
  -------
 1010001
  -------
  1111
```

Note the zero on the LSB as the numbers have been shifted to the left.

Again you can work this out by converting the binary to decimal to check your answer. In this case the first number is 27 (twenty-seven), the second number is 3 (three), so the answer is 81 (eighty-one).

Types of numbers

In computing terms it is important to distinguish between different types of numbers. So far in this chapter we have only looked at integers, which are whole numbers. More specifically we have only looked at unsigned integers, which means that all the numbers have been positive. Computers also need to handle real numbers – positive and negative numbers which may be shown to several decimal places. Many computer applications will require very large or very small numbers to be handled to a high degree of accuracy.

KEYWORD

Two's complement: a method of working with signed binary values.

Two's complement

Two's complement is a method used to represent signed integers in binary form. This means that it can be used to represent positive and negative integers. This method is very similar to the methods described in Chapter 24 and so it is assumed that you already have a good understanding of this before attempting this section. The purpose of this section is to show how two's complement can represent negative integers.

Assume we want to convert the binary code 10011100 into decimal using two's complement:

- Write out the denary equivalents as shown:

MSB LSB

–128	64	32	16	8	4	2	1

- You will notice that with two's complement, the most significant bit becomes negative. Using an 8-bit code, this means that the MSB represents a value of –128
- Now write in the binary code:

MSB LSB

–128	64	32	16	8	4	2	1
1	0	0	1	1	1	0	0

- Now add up the values:
  –128 + 16 + 8 + 4 = –100

Converting from denary to binary using two's complement can be slightly more difficult for negative numbers as you will be starting from a negative number and working forward. Remember that with two's complement, when the MSB is 1 it means that the number must be negative. You may find it easier to use the following method:

- To convert –102 into binary, first write out the binary equivalent of +102 as shown:

MSB LSB

128	64	32	16	8	4	2	1
0	1	1	0	0	1	1	0

- Starting at the LSB, write out the number again until you come to the first 1.
- Then reverse all the remaining bits, that is, 0 becomes 1 and 1 becomes 0.
- The number becomes 10011010.
- The number is now in two's complement.

MSB LSB

–128	64	32	16	8	4	2	1
1	0	0	1	1	0	1	0

To check, add these up: –128 + 16 + 8 + 2 = –102

With 8 bits it is possible to represent 256 different numbers. These range from –128 to 127. For an arbitrary number of bits (n), 2^n different values can be represented. If the n bits are used to store an unsigned binary value then the highest value that could be represented would be $2^n - 1$ as 0 is one of the values that can be represented. With two's complement for 8 bits the range is –128 through to 127 and for n bits it would be -2^{n-1} to $2^{n-1} - 1$.

Adding and subtracting using two's complement

Adding numbers together using two's complement is the same as adding numbers together in decimal: you add up the total and carry values across to the next column. For example, in decimal to add 48 to 83:

```
  48
  83+
 ---
 131
 ---
 11
```

Binary addition is the same. To add 01101100 to 10001000:

```
01101100
10001000+
--------
11110100
--------
    1
```

Remember that in binary 1 + 1 = 10 and that 1 + 1 + 1 = 11

The only true arithmetic that the computer can carry out is addition. In order to carry out subtractions, the method used is to convert the number to be subtracted to a negative number, and then to add the negative number. For example 20 – 13 in denary would actually be performed by adding 20 to –13 giving the answer of 7. To do this in binary:

- Calculate the binary equivalent of 20 which equals 00010100
- Calculate the binary equivalent of –13 which equals 11110011
- Add 20 to –13 in binary form:

```
00010100
11110011+
--------
00000111
--------
111
```

- Check your answer back by converting it to denary and the answer is 7 which is correct.
- You may notice that this calculation would have a final 1 to be carried. This is called an 'overflow bit' and is handled separately to the calculation. There is more on this later in the chapter.

Fixed point numbers

KEYWORD

Fixed point: where the decimal/binary point is fixed within a number.

In order to represent real decimal numbers, that is, numbers with decimal places or a fractional part, **fixed point** representation can be used. In the same way that decimal has a decimal point, binary has a binary point. The numbers after the binary point represent fractions. For example, if you had an 8-bit binary code, you may place the binary point after the fourth bit as shown:

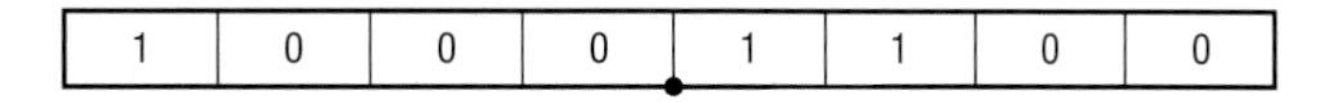

1	0	0	0	1	1	0	0

The binary point is not actually stored in the 8-bit code, its position is fixed by the programmer. It is shown here purely to aid understanding.

To convert this to a decimal number is a similar process to the other conversions we have done. This time, the digits after the binary point become fractions as follows:

8	4	2	1	$\frac{1}{2}$	$\frac{1}{4}$	$\frac{1}{8}$	$\frac{1}{16}$
0	1	1	0	0	1	1	0

The conversion of the bits before the binary point are handled in the same way as before with each value doubling as you move from right to left. The numbers after the binary point halve each time as you move from left to right as shown.

Therefore the number above is

$4 + 2 + \frac{1}{4} + \frac{1}{8}$

giving a total of $6\frac{3}{8}$ or 6.375.

The binary point can be placed anywhere within the byte but the position of the binary point restricts the size of the number that can be represented and also the accuracy of the number. With the binary point in the position shown in this example:

- The smallest number we could represent (apart from 0) is 0000.0001 which is $\frac{1}{16}$ or 0.0625.
- The next number we could represent is 0000.0010 which is $\frac{1}{8}$, $\frac{2}{16}$ or 0.125. It is not possible to represent any number between 0.0625 and 0.125
- The largest number we could represent is 1111.1111 which is $15\frac{15}{16}$ or 15.9375
- Moving the binary point to the left means that we can have more accurate decimals but reduces the range of whole numbers available.
- Moving the binary point to the right increases the range of whole numbers but reduces the accuracy.
- It remains the case that with an 8-bit code, we can represent 256 different combinations regardless of where we put the binary point.

The same techniques can also be used to represent negative numbers:

–8	4	2	1	$\frac{1}{2}$	$\frac{1}{4}$	$\frac{1}{8}$	$\frac{1}{16}$
1	1	1	0	1	1	0	0

This number would be:

$-8 + 4 + 2 + \frac{1}{2} + \frac{1}{4} = -1\frac{1}{4}$

Floating point numbers

A level only

The big problem with all the 8-bit systems we have investigated so far is that they can only store a very limited range of numbers. The biggest positive number we have been able to store so far is only 255 and the smallest positive number is 0.0625. There will be many scenarios when a program needs to cope with numbers that are larger or smaller than this. There are two ways round this problem:

- The first is to allocate more bits to store the number. For example, a 16-bit unsigned code would allow you to store all the integers from 0 to 65 535; a 24-bit code would allow you to cope with 16 777 216 different combinations; and so on.

- If you wanted to store negative and positive numbers you would need to use the two's complement system outlined above. Using 16 bits would allow you to store between –32 768 and 32 767.

KEYWORD

Floating point: where the decimal/binary point can move within a number.

The problem with allocating an ever-increasing number of bits to store large numbers is knowing when to stop. The solution to this is to use a **floating point** number representation.

In floating point, the binary point can be moved depending on the number that you are trying to represent. It 'floats' from left to right rather than being in a fixed position. In the previous example, the binary point was fixed after the fourth bit and this presented serious limitations on both the range and accuracy of numbers that can be represented. Floating point extends the fixed point technique described in the previous section and also involves two's complement so you should not attempt this section until you are happy with these two concepts.

A floating point number is made up of two parts – the mantissa and the exponent. In decimal, we often have to calculate large numbers on our calculators. Most calculators have an eight- or ten-digit display and often the numbers we are calculating need more digits. When this happens, an exponent and mantissa are used. For example, the number 450 000 000 000, which is 45 and ten zeros, would be shown as 4.5×10^{11}. 4.5 is the mantissa and 11 is the exponent meaning that the decimal place is moved 11 places to the right.

In binary, the exponent and mantissa are used to allow the binary point to float as in the previous example. Remember that the mantissa and/or the exponent may be negative as the two's complement method is also used on each part. Consider the following 12-bit code: 000011000011.

The code can be broken down as follows:

- the first eight bits are the mantissa which can be broken down further as:
  - the MSB is 0 which means that the number is positive
  - the next seven bits are the rest of the mantissa: 0001100
- the remaining four bits are the exponent: 0011.

It is common to show the mantissa and exponent more clearly as follows:

Mantissa

0	0	0	0	1	1	0	0

Exponent

0	0	1	1

- First, work out the exponent in the usual way, remembering that two's complement is being used.

  Therefore the exponent is +3.

–8	4	2	1
0	0	1	1

  This means that the binary point will 'float' three places to the right.

- Now calculate the mantissa. The binary point is always placed after the most significant bit as follows:

–1	$\frac{1}{2}$	$\frac{1}{4}$	$\frac{1}{8}$	$\frac{1}{16}$	$\frac{1}{32}$	$\frac{1}{64}$	$\frac{1}{128}$
0	0	0	0	1	1	0	0

- The point now floats three places to the right. The values for the conversion have changed because the binary point has now moved.

8	4	2	1	$\frac{1}{2}$	$\frac{1}{4}$	$\frac{1}{8}$	$\frac{1}{16}$
0	0	0	0	1	1	0	0

- Therefore, 00001100 0011 = 0.75

The following example shows the whole process of working out the floating point binary value of a decimal value. In this example, the value is 6.75:

- The first stage is to calculate the binary value using fixed point representation:

8	4	2	1	$\frac{1}{2}$	$\frac{1}{4}$	$\frac{1}{8}$	$\frac{1}{16}$
0	1	1	0	1	1	0	0

- The number needs to be normalised and there is more on this in a later section. At this stage, what this means is that for a positive number, the binary value must start with 01.
  This means that we need to move the floating point to the left 3 places as we convert to floating point:

0	1	1	0	1	1

- We now need to use the exponent to indicate that the floating point needs to be moved three places back to the right:

Mantissa

0	1	1	0	1	1

Exponent

0	1	1

- In this case therefore we have used a 6-bit mantissa and a 2-bit exponent to represent the value. It is possible to increase the number of bits used to both increase the accuracy and range of values. For example, a 3-bit exponent would enable the floating point to be moved up to six places.

Two's complement can also be used on the mantissa or exponent to represent negative values. The process of converting negatives is similar. A negative exponent would move the binary point to the left rather than the right.

For the value –10.5:

- First calculate the fixed point representation of the positive value, that is 10.5:

–16	8	4	2	1	$\frac{1}{2}$	$\frac{1}{4}$	$\frac{1}{8}$
0	1	0	1	0	1	0	0

- Starting at the LSB, write out the number again until you come to the first 1.
- Then reverse all the remaining bits, that is, 0 becomes 1 and 1 becomes 0.

–16	8	4	2	1	$\frac{1}{2}$	$\frac{1}{4}$	$\frac{1}{8}$
1	0	1	0	1	1	0	0

- The number is now in two's complement.

- Normalise the number, which for a negative value means you need to position the binary point so that the first bit of the mantissa after the binary point is the first 0 in the number. This means floating (moving) it four places to the left:

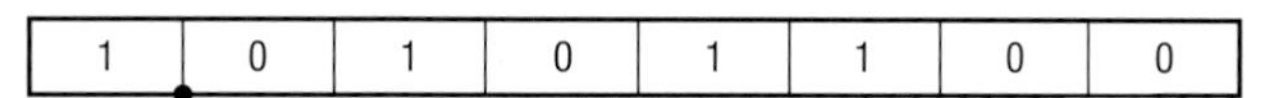

- We now need to use the exponent to indicate that the floating point needs to be moved four places back to the right:

Mantissa

1	0	1	0	1	1	0	0

Exponent

1	0	0

- In this case therefore we have used an 8-bit mantissa and a 3-bit exponent to represent the value.

This is a simplified example to show the concept. The real power of floating point is that the binary point can be moved several hundred places to the left or right as necessary to represent either very large or very small numbers.

An alternative way of representing the values is to use the same method that is also used to represent very large or very small numbers in base 10. For example, the value 3.345×10^3 means that the decimal point should be moved 3 places to the right. The value therefore is 3345. You can do the same in binary. For example with a 6-bit mantissa of 010011 and a 3-bit exponent of 100 this would be 19×2^4.

Fixed point compared to floating point

As you have seen, fixed point and floating point are two methods for representing values. Both systems have their advantages. The advantages of using floating point are:

- a much wider range of numbers can be produced with the same number of bits as the fixed point system
- consequently, floating point lends itself to applications where a wide range of values may need to be represented.

The advantages of using fixed point are:

- The values are handled in the same way as decimal values so any hardware configured to work with integers, can be used on reals. This also makes the processing of fixed point numbers faster than floating point values as there is no processing required to move the binary point.
- The absolute error will always be the same, whereas with floating point numbers the absolute error could vary. This means that precision is retained albeit within a more limited range than floating point representation.
- It is suited to applications where speed is more important that precision, e.g. some aspects of gaming or digital signal processing.
- It is suited to applications where an absolute level of precision is required, e.g. currency, where the number of places after the binary point does not need to change.

KEYWORDS

Signed binary: binary with a positive or negative sign.

Overflow: when a number is too large to be represented with the number of bits allocated.

Underflow: when a number is too small to be represented with the number of bits allocated.

Underflow and overflow

It is possible when using **signed binary** that you will generate a number that is either too large or too small to be represented by the number of bits that are available to store it. When the number is too large, we get an **overflow** and when it is too small we get an **underflow**.

Looking at an overflow first, as we have seen, the most significant bit (MSB) is used to indicate whether the number is positive or negative. 0 means positive and 1 means negative.

To take a simple example, using two's complement with just 8-bits, we could represent the numbers –128 to +127. What would happen therefore if we asked the computer to calculate a number that returned the value of +128. If it used the MSB, we would actually get the answer –128.

For example, if 00000001 (1 in decimal) was added to 01111111 (127 in decimal), it would generate the answer 1000000 (128 in decimal). However, using two's complement, we would actually get the answer –128 as shown below:

–128	64	32	16	8	4	2	1
1	0	0	0	0	0	0	0

As another example, if we added 78 to 75 we would expect to get 163. In binary with 8 bits this would be 128 + 32 + 2 + 1:

128	64	32	16	8	4	2	1
1	0	1	0	0	0	1	1

In two's complement we would actually get the answer –128 + 32 + 2 + 1 = 93:

–128	64	32	16	8	4	2	1
1	0	1	0	0	0	1	1

Underflow is a similar concept but occurs when the number is too small to be represented with the number of bits allocated.

If a calculation generated a small positive or negative number (i.e. one that is close to zero), then there may not be enough bits to represent it. The example below has 6 bits and a fixed point. If we tried to represent the value $\frac{1}{64}$ or 0.015625 it would create an underflow error as the smallest number we can represent with 6 bits and a fixed point is $\frac{1}{32}$ or 0.03125.

1	$\frac{1}{2}$	$\frac{1}{4}$	$\frac{1}{8}$	$\frac{1}{16}$	$\frac{1}{32}$
0	0	0	0	0	1

As you can see, an underflow or overflow could cause serious errors in a program. It could generate erroneous results or even cause the program to crash. There are various methods for dealing with overflows and underflows. A common method is the use of a flag to indicate where an overflow or underflow has occurred and to 'carry' the additional bits in the same way that you carry digits when adding up in decimal. Overflow can also be represented as ∞.

Normalisation and precision

KEYWORDS

Normalisation: a process for adjusting numbers onto a common scale.

Precision: how accurate a number is.

Normalisation is a technique used to ensure that when numbers are represented they are as precise as possible in relation to how many bits are being used. Another benefit is that normalisation ensures that only one representation of a number is possible. The easiest way to think about this is to consider how many decimal places you would choose to use when representing a decimal number.

Assume that you are creating a system to record the results of various athletics events: 100 m, 400 m and 1500 m, and that you allow six digits to store the time taken for each race.

- The winner of the 100 m event ran it in 10.4357 s. Four numbers have been used after the decimal place to provide a precise number.
- The winner of the 400 m event took 47.3453 s. Again four numbers are used after the decimal place.
- The winner of the 1500 m race took 150.435 s. This time, the number is only precise to three decimal places as three digits are needed for the integer. The decimal point has floated here in order to represent the number. However, the result is not as precise as the results for the 100 m and 400 m races.
- The result of the 100 m race could be stored as 010.436. However this would not be sensible as the number is not as precise as it could be with six digits available.

The same thing happens when using a mantissa and exponent. With a fixed number of bits that can be used to represent the mantissa, the **precision** of the number can be affected by where the binary point is positioned. The exponent is used to ensure that the floating point is placed to optimise the precision of the number.

For example 234 000 can be represented as:

- 23400×10^{1}
- 2.34×10^{5}
- $0.00000234 \times 10^{11}$

The second option is the best way to represent the number as it uses the least number of digits yet provides a precise result. This number is referred to as being in 'normal form' or 'normalised'.

With binary codes, normalisation is equally important. In order to be 'normalised' the first bit of the mantissa, after the binary point, should always be a 1 for positive numbers, or a 0 for negative numbers and the bit before the binary point should be the opposite. So a normalised positive floating point number must start 0.1 and a normalised negative floating point number must start 1.0. For example, the binary equivalent of 108 in decimal using an 8-bit code would be 01101100:

- The normalised mantissa would be 0.1101100
- The binary point will have to be moved seven places to the right in order to convert it back to the original number
- Therefore the exponent must be 7
- Two's complement for 7 is 0111.
- Therefore the normalised representation of 108 is 0.11011000111.

You might be wondering why it is worth showing 108 in this way when it could be shown more simply as 01101100. The reason for this is that with an 8-bit mantissa and a 4-bit exponent it is possible to represent a much wider range of positive and negative numbers than using eight bits alone.

For example, if you had 8 bits and a fixed point you could represent:

- Lowest positive value: $\frac{1}{16}$ or 0.0625

8	4	2	1	$\frac{1}{2}$	$\frac{1}{4}$	$\frac{1}{8}$	$\frac{1}{16}$
0	0	0	0	0	0	0	1

- Highest value: 15.9375 or $15\frac{15}{16}$

8	4	2	1	$\frac{1}{2}$	$\frac{1}{4}$	$\frac{1}{8}$	$\frac{1}{16}$
1	1	1	1	1	1	1	1

With a floating point and 8 bits for the mantissa and 4 bits for the exponent and assuming that two's complement is being used on both, the range of values would be:

- +127 for the mantissa and +7 for the exponent
- –128 for the mantissa and –8 for the exponent

Rounding errors

When working with decimal numbers, we are used to the idea of rounding numbers up or down. As a consequence, we will get rounding errors. For example, $\frac{1}{3}$ in decimal is 0.3 recurring. Consequently we are comfortable with using 0.33 or perhaps 0.333 to represent $\frac{1}{3}$. Obviously there is a degree of error in this calculation. Whether it is acceptable or not depends on what our program was doing. If it was taking exact scientific measurements we might want a greater degree of accuracy.

A similar phenomenon occurs with binary representation. For example, if you try to convert 0.1 in decimal into binary you will find that you get a recurring number, so it is not possible to exactly represent it. If we continue the binary examples we used before, if you try to represent 1.95 with 8-bits and a fixed point:

1.1111010 would give us 1.953125:

1	$\frac{1}{2}$	$\frac{1}{4}$	$\frac{1}{8}$	$\frac{1}{16}$	$\frac{1}{32}$	$\frac{1}{64}$	$\frac{1}{128}$
1	1	1	1	1	0	1	0

1.1111001 would give us 1.9453125:

1	$\frac{1}{2}$	$\frac{1}{4}$	$\frac{1}{8}$	$\frac{1}{16}$	$\frac{1}{32}$	$\frac{1}{64}$	$\frac{1}{128}$
1	1	1	1	1	0	0	1

With 8 bits the nearest we can get is 0.003125 out. We could extend the number of bits that we use to try and get an answer that is closer to 1.95.

It is the job of the programmer to decide on what is an accurate enough number and to allocate an appropriate number of bits to store the data to the required level of precision.

Absolute and relative errors

There are two main methods for calculating the degree of error in numbers that we use within a program. The absolute error is the actual mathematical difference between the answer and the approximation of it that you can store. For example, if a calculation requires 8 decimal places, but we only allocate 8 digits, we would have to either round or truncate the number. So the number 1.65746552 would become 1.6574655. In this case, to work out the absolute error we would subtract the two values and that would give us an absolute error of 0.00000002. Note that the absolute error is always a positive number.

You could apply a margin of error when deciding whether a number is accurate enough. For example, you could apply an absolute measure. In the example above you could identify ±0.00005 as an acceptable margin of error.

You can see that for a number that is around 1, this degree of accuracy is probably sufficient. If the number we were storing was larger, for example 111 001.65746552, you would probably not need to store it to eight decimal places anyway. Perhaps one or two decimal places would be sufficient.

However, if the number was much smaller, e.g. a microscopic measurement, you might need a much larger number of decimal places.

This is where the idea of a relative error comes in. Rather than applying a rigid margin of error, you would look at the value that was being stored and then decide on a relative margin of error. In this way you are comparing the actual result to the expected result. For example you might decide that ±5% would be sufficient. This would mean that for a number in the thousands, you would not need any decimal places and you could store it as an integer. With very small numbers, you might need to allocate 10 decimal places with no whole number at all in order to record the result as accurately as needed.

KEYWORDS

Mantissa: the significant digits that make up a number.

Exponent: the 'power of' part of a number indicating how far a binary point should be shifted left or right.

Relative error can be calculated using the formula:

Relative error = Absolute error / Number intended to be stored

For example, if trying to represent the value 6.95 using floating point with an 8-bit **mantissa** and a 3-bit **exponent**, it could be shown as 0.1101111011. This works out to be 6.9375. Therefore:

Absolute error = 6.95 – 6.9375 = 0.0125

Relative error = 0.0125 / 6.95 = 0.001798561151

Practice questions can be found at the end of the section on page 228.

KEY POINTS

- Binary addition and multiplication of positive integers use the same methodology as decimal addition and multiplication.
- Two's complement is a method used to represent signed integers in binary form.
- Fixed point binary numbers are used to represent fractions. The binary point is in a fixed position.
- Floating point binary numbers are used to represent fractions. The binary point can move position.
- A floating point number is made up of two parts – the mantissa and the exponent.
- It is possible that when using signed binary you will generate a number that is either too large or too small to be represented by the number of bits that are available to store it.
- Normalisation is a technique used to ensure that when numbers are represented they are as precise as possible in relation to how many bits are being used.
- When working with decimal numbers, we are used to the idea of rounding numbers up or down. As a consequence, we will get rounding errors, which can be quantified as absolute or relative.
- Precision refers to how accurate the number needs to be in the context that it is being used.

TASKS

1. Explain the term 'unsigned integer'.
2. Show how an 8-bit two's complement integer would be used to store:
   a) +64 **b)** –64 **c)** +100 **d)** –100
3. Add the binary numbers 1001 and 1100. Leave your answer as a binary number.
4. Use two's complement to carry out the following calculations.
   a) 12 + 8 **b)** 25 – 17
5. Describe the benefits of using a fixed point number system over floating point.
6. Convert $3\frac{1}{4}$ into a binary number.
7. Convert the binary number 10.11 into its decimal equivalent.
8. A computer uses 12 bits to store numbers in floating point format. Seven bits are used to store the mantissa and the other five bits the exponent. Both parts use two's complement. The following numbers are stored using this system. Work out their decimal equivalents.
   a) 011100001111 **b)** 101000011111
9. Represent the following decimal values as floating point binary values.
   a) 13.5 **b)** –3.75
10. Explain why floating point numbers should always be normalised.
11. 011.111010000 is used to represent the decimal value 3.9.
    a) What is actual decimal value represented?
    b) Calculate the relative and absolute error.
    c) Explain how the floating point system could create a more accurate representation of 3.9.

26 Coding systems

SPECIFICATION COVERAGE

3.5.5.1 Character form of a decimal digit

3.5.5.2 ASCII and Unicode

3.5.5.3 Error checking and correction

3.5.6 Representing images, sound and other data (all subsections except 3.5.6.10 Encryption)

LEARNING OBJECTIVES

In this chapter you will learn:

- how all forms of data are represented as 0s and 1s
- that ASCII and Unicode are coding systems for characters and numbers
- how to check for and correct errors in data
- how bit-mapped graphics are created
- how analogue and digital data are transmitted and converted
- how sound is sampled and digitised
- how to compress data using lossy and lossless techniques.

A-level students will learn:

- how vector graphics are created.

KEYWORD

Character code: a binary representation of a particular letter, number or special character.

INTRODUCTION

In addition to the various number systems described in Chapter 23, binary codes can also be used to represent text and characters. The term character is used in the widest sense to include all the keyboard characters, control and special characters. Each character has a **character code**, which is its binary representation.

There is an important distinction to be made here about the way in which binary can be used to represent numbers either as true numbers, or as characters. So far, we have looked at the pure binary representation of a number. This means that we are able to carry out calculations using that number. Sometimes numbers are not used as part of mathematical calculations and the character code for the number will be used instead. For example, a house number in an address or a telephone number uses numerical characters. In these cases, different coding systems are used.

ASCII and Unicode

In the early days of computing, programmers would combine groups (sequences) of 0s and 1s to represent different things. For example, they might decide that 00000000 could be used to represent an A and 00000001 could be used to represent a B and so on. The problem was that different programmers used their own coding systems so the sequences meant different things to different people.

KEYWORD

ASCII: a standard binary coding system for characters and numbers.

As a result of the confusion this caused, a standard was agreed upon for the representation of all the keyboard characters, including the numbers, and other commonly used functions. This standard is called **ASCII** or the American Standard Code for Information Interchange. In fact, a 7-bit code was agreed upon as 7 bits gives 128 permutations, which is enough for the most commonly used characters. More recently, extended ASCII was introduced which is an 8-bit code allowing for 256 characters.

Table 26.1 shows an extract of the keyboard characters and the string of bits that represent them. You don't need to remember any of these but it is useful to understand the principle behind them.

Table 26.1 ASCII look-up table

Char	Decimal	Binary	Char	Decimal	Binary	Char	Decimal	Binary
!	33	00100001	0	48	00110000	F	70	01000110
"	34	00100010	1	49	00110001	G	71	01000111
#	35	00100011	2	50	00110010	H	72	01001000
$	36	00100100	3	51	00110011	a	97	01100001
%	37	00100101	4	52	00110100	b	98	01100010
&	38	00100110	5	53	00110101	c	99	01100011
'	39	00100111	6	54	00110110	d	100	01100100
(	40	00101000	7	55	00110111	e	101	01100101
)	41	00101001	8	56	00111000	f	102	01100110
*	42	00101010	9	57	00111001	g	103	01100111
+	43	00101011	A	65	01000001	h	104	01101000
,	44	00101100	B	66	01000010	i	105	01101001
-	45	00101101	C	67	01000011	j	106	01101010
.	46	00101110	D	68	01000100	k	107	01101011
/	47	00101111	E	69	01000101	l	108	01101100

You will notice that all the keyboard characters have a code covering upper and lower case letters, numbers on the keypad, special characters (%, @, /, #, etc.) and non-printing characters (ACK, BS, etc.). Non-printing codes mainly cover the communication codes that are used to allow devices to be understood by the processor.

ASCII was until recently the standard method of converting keyboard and other characters into binary codes. However, ASCII does have certain limitations:

- 256 characters are not sufficient to represent all of the possible characters, numbers and symbols.
- It was initially developed in English and therefore did not represent all of the other languages and scripts in the world.
- Widespread use of the web made it more important to have a universal international coding system.
- The range of platforms and programs has increased dramatically with more developers from around the world using a much wider range of characters.

KEYWORD

Unicode: a standard binary coding system that has superseded ASCII.

As a result, a new standard called **Unicode** has emerged which follows the same basic principles as ASCII in that in one of its forms it has a unique 8-bit code for every keyboard character on a standard English keyboard. ASCII codes have been subsumed within Unicode meaning that the ASCII code for a capital letter A is 65 and so is the Unicode code for the same character. Unicode also includes international characters for over 20 countries and even includes conversions of classical and ancient characters.

To represent these extra characters it is obviously necessary to use more than 8 bits per character and there are two common encodings of Unicode in use today (UTF-8 and UTF-16). As the name suggests the latter is a 16-bit code.

Unicode is being constantly developed and updated to include more of the 'diverse languages of the modern world', for example, the Arabic and Chinese alphabets are significantly different to English and even alphabets that are similar to English, such as French and German, still contain specific characters not found on standard English keyboards.

KEYWORD

Parity bit: a method of checking binary codes by counting the number of 0s and 1s in the code.

The importance of any standard is that it is universally adopted, which in this case involves everyone in the computing industry throughout the world. The increased use of the Internet has meant that much more data is being passed around global networks. If different encoding systems are used it means that data can be corrupted when used on any system other than that on which it was created. Unicode aims to cover every platform in terms of hardware and operating systems, every foreign language and every program.

Error checking and correction

Data is being transmitted around the computer all the time. All of this data is made up of strings of 0s and 1s. It is possible that the data can get corrupted at any point either when it is being processed or transmitted. There are various methods for checking and correcting errors in data.

A **parity bit** is a method of detecting errors in data during transmission. The way it works is quite simple, but it will not identify all errors in transmission. When you send data, it is being sent as a series of 0s and 1s. In Figure 26.1, a Unicode character is transmitted as the binary code 01101111. It is quite possible that this code could get corrupted as it is passed around either inside the computer or across a network.

As you will read in Section Nine these codes are being sent on carrier waves. Any slight variation in the frequency could mean that a 0 is misinterpreted as a 1. This would make the data very unreliable. Depending on the nature of the data, this could be critical. In the top example the parity bit is set to a 0 to maintain an even number of ones. The bottom example shows another binary code where the parity bit is set to 1 in order to ensure an even number of ones.

0 1 1 0 1 1 1 1 0

0	1	1	0	1	1	1	1	0

Parity bit

Parity bit = 0 to ensure number of ones is even

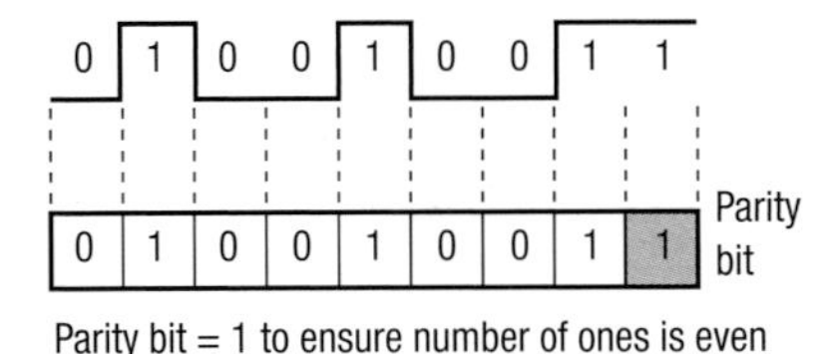

Parity bit = 1 to ensure number of ones is even

Figure 26.1 Even parity

One method for detecting errors is to count the number of ones in each byte before the data is sent to see whether there is an even or odd number. At the receiving end, the code can be checked to see whether the number is still odd or even.

- Even parity: the number of 1s in the code are counted. If there are an odd number of 1s, the parity bit is set to one to make the total number of 1s even. When the data is received, it is checked to ensure that there are still an even number of 1s. If there are, then the data is assumed to be correct.
- Odd parity: the number of 1s in the code are counted. If there are an even number of 1s, the parity bit is set to one to make the total number of 1s odd. When the data is received, it is checked to ensure that there are still an odd number of 1s. If there are, the data is assumed to be correct.

Majority voting

KEYWORD

Majority voting: a method of checking for errors by producing the same data several times and checking it is the same each time.

Majority voting is another method of identifying errors in transmitted data. In this case, each bit is sent three times. So the binary code 1001 would be sent as:

111000000111

When the data is checked, you would expect to see patterns of three bits. In this case, it is 111 for the first bit, then 000 and so on. Where there is a discrepancy, you can use majority voting to see which bit occurs the most frequently. For example, if the same code 1001 was received as:

101010000111

you could assume that the first bit should be 1 as two out of three of the three bits are 1. You would assume that the second bit is 0 as two of the three bits are 0. The last two bits are 0 and 1 as there appears to be no errors in this part of the code.

This same principle can also be applied on a larger scale. For example, on the Space Shuttle missions, the Columbia had at least four computers processing the same data and comparing each other's results. Where there was a discrepancy between results, the majority voting method was used to identify what was considered to be the correct course of action.

Check digits

KEYWORD

Check digit: a digit added to the end of binary data to check the data is accurate.

Check digits are a common way of checking data, often when they are being entered into a computer. For example, check digits are used on the barcodes printed on goods at a supermarket. Like a parity bit, a check digit is a value that is added to the end of a number to try and ensure that the number is not corrupted in any way.

The check digit is created by taking the digits that make up the number itself and using them in some way to create a single digit. The simplest, but most error-prone method is to add the digits of the number together, and keep on adding the digits until you have only a single digit remaining. So the digits of 123456 add up to 21 and 2 and 1 in turn add up to 3, so the number with its check digit becomes 1234563. When the data is being processed the check digit is recalculated and compared with the digit that has been transmitted. Where the check digit is the same then it is assumed that the data is correct. Where there is a discrepancy, an error message is generated.

The problem with this system is that if two numbers are transposed (swapped round) the check digit will be the same. For example, the numbers 1234 and 4321 both add up to 10 and therefore produce a check digit of 1. In order to overcome this, each number in the pattern is given a weighting. This means that each number is multiplied by a different weight or scaling factor. A common method for calculating a check digit is known as a modulo-11 and is shown below.

- Original number 2 3 0 4 5
- Weighting 6 5 4 3 2
  (this starts from 2, not 1)
- Multiply by weight 12 15 0 12 10
- Add together 12 +15 +0 +12 +10 = 49
- Divide by 11 49 ÷ 11 = 4 remainder 5
- Subtract the remainder from 11 11 – 5 = 6
- So the check digit is 6 and this makes the number 230456

Bit-mapped graphics

KEYWORDS

Bit-mapped graphic: an image made up of individual pixels.

Pixel: an individual picture element.

Resolution: width × height or pixels per inch.

Up to now we have concentrated on how binary can represent text and numbers. It is also used to represent sound and graphics. Graphics are the display of pictures on your computer and can range in complexity from simple line drawings through to full animations. All computer graphics are represented using sequences of binary digits (bits).

The display on a monitor is made up of thousands of tiny dots or picture elements called **pixels**. A typical monitor might have a grid of 1366 by 768 pixels. This is known as the **resolution**. This term is also used on individual picture files. So the formula for the resolution of a file in pixels is:

resolution = width × height

You can also define resolution in terms of the number of pixels per inch (PPI). For example, a monitor that is 12 × 9 inches, which is 1366 by 768, will have 1366 divided by 12 pixels per inch, or around 114 PPI on the horizontal axis and 768 divided by 9, which is around 85 PPI on the vertical axis.

Each of these pixels can be controlled to display different colours. By combining the pixels, a picture is created on the screen. At a very simple level, each pixel could be controlled by one bit. This means that each pixel is mapped to one bit in memory. The bit could be set to either 0 or 1 representing off or on which in this case would be black or white.

Data stored in memory

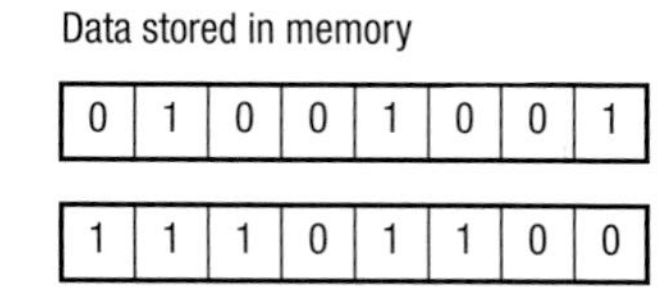

Data displayed on screen

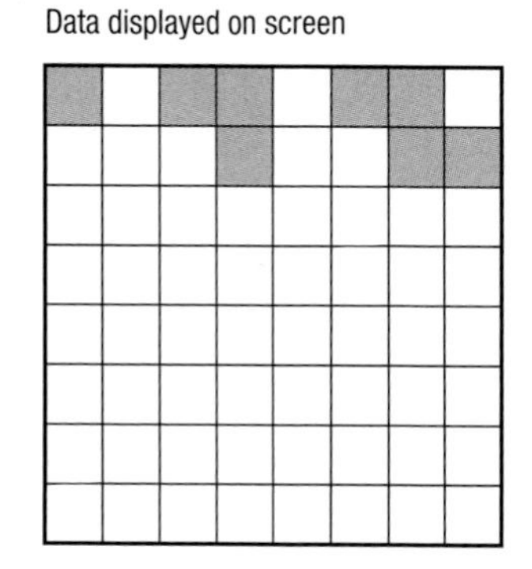

Figure 26.2 Bit-mapped display

KEYWORD

Colour depth: the number of bits or bytes allocated to represent the colour of a pixel in a bit-mapped graphic.

To create colour graphics, each pixel is mapped to more than one bit. For example, a pixel might be represented by a byte (8 bits) in memory. This means that each pixel could be any one of 2^8 or 256 different colours. The amount of memory allocated to each pixel is referred to as the **colour depth**. Your computer will contain a graphics card for controlling graphics. The amount of memory allocated for bit-mapping depends on the amount of memory on this card:

storage = resolution × colour depth

If 24 bits were allocated to each pixel this would give 2^{24} combinations or 16 777 216 different colours. Twenty-four bits are typically used as eight bits are allocated to each of the three primary colours: red, green and blue (RGB) from which all other colours are created. This means that with a 1024 × 768 display with 24 bits per pixel you get 18 874 368 bits or 2.35 MB of memory to make one picture as calculated as follows:

- screen resolution of 1024 × 768 = 786 432 pixels
- 24 bits are allocated to each pixel to give 786 432 × 24 = 18 874 368 bits
- divide by 8 to get the answer in bytes = 2 359 296 bytes
- convert to megabytes to get 2.36 MB.

Note that bitmaps may also contain metadata, which means that the bitmap file will be storing information about itself. Metadata is normally found at the beginning of the file in a header, and might include information such as the file type, the width and height in pixels and the colour depth.

Vector graphics

A level only

KEYWORD

Vector graphic: an image made up of objects and coordinates.

Vector graphics are created using objects and coordinates. A vector is a measure of quantity and direction. It is easier to think of vector graphics as geometric shapes. For example, if we had a vector graphic of a square it would be made up of four coordinates with lines drawn between them. To rescale the object requires an adjustment of the coordinates. Therefore, the graphics are being controlled mathematically rather than being completely regenerated as with a bit map. An image created on the screen will be made up of lines and the scale and position of the lines will be adjusted as the screen display changes to create an image.

Consider the two images in Figure 26.3.

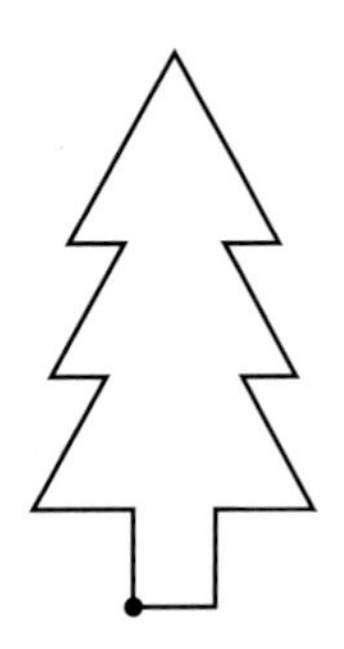

• Start coordinate

Figure 26.3 Scaled vector graphic

The first image is a series of lines, the second image is the same series of lines but the dimensions are different. If this second image was stored as a vector graphic, the file would contain the new dimensions. One of the advantages of this method is that the file would be much smaller than a bit-mapped file containing the same image because vector graphics files contain the mathematical description required to create an image rather than storing the actual image as with a bit-mapped graphic. They also facilitate perfect rescaling as, when drawn at a different scale, it is done using the mathematical description. They are not practical for every scenario where graphics are needed such as scanning and digital photography.

CAD/CAM packages make use of vector graphics as these packages tend to use line-based drawings. Some two- and three-dimensional animation programs also use vector graphics. This is because an animated image is a series of still images combined together. Once the still image has been created, the vectors can be manipulated to create the various frames within the animation.

Vector images are made up of primitives, which are the basic pieces of data needed to create an image. Typically this will be points, lines, curves and polygons. This includes common shapes and will also include letters. The colour gradient may also be contained as a primitive.

Analogue and digital signals

All the processing carried out by a computer is digital, yet there are occasions when either the input or output required are analogue. For example, some of the data sent around the Internet are sent in analogue form over the telephone network. This is because the telephone lines were originally designed to carry voice data which are analogue. A microphone takes speech input which is analogue, or a musical instrument digital interface (MIDI) takes in data from a musical instrument which may be analogue.

Analogue data are data that are infinitely variable and are often represented in the form of a wave. Figure 26.4 shows a typical sound wave.

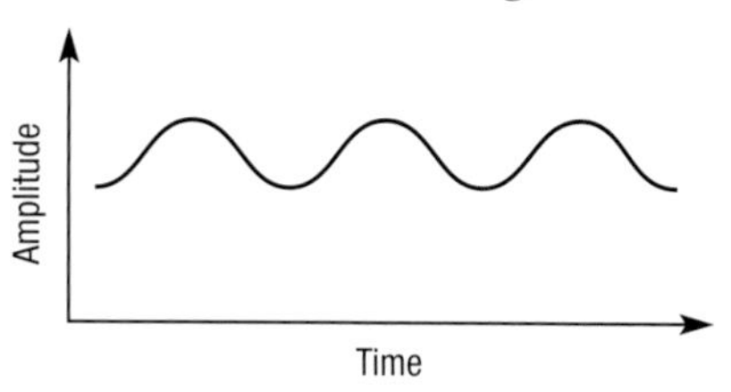

Figure 26.4 An analogue wave

Digital data are often represented as discrete values shown in Figure 26.5 with the ons and offs shown as set peaks and troughs. As we have seen in this section, digital data are often represented as a sequence of 0s and 1s.

Figure 26.5 A digital signal

Analogue to digital conversions

The problem arises when we need to input analogue data into the computer or when we want to output digital data from the computer in analogue form. In order to do this, a converter is needed, which could be either an analogue to digital converter (ADC) or a digital to analogue convertor (DAC).

One example where an ADC is used is between a microphone and a computer. The microphone inputs sound in the form of changes in air pressure and then converts them into electrical signals. These analogue electrical signals are then converted by the ADC into digital signals that the computer can process.

Another example is a MIDI device for an acoustic guitar. This device fits beneath the strings on the guitar and when the strings are played they generate an analogue sound wave. The sound waves are picked up and converted to digital form.

MIDI uses event messages to control various properties of the sound. These messages are typically encoded in binary and provide communication

Figure 26.6 A MIDI keyboard

between MIDI devices or between a MIDI device and the processor. For example, on a MIDI keyboard, an event message may contain data on:

- when to play a note
- when to stop playing the note
- timing a note to play with other notes or sounds
- timing a note to play with other MIDI-enabled devices
- what pitch a note is
- how loud to play it
- what effect to use when playing it.

The advantages of using midi files over other digital audio formats are:

- MIDI files tend to be much smaller. This means they require less memory and also load faster, which is particularly advantageous if the MIDI file is embedded in a web page
- MIDI files are completely editable as individual instruments can be selected and modified
- MIDI supports a very wide range of instruments providing more choices for music production
- MIDI files can produce very high quality and authentic reproduction of the instrument.

Sound sampling and synthesis

Sampling is the process of converting analogue sound waves into digital form to create what is commonly known as digitised or digital sound. This is sometimes referred to as analogue to digital (ADC) conversion. An analogue sound wave is infinitely variable so in order to store this digitally, a series of readings at fixed intervals are taken from the wave in order to create the discrete data values that are a defining feature of binary data. These readings are then stored as binary codes. It is called sampling because you do not record every single change in amplitude of the waveform. Instead, you choose set points at which a reading (or sample) will be taken.

Figure 26.7 shows the points at which the sample readings are taken.

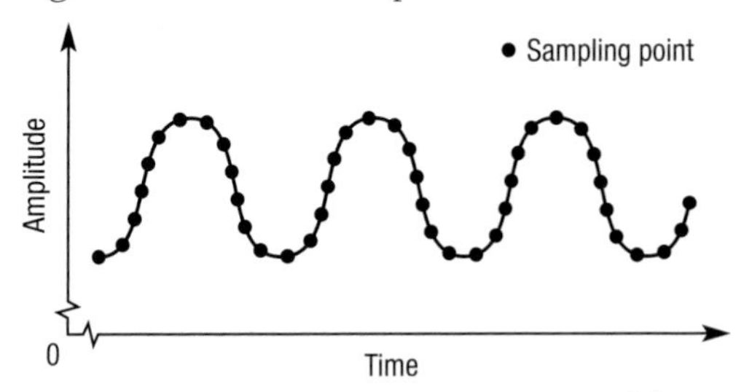

Figure 26.7 Sampling an analogue wave (1)

The amplitude of the wave is only recorded at the point where each sample is taken. Other variations in the amplitude are not recorded. Therefore, to create an exact replica of the analogue sound would require a sample to be taken every time the amplitude of the wave changed even by a small amount. However, the human ear doesn't notice very small changes, so sound can be faithfully created with fewer samples. Taking more samples

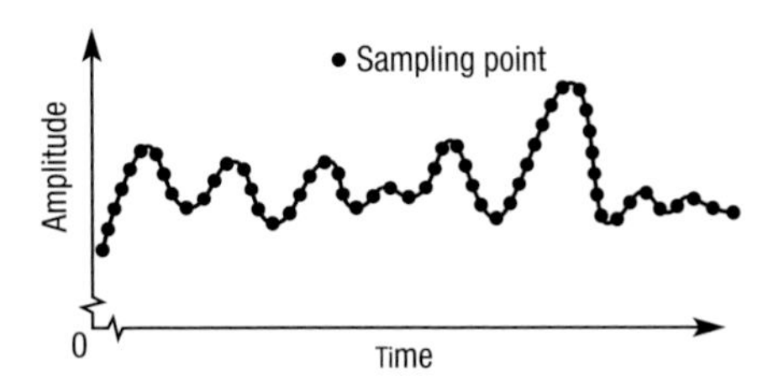

Figure 26.8 Sampling an analogue wave (2)

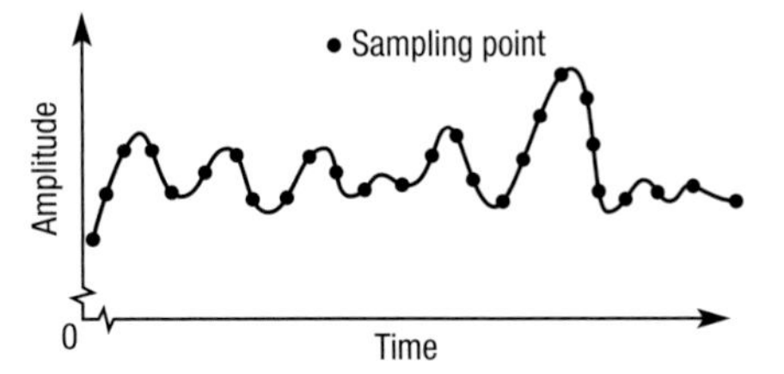

Figure 26.9 Sampling an analogue wave (3)

allows more accurate reproduction of the original analogue sound. Consider the following example:

Imagine that the sound wave shown in Figures 26.8 and 26.9 represents part of a song. The first sample would create a more accurate recording as it involves more samples. However, it would require much more memory and storage space than the sample in the second diagram where fewer samples have been used.

To calculate how large the file size will become you can use the following calculation:

sample rate (Hz) × length of recording (seconds) × sampling resolution (bits)

Sample rate represents the number of samples that will be taken per second. The length of the recording is simply measured in seconds and the sampling resolution refers to the number of bits allocated to representing the sound.

For example the following three samples are recorded at a resolution of 16 bits (two bytes).

0011 1100 0000 0011
1111 0101 0110 0101
0110 0110 0011 1110

Assuming a sample rate of 44 000 Hz, with a recording lasted 60 seconds, the file size would be:

44 000 × 60 × 16 = 42 240 000 bits = 5 280 000 bytes or 5.28 MB

When deciding on the optimum sampling rate, many programmers refer to Nyquist's Theorem, which states that to faithfully recreate the analogue signal, you should sample at least twice the highest frequency. For example, if the human ear can cope with frequencies of 20 Hz – 20 000 Hz, then the analogue frequency must be sampled at at least 40 000 Hz. The reason for doubling the frequency is to ensure that the sample covers the complete range of peaks and troughs in the analogue signal, which then allows a faithful reproduction of the sound.

Samples can be edited to remove any background noise or interference from the original sound wave. Some people argue that CDs produce better quality sound than vinyl disks for this reason.

Sound synthesis is another term that is used to refer to sound that is produced digitally rather than in analogue format. It means that the sound is synthesised or manufactured rather than being in its original analogue format. By definition, all sounds created by a computer are digital.

After sound has been digitally recorded, in order to hear it, the user will use either earphones or speakers. These devices are driven by audio signals yet the data is stored as digital signals. In order to convert it so that it can be amplified and played, a digital to analogue convertor (DAC) is required. Typically, the DAC is embedded in the device that plays the audio data and the signal is passed in analogue form to the loudspeakers or headphones.

Data compression

There are many scenarios where the files used to store data can get very large. For example, high resolution photographs or music sampled at a high frequency will result in files that could be several megabytes. A whole movie will take up several gigabytes. In order to reduce storage requirements, and make it quicker to transmit these files, they are often compressed.

> **KEYWORD**
>
> **Compression**: the process of reducing the number of bits required to represent data.

Compression is the process of encoding information with fewer bits, so that the files take up less memory. There are several methods for doing it, depending on the type of data being encoded. You are probably familiar with the concept of using zip files, or reducing a high resolution image to low resolution. Many familiar file types such as jpeg, mpeg and mp3 are compressed files.

Any type of data can be compressed and different techniques are used depending on the data type. These techniques lead to either lossless or lossy compression.

- Lossless means that the compressed file is as accurate as it was before compression, i.e. no data is lost.
- Lossy means that there will be some degradation in the data, for example, a grainier image might be produced.

Lossless compression

> **KEYWORDS**
>
> **Run-length encoding**: a method of compressing data by eliminating repeated data.
>
> **Dictionary-based encoding**: a method of compressing text files.

Imagine a picture made up of millions of pixels. The picture file will contain data about each pixel, for example its colour. So part of the file might simply read: blue, blue, blue, dozens of times where there is a run of blue pixels. Rather than storing this same data over and over, you can use '**run-length encoding**', which states that the next *x* pixels will be blue. So B,B,B,B,B,B,B,B becomes 8B. This is a simple example but you can see that only two encoded digits are needed to represent eight uncompressed ones, and there is no loss of data accuracy.

When compressing text files, **dictionary-based** compression techniques can be used. These work on the basis that within the text, there will be common strings of characters. Rather than rewriting these same strings, they can be coded in some way.

For example, the characters 'tion' are commonly found at the end of many words, such as 'station', 'nation' and 'creation'. Rather than storing the words individually, 'tion' can be encoded to the dictionary and then used in combination with other prefixes to form words. At the same time, 'sta', 'na' and 'crea' can be encoded to the dictionary as they too can be made into other words. Now when you need to encode any words that contain those strings of text, you can use the dictionary entry rather than writing the whole strings out again in the file.

For example, you could have a dictionary file set up with tokens or codes that represent different words or parts of words. Following on from the example above, we could assign numeric tokens as follows:

Data	Token
tion	1
sta	2
na	3
crea	4

A data file containing the strings 21, 31, 41 would result in the words 'station', 'nation' and 'creation' but use six characters instead of 21 characters.

Dictionary-based techniques can be used on non-text data if it is considered as a sequence of 0s and 1s.

Lossy compression

There are cases where lossless compression still results in a large file as there is a limit to how small it can get, while still maintaining accuracy. In some cases where the amount of memory is an issue, or where data is being transmitted across a network and the speed of transmission is vital, it might be necessary to make these files even smaller.

Figure 26.10 Lower resolution images resulting in pixellation

This is often the case with streaming audio or video. In these cases, a compression technique that leads to some degradation in data quality may be acceptable. For example, if you were streaming a movie, you might not expect the picture quality to be as good as that on a DVD.

Programmers must take into account issues such as memory and transmission requirements when deciding on the most appropriate compression techniques. The widespread use of mobile data on various portable devices means that lossy compression is commonly used, even though it leads to a degradation of the original data.

Lossy compression techniques work by identifying data that can be removed, while still creating an acceptable representation. In the case of audio, graphics and video, the user will have some control over the level of compression and therefore the quality of the compressed file. For example, a low resolution JPEG file will have more of the original data removed and therefore produce a pixellated image.

JPEG compression works by breaking an image up into blocks of 8 × 8 pixels. In each block, the data is converted into frequencies. Some frequencies are considered to be more important than others.

Figure 26.11 JPEG image broken into 8 × 8 pixel grids with high and low frequencies

This is because high frequency data is more difficult for the human eye to perceive so changes made to high frequency data will be less noticeable. Low frequency data on the other hand is more noticeable.

For example, in Figure 26.11, the 8 × 8 pixel grid is of a grey road. This is a high frequency image as the human eye will not notice slight changes in the grey. However, if you look at the red box that represents another 8 × 8 grid, there is a large contrast between the dark building and the light sky. When compressing the image it is important to maintain this contrast. The slight contrasts in grey on the road are much less important.

Therefore JPEG analyses the pixel data within each 8 × 8 block and removes data that is least likely to affect the human perception of the image. It then uses the same run-time encoding methods described above to eliminate any repeating data to reduce the file size.

Practice questions can be found at the end of the section on page 228.

TASKS

1. The number 5 could be represented in binary as 00000101. Using ASCII the binary code for 5 is 00110101. Explain why there are two different binary codes for the same number.
2. Describe how the following methods work and explain why none of them are guaranteed to spot all errors in data.
   a) parity bit **b)** majority voting **c)** check digits
3. What is a bit-mapped graphic?
4. If eight bits of memory are allocated to each colour (red, green. blue) within a pixel, how many megabytes of memory are needed to store a 1024 × 768 display?
5. What is a vector graphic?
6. Give two examples where analogue data needs to be converted to digital data.
7. Give two examples where digital data needs to be converted to analogue data.
8. 'Analogue produces a purer sound than digital.' Give one reason why this statement could be true.
9. Explain why the storage of video with sound requires a large amount of disk space.
10. Explain two situations where you might use:
    a) lossy compression **b)** lossless compression.
11. Explain the difference between odd and even parity, giving an example of each.

KEY POINTS

- Binary codes can be used to represent text, characters, numbers, graphics, video and audio.
- ASCII and Unicode are systems for representing characters.
- It is possible that the data can get corrupted at any point when it is being either processed or transmitted.
- Error detection and correction methods include check digits and majority voting.
- Bit-mapped graphics are made up of individual pixels (picture elements).
- Vector graphics are composed of objects.
- Resolution is the measure of the height and width of an image.
- Analogue signals such as sound waves need to be converted into digital form so they can be processed by the computer by sampling.
- Data is compressed to make file sizes more manageable.
- Compression can either be lossless, which means no degradation of the data after compression, or lossy, which means there will be degradation of the data.

STUDY / RESEARCH TASKS

1 Identify one application where it would be more appropriate to use vector graphics rather than bit-mapped graphics.
2 Identify one application where it would be more appropriate to use bit-mapped graphics rather than vector graphics.
3 What is the standard sampling rate for CD audio quality?
4 Video messages sent via mobile phone often appear slightly jagged and jerky. Explain why this is the case.
5 What is a CODEC?
6 Research one of the better known compressed file types and find out what compression algorithm is used.
7 Research the LZ77, LZ78 and LZW compression methods.

27 Encryption

SPECIFICATION COVERAGE

3.5.6.8 Encryption in AS Level

3.5.6.10 Encryption in A Level

LEARNING OBJECTIVES

In this chapter you will learn:

- what encryption is and why it is important
- the basics of encryption techniques
- how to encrypt a message using the Caesar cipher
- how to encrypt a message using a transposition cipher
- how to encrypt a message using the Vernam cipher
- how to use frequency analysis to help decrypt ciphers
- what computational security is.

INTRODUCTION

Encryption is the process of scrambling data so that it cannot be understood by another person unless they know the encryption method and key used. **Decryption** is the process of turning the scrambled data back into data that can be understood. Data is encrypted before it is transmitted and decrypted when it is received. Therefore encryption keeps data secure during transmission.

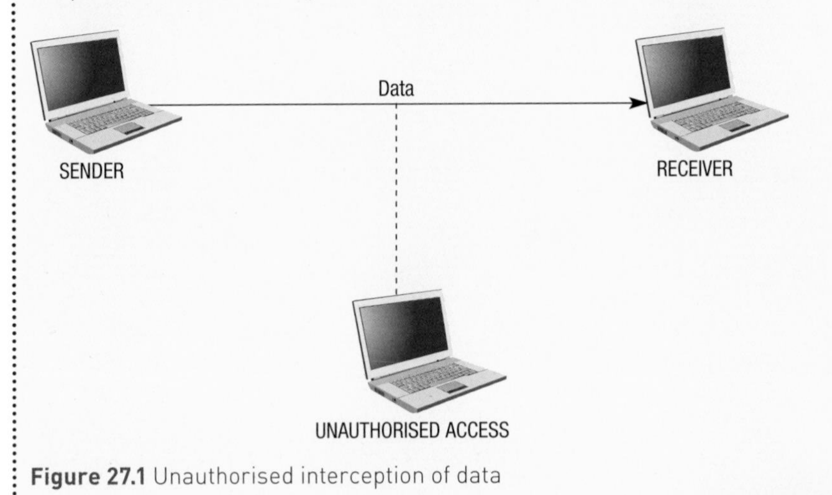

Figure 27.1 Unauthorised interception of data

KEYWORDS

Encryption: the process of turning plaintext into scrambled ciphertext, which can only be understood if it is decrypted.

Decryption: the process of deciphering encrypted data or messages.

Data are vulnerable when they are accessed by a third party who should not have access to them. This might occur if someone gains access to saved files on a computer, or when data are transmitted across networks. Often the nature of the data being sent is personal or sensitive and there is a risk if someone were to intercept them. For example:

- online banking transactions need to be kept secure to prevent theft and fraud
- health information needs to be kept secure as it contains very sensitive personal information
- government security data needs to be kept secure as they could have implications for national security.

In fact the use of encryption is closely linked to matters of government and military security as many techniques were developed and used during conflicts. With the widespread use of the Internet over recent years, encryption has become a vital mechanism for securing data sent across local and wide area networks.

Encryption basics

KEYWORDS

Plaintext: data in human-readable form.

Ciphertext: data that has been encrypted.

All encryption works on the basis of turning **plaintext** into **ciphertext** as shown in Figure 27.2.

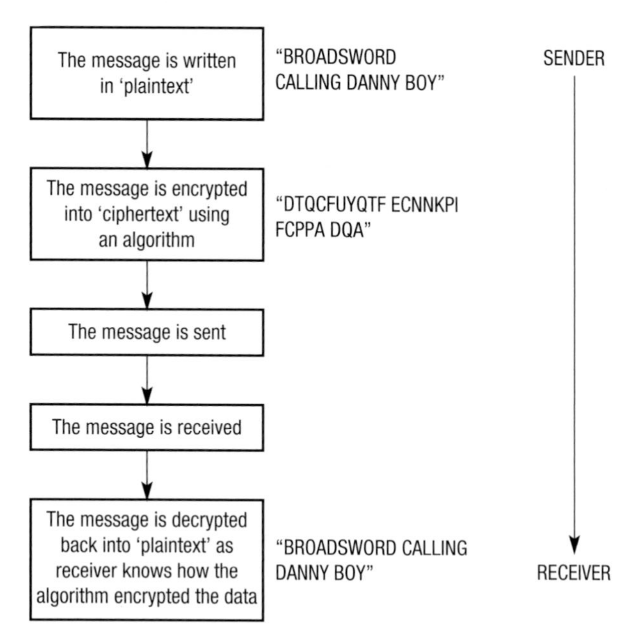

Figure 27.2 The encryption–decryption process

Plaintext is the original data in a format that can be understood. The data are then encrypted or encoded using an algorithm that turns the data into ciphertext. The message can then be sent. Encryption does not prevent the data from being intercepted, it just makes them meaningless to the person who intercepted them unless they know how the data have been encrypted.

It is common practice therefore for people to try and work out what algorithm has been used, that is, to try and crack the code. There are many famous examples of code-breakers throughout history, not least the work done on the German Enigma machine during WWII by both Polish and British mathematicians. Many people believe that their ability to crack this code

KEYWORDS

Caesar cipher: a substitution cipher where one character of plaintext is substituted for another, which becomes the ciphertext.

Vernam cipher: a method of encryption that uses a one-time pad (key) to create ciphertext that is mathematically impossible to decrypt without the key.

Transposition cipher: a method of encryption where the characters are rearranged to form an anagram.

Key: in cryptography it is the data that is used to encrypt and decrypt the data.

Substitution cipher: a method of encryption where one character is substituted for another to create ciphertext.

enabled the Allies to win the war as they were able to intercept and read messages about the German war effort.

There are various techniques for encrypting data and in this chapter we will be considering two of the common ones: the **Caesar cipher** and the **Vernam cipher**. We will also be considering **transposition ciphers**. A-level students also need to be aware of public and private key encryption and this is covered in Chapter 44.

A **key** is a piece of data used in encryption that defines the way in which plaintext is turned into ciphertext. You will see some examples in this chapter. The key that is used by the sender to encrypt the message must be known by the receiver so that they can decrypt the message. Some keys, like those used with a Caesar cipher are too easy to work out, whereas other keys like those used on a Vernam cipher are theoretically impossible to crack.

The Caesar cipher

Named after the Roman Emperor who used it for all of his personal correspondence, the Caesar cipher is an example of a shift or **substitution cipher**. This method substitutes each letter of the alphabet for another character by simply shifting the letters forwards or backwards. A variation on this would be to shift letters on a random basis.

The diagram below shows a substitution cipher with a two-letter shift. In this case, the message: "BROADSWORD CALLING DANNY BOY" becomes "DTQCFVYQTF ECNNJPI FCPPA DQA".

A	B	C	D	E	F	G	H	I	J	K	L	M	N	O	P	Q	R	S	T	U	V	W	X	Y	Z
C	D	E	F	G	H	I	J	K	L	M	N	O	P	Q	R	S	T	U	V	W	X	Y	Z	A	B

A random substitution might look like this:

A	B	C	D	E	F	G	H	I	J	K	L	M	N	O	P	Q	R	S	T	U	V	W	X	Y	Z
O	Z	P	C	Y	D	B	Q	X	K	L	A	V	W	R	I	S	M	J	E	G	N	F	T	U	H

Our message when encrypted with this random substitution becomes: "ZMROCJFRMC POAAXWB COWWU ZAU".

Both of these methods would be fairly easy to work out, even without the key. It could be made more secure by adding a keyword, for example, you might select the word "BEESWAX". First, you need to delete any repeated letters in the word, leaving you with "BESWAX". Then add this word to the start of the alphabet. In other words, the first seven letters of the alphabet are substituted for the keyword. Then add the remaining letters in alphabetical order:

A	B	C	D	E	F	G	H	I	J	K	L	M	N	O	P	Q	R	S	T	U	V	W	X	Y	Z
B	E	S	W	A	X	C	D	F	G	H	I	J	K	L	M	N	O	P	Q	R	T	U	V	Y	Z

The message can then be encrypted. With any of these three examples, the receiver would need to know what key is being used. If it is a two-letter shift then they only need to shift +2 to decrypt the message. If a keyword is being used then they need to know the keyword. In the case of the random substitution they would need the entire look-up table. Therefore to decipher the code, the recipient would need the key, which in this case is the keyword, and the encryption method.

KEYWORD

Polyalphabetic: using more than one alphabet.

A further level of complexity can be added to a substitution cipher with the use of more than one alphabet. This is known as a **polyalphabetic** cipher and rather than having one set of substitutions it could have any number of alphabets or look-up tables to work with. For example, there might be ten different versions of the look-up table above. Each letter that is encrypted is passed through all of them to produce a final encrypted letter. For example, using four different alphabets an example might look like this:

- The letter 'A' is encrypted using the first alphabet to the letter 'W'
- The letter 'W' is encrypted using the second alphabet to the letter 'S'
- The letter 'S' is encrypted using the third alphabet to the letter 'P'
- The letter 'P' is encrypted using the fourth alphabet to the letter 'D'.

The letter "D" is then used in the encrypted message. All four alphabets will be needed to decrypt the message.

This concept goes back to the 15th century and is known as the Alberti cipher. This is also one of the underlying principles used in the Enigma machine, where several randomised alphabets were used to encrypt a message.

Frequency analysis

The Caesar cipher is one of the easiest to crack because of the nature of language. In English, for example, there are certain letters that are used more frequently than others and certain combinations of letters that are also common. By examining ciphertext for frequently used letters and patterns of letters it is possible to work out what letter has been substituted for which other letter.

KEYWORD

Frequency analysis: in cryptography it is the study of how often different letters or phrases are used.

For example, a **frequency analysis** of letters used in most English writing might look like Figure 27.3.

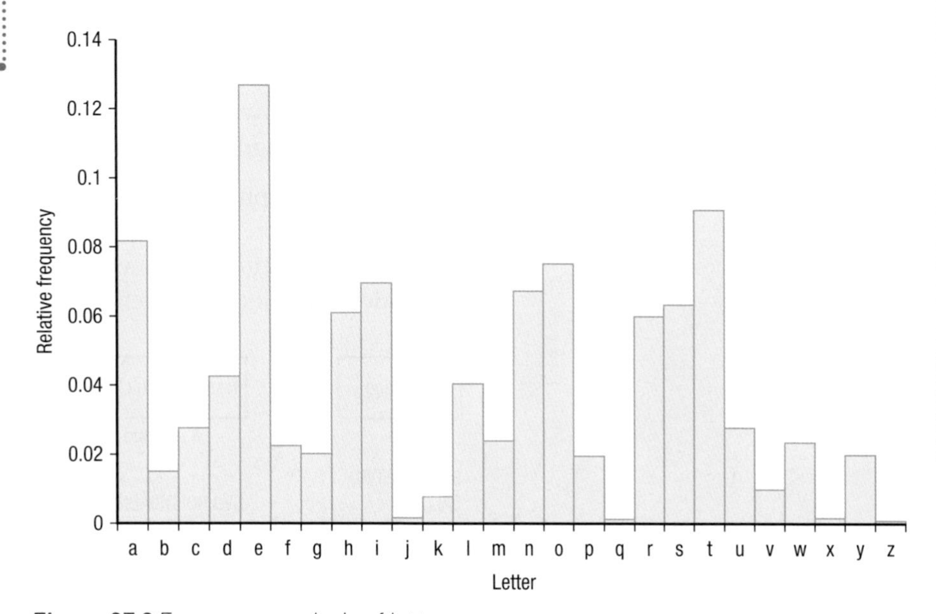

Figure 27.3 Frequency analysis of letter usage

Notice that 'e' is the most frequently used letter followed by 't'. 'j' and 'x' are used infrequently. If you took a large block of ciphertext and carried out the same analysis you would find a pattern. For example, if the 'e' was replaced by a 'p', then 'p' would be the most frequently used letter in the ciphertext and you can therefore assume that e = p. You could do this for all of the other letters by matching their frequencies.

In modern computing, cryptography like this could be cracked instantly, so more sophisticated methods have been developed.

Transposition cipher

KEYWORDS

Railfence cipher: a type of transposition cipher that encodes the message by splitting it over rows.

Route cipher: a type of transposition cipher that encodes the message by placing it into a grid.

With this type of cipher, the letters of the message are transposed, or rearranged to form an anagram. You must rearrange the letters according to a set pattern to make it possible to decrypt the message. One way of doing this is called the **railfence** method where the message is split across several lines. For example:

B		O		D		W		R		C		L		I		G
	R		A		S		O		D		A		L		N	

If you now read the message off line by line it becomes: "BODWRCLIGRASODALN". If you were decrypting this message you would need to know that the key is that it has been split over two lines.

To decrypt it, you simply read it back by moving down and then up. You could use any combination of lines and line lengths. For example, you could put the message across three lines:

B				D				R				L				G
	R		A		S		O		D		A		L		N	
		O				W				C				I		

In this case, the message becomes: "BDRLGRASODALNOWCI".

B	W	L
R	O	L
O	R	I
A	D	N
D	C	G
S	A	A

A variation is to put the message into a grid. This is called a **route cipher**. For example, the same message is placed into a 6 × 3 grid.

Reading down in columns from left to right will decrypt the message. Notice that you can add null or meaningless values if you have spare cells in your grid. In this case, the letter 'A' has been added to the bottom right-hand cell.

As with Caesar ciphers, complexity can be added to the key with the addition of a keyword before the main message. It is also common practice to combine then apply a substitution cipher to the transposed message so that there are two codes to crack to decipher the original message as shown in Figure 27.4.

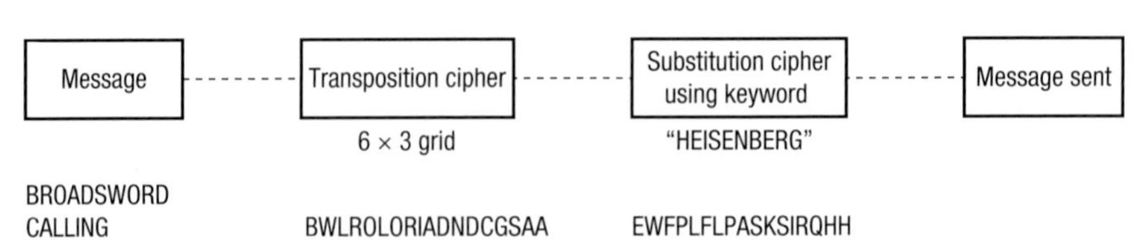

Figure 27.4 A substitution cipher

Although more complex than a straight transposition cipher, it is still possible to decrypt this message without knowing the key using a combination of frequency analysis and anagram deciphering.

Vernam cipher

Gilbert Vernam invented this cipher around 100 years ago as a means of keeping data secure whilst it was being transmitted using telex machines. The Vernam cipher is an example of a class of encryption techniques known as **one-time pad** techniques. The key that is used is a sequence of letters that should be as long as the plaintext that is being encoded. The key can be recorded on a pad, although in the Vernam case the key was written on a punched taped for input into the telex device. For maximum security, a particular key should only be used to encrypt one message, hence the name one-time pad.

KEYWORDS

One-time pad: a key that is only used once to encrypt and decrypt a message and is then discarded.

Baudot code: a five-digit character code that predates ASCII and Unicode.

To encrypt a message, each character in the plaintext is combined with the character at the corresponding position in the key by converting the corresponding plaintext and key characters into a binary code (originally a 5-bit **Baudot code**) and using a logical XOR on these binary representations to produce a new binary code, which in turn maps back to a character. Once the ciphertext is created the key is never used again for encryption, although it will be used once more to decrypt the ciphertext.

A message is encrypted and decrypted as follows:

- The key is created, which is a completely random sequence of characters. For example, each letter in the plaintext message "BROADSWORD" is combined with the letter at the corresponding position in the key that is written on the pad or tape:

Plaintext message	B	R	O	A	D	S	W	O	R	D
Key	H	E	L	K	K	J	V	T	U	I

- For each character in the plaintext and the key, the 5-bit Baudot representation is identified. For example, B = 11001, H = 10100. A logical XOR is then performed on the two values, which is a bitwise operation that results in a 1 only if the two bits being compared are different. In this case:

Baudot code for plaintext B	1	1	0	0	1
Baudot code for one-time pad H	1	0	1	0	0
XOR the two codes to produce ciphertext	0	1	1	0	1

- The Baudot table is then used to find the corresponding character that is represented by 01101, which is F. Therefore the first character in the ciphertext becomes F. This process is repeated for each letter in the plaintext.
- On receipt of the ciphertext, assuming the receiver has access to the key, an XOR can be performed on the ciphertext with the key to find the original plaintext:

Baudot code for ciphertext F	0	1	1	0	1
Baudot code for key H	1	0	1	0	0
XOR the two codes to reproduce plaintext	1	1	0	0	1

- You can see that the result of the XOR operation is 11001, which is the original plaintext character of B. This process is repeated on every character in the ciphertext.

Nowadays, it is more likely that a character code such as ASCII or Unicode would be used to convert the characters to binary values instead of the Baudot code.

Once the entire message has been encrypted and decrypted the key is destroyed and a new random key is created. As long as the key is completely random, and is kept secret and only used once, then it is mathematically impossible to crack the code.

This is because the key is entirely random and therefore creates outcomes that are random. If you don't know the key, then ciphertext letter A is just as likely to be plaintext H as it is L, or any other letter. Indeed, if the letter A occurred at more than one position in the ciphertext then it is likely that each occurrence of an A will correspond to a different letter in the plaintext. If someone intercepted the message, the best they could do is to try all possible keys to work out every possible plaintext, but there would be a huge number of these and no way to know which the correct one was.

This is known as perfect security and it means that however much time and ciphertext a cryptographer has to work with, they will never be able to crack the code.

In practice, complete security is more difficult to achieve as:

- generating completely random numbers is complex as any algorithm is likely to contain some element of predictability
- letting the receiver know what the key is (key exchange) is difficult as this information itself could be intercepted
- there is no way of authenticating the sender and receiver so if the key was intercepted, messages could be sent from unauthorised sources.

Computational security

KEYWORDS

Computational security: a concept of how secure data encryption is.

Computational hardness: the degree of difficulty in cracking a cipher.

The Vernam or one-time pad cipher is the only cipher that is considered to be 100% mathematically secure. All other ciphers can be cracked given enough time and enough ciphertext to work on. This leads to the concept of **computational security** or **computational hardness**. A cipher that is computationally secure is theoretically breakable but not when using current technology in a timeframe that would be useful. This recognises the fact that although most encryption can theoretically be cracked, in practice it will be secure enough to withstand most threats.

When devising encryption algorithms it means that programmers need to be aware that some levels of encryption are harder to crack than others and that the level of security they use needs to be commensurate with the level of risk of the data being intercepted. For example, you would expect a much more sophisticated level of encryption on a data file being used to store data on military movements than one used to store a school project.

Computational security means that cryptographers need to be aware of the ways in which their encryption could be cracked. In addition to frequency analysis discussed earlier, there are several different methods for cracking codes:

- Identifying commonly used techniques: Many ciphers are based on substitution or transposition. Experienced cryptographers are able to recognise patterns in data that has been encrypted using these methods.

- Reverse engineering: This is the process of going back step by step until you work out how something has been put together.
- Dictionary attacks: This is the process of using a dictionary that contains common words and phrases. After each attempt to decrypt text is made, the text can be compared to the dictionary to see if it matches.
- Brute force: This is similar to a dictionary attack but takes much longer as rather than looking at common words and phrases it looks at every single permutation of characters that can be created and then compares the decrypted text to these permutations.

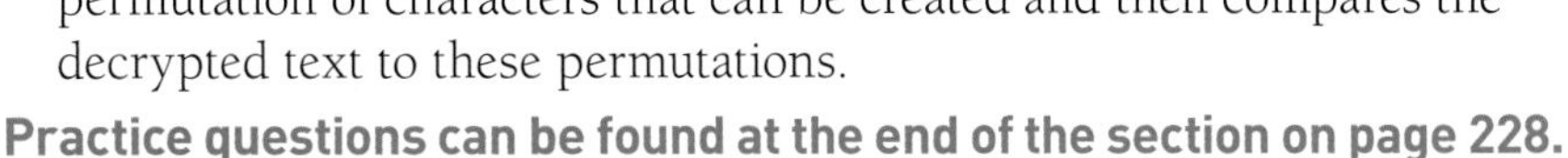

Practice questions can be found at the end of the section on page 228.

TASKS

1. What is encryption and why is it used?
2. Explain the use of a key in encrypting and decrypting data.
3. Encrypt a message using:
   a) a simple shift cipher
   b) a simple substitution cipher
   c) a substitution cipher with a keyword.
   You could do this on paper or create a programmed solution.
4. What is a polyalphabetic substitution cipher?
5. Why are substitution ciphers relatively easy to crack?
6. Encrypt a message using a transposition cipher. You could do this on paper or create a programmed solution.
7. Explain how the Vernam (one-time pad) cipher works with the aid of an example. You could do this on paper or create a programmed solution.
8. Why is the Vernam cipher said to be mathematically unbreakable?
9. What conditions must be met for a Vernam cipher to be 100% secure?
10. How are most codes cracked?
11. What is computational security?

KEY POINTS

- Data are vulnerable to interception whenever they are being transmitted.
- Encryption is the process of scrambling data so that they can only be understood if they are decrypted.
- Decryption is the process of turning encrypted data back into meaningful data.
- A cipher is a code or key applied to plaintext to turn it into ciphertext.
- There are three main types of cipher you need to be aware of: Caesar, Vernam and transposition.
- As a programmer you need to be conscious of security issues relating to your own programs and data.

STUDY / RESEARCH TASKS

1. Create a timeline of famous codes and ciphers that have been used throughout history.
2. Research the Enigma machine and find out how messages sent using it were encoded and how the code was eventually cracked.
3. There are many examples of ciphers and codes that have never been cracked. Find out about them. Why has no-one ever managed to decipher them?
4. According to the government, as many as 80% of large business experience security breaches every year. What are the implications for large organisations if their data is compromised?

Section Five: Practice questions

1 Represent the decimal value 6.125 as an unsigned fixed point number, with 4 bits before and 4 bits after the binary point.

2 Represent the decimal value –57 as an 8-bit two's complement binary integer.

3 Describe what happens when using 8-bit two's complement to perform the calculation 56 + 87, by performing the addition and commenting on the result.

4 Calculate the decimal equivalent of the following floating point numbers, which are stored using two's complement. Note there is an 8-bit mantissa and a 4-bit exponent:

a) 0.0110000 0010

b) 1.0010100 0100

5 Write the normalised floating point representation of –7.5 using an 8-bit mantissa and a 4-bit exponent. Explain how you arrived at your answer.

6 Hexadecimal numbers are used widely in computing.

a) Give one example of where hexadecimal numbers are used, and explain why they are used here rather than binary numbers.

b) Convert the binary data 10110111 00111110 into hexadecimal.

c) What is the decimal equivalent of the hexadecimal number E4?

7 For digital audio systems, signals received from a microphone are sampled and the measurement of the amplitude of the waveform can be stored as digital data. To reproduce the sound, the digital data are fed through a digital-to-analogue converter.

a) Explain how the sample rate affects the quality of sound.

b) Why does the digital data need to be passed through a digital-to-analogue convertor?

c) Explain how Nyquist's Theorem applies to this scenario.

8 A secure message needs to be sent across an open network. In order to keep it secure it must be encrypted. The message is: "Head South".

a) Describe what is meant by encryption.

b) Encrypt the message using a Caesar cipher. Explain your method.

c) The Vernam cipher is said to be impossible to crack. Identify two conditions that must be met to ensure the security of plaintext encrypted with a Vernam cipher.

d) Identify one protocol that could be used to increase security when sending data around a wide area network.

e) Some governments can insist on being provided with a 'back door' into encrypted data. Give two reasons why organisations might object to providing governments with this access.

9 The 8 × 8 grid represents how a mono bitmap image is stored. Each grid represents a pixel.

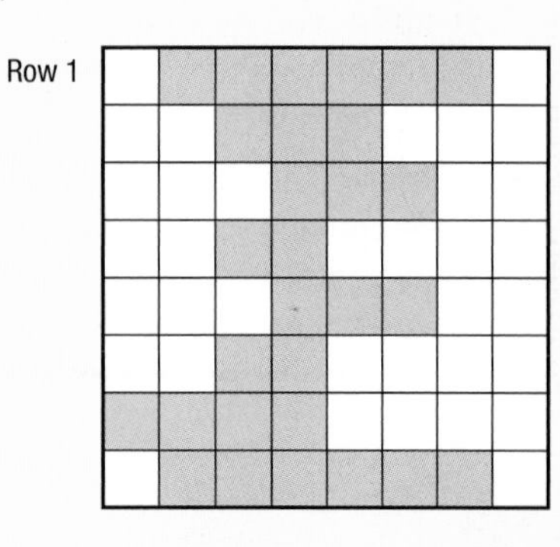

a) If one bit is used to represent each pixel, write out a possible bit string for row 1.

b) How could run-length encoding be used to compress the file?

c) Explain how a bitmap could be used to store a colour image.

d) Explain the difference between a bitmap image and a vector image.

10 Explain in detail how digital cameras capture an image.

11 Describe one example of when it might be necessary to convert analogue signals to digital and one example when it might be necessary to convert digital signals to analogue.

Section Six: Fundamentals of computer systems

28 Hardware and software

SPECIFICATION COVERAGE

3.6.1 Hardware and software

LEARNING OBJECTIVES

In this chapter you will learn:

- the definition of hardware and software
- that hardware is classified into internal and external components
- that there are many different types of software
- many programs are created to make a computer system work, such as operating systems, library and utility programs
- specific software exists to convert programming code into machine code
- how the operating system manages resources.

INTRODUCTION

The term 'system' is often used to refer to the various physical components of your computer – the monitor, keyboard, etc. A computer system has two main elements: **hardware** and **software**. It is only when the two are combined that you create a fully working system. There is no point in having one without the other. Some definitions of a computer system add in a third vital element – the user. In this chapter we will look at the main elements of hardware and software that make up a computer system.

KEYWORDS

Hardware: a generic term for the physical parts of the computer, both internal and external.

Software: a generic term for any program that can be run on a computer.

Hardware

Computer hardware is the physical components of the computer. Sometimes this is described as 'the parts you can touch'. This is not particularly helpful as many elements of hardware are contained inside your computer and can only be seen (or touched) by taking off the case. Therefore, it is important to distinguish between the internal components, which are the processing and storage devices, and external components, normally referred to as peripherals.

- **External components (peripherals):** The external components of hardware are the parts that you can touch, for example the monitor, mouse, keyboard, and printer. The external components are used either to get data into or out of the system. Consequently, they are referred to as input and output (I/O) devices. Some storage devices are also external, for example DVD and flash drives may be added as peripherals. Input devices include the keyboard, mouse, scanner, digital camera and microphone. Output devices include the monitor, printer and speakers.

Figure 28.1 External hardware devices

- **Internal components (processing and storage):** the internal hardware components are housed within the casing of the computer and include the processor, the hard disk, memory chips, sound cards, graphics cards and the circuitry required to connect all of these devices to each other and to the I/O devices.

Figure 28.2 Internal hardware components

Software

Hardware is useless on its own unless we have some programs to run on it. Software is the general term used to describe all of the programs that we run on our computers. These programs contain instructions that the processor will carry out in order to complete various tasks.

This covers an enormous range of possibilities from standard applications, such as word processors, spreadsheets and databases, to more specific applications, such as web-authoring software and games. It also includes programs that the computer needs in order to manage all of its resources, such as file management and virus-checking software. As users, we tend to be aware of the software that we use on a regular basis, yet this is only one part of the software that is on our computers.

The range of software is so great now that some classification is needed in order to make sense of it all. A first level of distinction is made between application software and system software.

KEYWORD

Application software: programs that perform specific tasks that would need doing even if computers didn't exist, e.g. editing text, carrying out calculations.

Application software

Application software refers to all of the programs that the user uses in order to complete a particular task. In effect, it is what users use their computers for. People do not buy computers for the sake of it, they buy them because they have a need to do something: write essays, email, manage a business, create web pages, etc. To carry out any of these, application software is needed that has the necessary features.

There is a wide range of application software available and, in most cases, a number of different applications to choose from that complete the same task. For example, you need application software to access web pages on the Internet – it is called a browser. There are three main ones to choose from: Internet Explorer, Google Chrome and Mozilla Firefox; and many more less well known ones. All of them do the same thing although there are subtle differences between them.

System software

Whereas application software is what we use our computers for, system software covers a range of programs that are concerned with the more technical aspects of setting up and running the computer. Many aspects of system software are invisible to the user which means that they will not even realise that they have system software on their computer. System software exists to support the applications software.

There are four main types: utility programs; library programs; compilers, assemblers and interpreters; and operating system software.

KEYWORD

Utility programs: programs that perform specific common task related to running the computer, e.g. zipping files.

Utility programs

This covers software that is written to carry out certain housekeeping tasks on your computer. They are normally supplied with the operating system though they can be purchased separately. **Utility programs** are often made available as free downloads. Utility programs are designed to enhance the use of your computer and programs though your computer will still work without them.

A common example of a utility program is compression software which allows you to compress files, making them much smaller. Other examples include anti-virus software, back-up software and registry cleaners.

> **KEYWORD**
>
> **Library programs**: code, data and resources that can be called by other programs.

Library programs

Library programs are similar to utility programs in that they are written to carry out common tasks. The word library indicates that there will be a number of software tools available to the users of the system.

Whereas some utility programs are non-essential, library programs tend to be critical for the applications for which they were built. For example, the Windows operating system uses Dynamically Linked Library (DLL) files, which contain code, data and resources. These are similar to executable files and are loaded dynamically by Windows as they are required. There are hundreds of DLL files that carry out a wide range of actions including controlling dialog boxes, managing memory, displaying text and graphics and configuring device drivers.

The Python programming languages also has an extensive library that contains built-in modules that provide various standard system functions and solutions. For example, the library contains code modules for handling common data types, displaying fonts and graphics and performing mathematical and functional operations.

> **KEYWORDS**
>
> **Translators**: software that converts programming language instructions into 0s and 1s (machine code). There are three types – compilers, assemblers and interpreters.
>
> **Compiler**: a program that translates a high-level language into machine code by translating all of the code.
>
> **Assembler**: a program that translates a program written in assembly language into machine code.
>
> **Interpreter**: a program for translating a high-level language by reading each statement in the source code and immediately performing the action.
>
> **Operating system software**: a suite of programs designed to control the operations of the computer.
>
> **Virtual machine**: the concept that all of the complexities of using a computer are hidden from the user and other software by the operating system.

Translators: compilers, assemblers and interpreters

Translators are software used by programmers to convert programs from one language to another. There are three types: **compilers**, **assemblers** and **interpreters**. At some point, every piece of software, whether it is application software or system software, has to be written by a programmer. A program is simply a series of instructions written by a programmer that the computer's processor must carry out.

In order to write software, programmers use programming languages which allow them to write code in a way that is user-friendly for the programmer. However, the processor will not understand the programmers' code, so it has to be translated into machine code, that is, 0s and 1s. Compilers, assemblers and interpreters are used to carry out this translation process. There is more detail on how these work in Chapter 29.

Operating system software

An operating system is a collection of software designed to act as an interface between the user and the computer and manages the overall operation of the computer. It links together the hardware, the applications and the user, but hides the true complexity of the computer from the user and other software – a so-called **virtual machine**.

When you are using a computer you are obviously aware of the applications software you are using, whether it is an Internet browser, a spreadsheet or a game of some sort, but you are much less aware of the software that is running in the background. The systems software is dominated by the operating system (OS).

The OS in a modern computer is very large. For example, Microsoft Windows 8 needs a minimum of 1 GB of RAM and 16 GB of hard disk space when the computer is in use. This gives you some idea of the complexity of the OS. This is because the OS carries out many tasks. For example it:

- controls the start-up configuration of your computer, including what icons to put on your desktop and what backdrop to use
- recognises when you have pressed a mouse button and then decides what, if any, action to take
- sends signals to the hard disk controller, telling it what program to transfer to memory
- decides which sections of memory to allocate to the program you are intending to use and manages memory to ensure all of the programs you want to run are allocated the space they need
- attempts to cope with errors as and when they occur. For example, if a printer sends an 'out of paper error' or you fail to save a file correctly, it is the operating system that displays the appropriate message
- makes sure that your computer shuts down properly when you have finished
- controls print queues
- manages the users on a network – it maintains the lists of usernames and passwords and controls which files and resources users have access to.

Resource management

Computers are capable of running many programs, seemingly at the same time. It is the job of the operating system to make sure that each program is allocated enough memory to operate efficiently.

In a computer with only one **processor**, only one program can actually be live at any one moment in time. In order to allow more than one program to appear to run simultaneously, the operating system has to allocate access to the processor and other resources such as peripherals and memory. One of the main tasks that an operating system has to do is to make sure that all these allocations make the best possible use of the available resources.

Usually the most heavily used resource in a computer is the processor. The process of allocating access to the processor and other resources is called **scheduling**. The simplest way that an operating system can schedule access to the processor is to allocate each task a time slice. This means that each task is given an equal amount of processor time. This process of passing access to the processor from one task to the next is also known as 'round-robin' scheduling.

Time slicing is a crude system because a particular task might not need all or even any of its allotted time slice with the processor. For example, a word processor might be waiting for the slowest part of any computer system, the user, to press a key. Waiting for this to happen is a waste of processor time and so a more sophisticated scheduler might pass the time slice on to the next task before the word processor's time slice has expired.

KEYWORDS

Resource management: how an operating system manages hardware and software to optimise the performance of the computer.

Processor: a device that carries out computation on data by following instructions, in order to produce an output.

Scheduling: a technique to ensure that different users or different programs are able to work on the same computer system at the same time.

Managing input/output devices

The operating system also controls the way in which the various input and output devices are allocated, controlled and used by the programs that are using them. Common examples would include:

- allocating print jobs to printers
- rendering the windows, frames and dialog boxes to the screen
- controlling the read/write access to the hard disk.

Accessing some devices is a relatively slow process compared to the speed at which the processor can handle requests. At the same time there are likely to be competing requests where several processes are waiting for the same device. For example, reading and writing to a standard hard disk is relatively slow. Rather than wait for each process to end before it can continue, the OS can effectively create a queue of commands that are waiting for the device and then handle each request in sequence or based on priority.

Every input/output device has a device driver, which is a piece of software that enables the device to communicate with the OS. Device drivers are often built in to the OS or installed when new devices are attached. When the OS starts up it loads the various drivers for all of the input/output devices it detects.

There is more on how input/output devices connect to the computer in Chapter 32.

Memory management

In the same way that the operating system of a computer controls the way files are stored on a secondary storage system, such as a hard disk, the operating system also controls how the primary memory or RAM is used.

The operating system stores details of all the unallocated locations in a section of memory known as the heap. When an application needs some memory, this is allocated from the heap, and once an application has finished with a memory location or perhaps an application is closed, the now unneeded memory locations are returned to the heap.

> **KEYWORD**
> **Memory management**: how the operating system uses RAM to optimise the performance of the computer.

When a user asks for a file to be loaded, it is the job of the **memory management** routines to check to see if enough memory is available and then allocate the appropriate memory and load the file in to those locations.

The operating system controls the use of main memory by creating a memory map, which shows which blocks of memory have been allocated to each task. This way an operating system can control more than one task in the RAM at any one time. The amount of memory needed by each task is dependent on the size of the program itself, the variables that will be needed and any files that might be generated by the task, but it is up to the operating system to decide how much space can be allocated.

When the operating system processes a request to load an application or file from the hard disk it is the job of the memory management system to decide whereabouts in the RAM the file will be stored.

As you add to a file it is highly likely that the work will need more than one block of RAM. In this case, a type of linked list is used to show where each subsequent block is stored.

App1	File1	unused

In this stylised example of a memory map, the application, **App1** was loaded first and work started on **File1**.

App1	File1	App2	File1	unused

After a while the application **App2** is loaded, then work carried on with **File1**. Because **App2** is now in the memory, **File1** is now spread across two sections of RAM.

App1	File1	App2	File1	File2	File1	unused

A little work is done using **App2** which generates **File2**, and then work recommences on **File1** which is now spread across three sections of RAM and so on.

App1	unused	App2	unused	File2	unused

File1 is now saved and then deleted from RAM, the memory map now looks something like this.

The operating system might now decide to rationalise this and move the applications and existing files so that all the unused space is put together. Even with this simple example you can see that managing the memory can be a complex task for the operating system to control.

Virtual memory and paging

In some cases, the application or file you are trying to work with will be too big to fit in the available RAM. In this case a process called virtual memory can be used. This involves using secondary storage such as a hard disk to store code or files that would normally be held in RAM. The operating system then treats that part of the hard disk as if it is part of the RAM, hence the name virtual memory.

An alternative method is to hold a kernel or central block of the code in RAM. Other sections of code known as 'pages' are loaded from the secondary storage as and when they are needed. Using this method allows very large applications to run in a small section of RAM. This in turn frees up memory for other applications to use.

A word processor will hold instructions about how to cope with the text itself in the kernel but if the user selects a less commonly used task, such as spell checking or mail merging, the appropriate page of the program will be loaded into RAM from the hard disk. This downloading often causes a noticeable delay in the operation of the application.

File management

KEYWORD

File management: how an operating system stores and retrieves files.

One of the many tasks the operating system has to deal with is managing files – this includes controlling the structures that are used to store the files.

The hard disk on a home computer is likely to contain many thousands of files and without some sort of logical storage structure it would be impossible to find a specific file amongst so many. The operating system on a home computer uses folders or directories. These allow the user to group similar files together so that, for example, all the files for your Computing course might be kept in one folder whilst all the photos from your digital camera would be kept in another. In the case of the photos, it is highly likely that there will be folders within folders. The way folders like these are arranged is known as a hierarchical structure – it looks rather like an inverted tree and indeed the start or base folder is normally referred to as the root folder.

Each file has a filename but because of the folder structure it is possible to use the same filename for different files. In the coursework/photo example there might be a folder called **miscellaneous** and a file called **latest** in each area. Obviously this is not particularly good practice, especially if you want to move files between folders and you lose track of which folders you are working with. It is a good practice to use folder and file names that indicate what they are being used for.

As hard disks get larger and larger, it is becoming increasingly common to split up or partition a hard disk. This means that although you actually only have one hard disk in your computer, the operating system splits it up into a number of partitions or logical drives. This means your computer seems to be fitted with more than one hard drive. You might use this system to store your applications on one logical drive and your data on another.

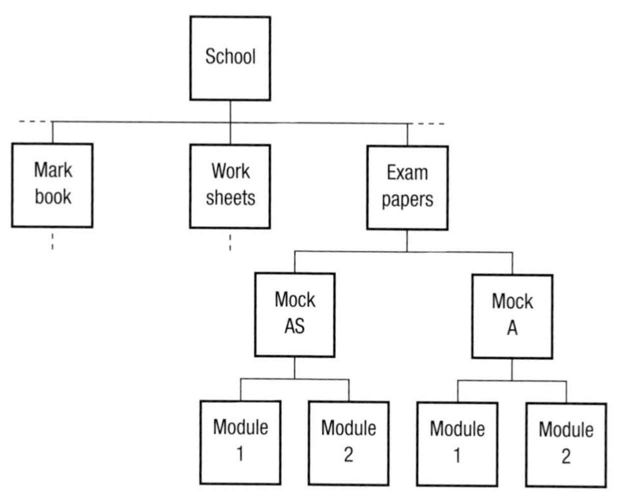

Figure 28.3 Hierarchical file structure

Practice questions can be found at the end of the section on page 264.

KEY POINTS

- A computer system is made up of hardware and software.
- Hardware is usually classified in terms of internal and external components.
- System software includes the operating system, library and utility programs.
- Compilers, interpreters and assemblers are programs that convert high-level programming language into executable instructions.
- The operating system plays a critical role in managing resources.

TASKS

1. Explain the terms hardware and software, differentiating between different types of hardware and software.
2. Identify two utility programs that are not normally part of an operating system.
3. Explain the term virtual machine in the context of an operating system. This should include details of the main tasks performed by the OS.
4. Explain how it is possible for two programs to apparently be running at the same time.
5. Explain how it is possible for two files with the same name to be stored in a file structure.
6. Explain the difference between resource and file management.

STUDY / RESEARCH TASKS

1. What type of software is a DLL and what is its purpose?
2. What are the most common operating systems in use for mobile devices?
3. Compare two commonly used operating systems. Why would you choose to use one over the other?
4. Open source operating systems such as Linux are not as widely used as Windows or Mac OS. Why not?
5. In a computer system with multiple processors, what implications does this have for how the OS handles resources?

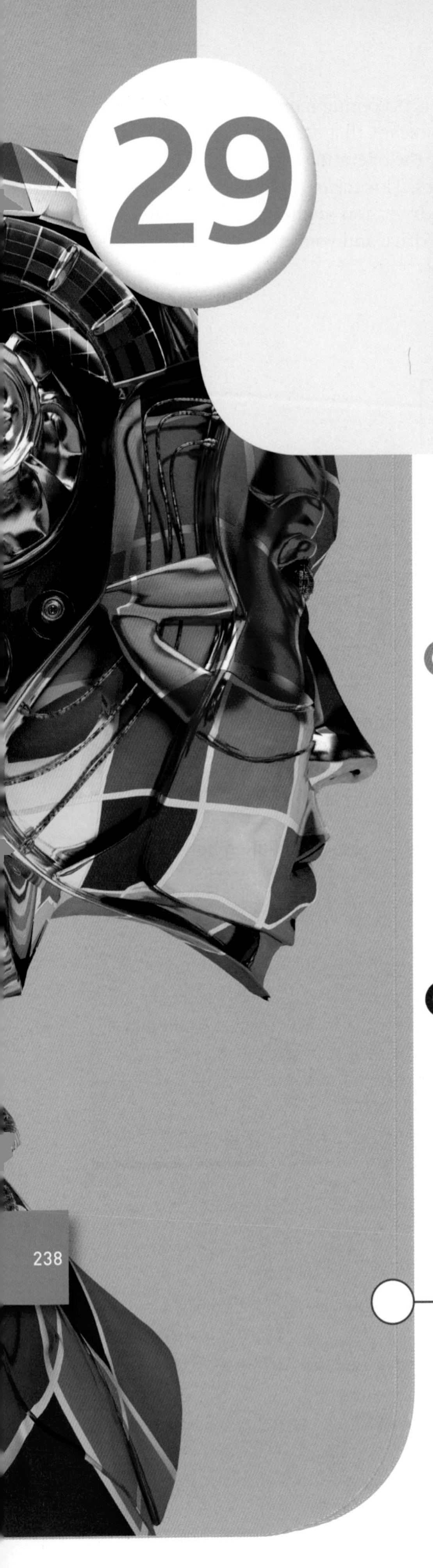

29 Classification of programming languages and translation

SPECIFICATION COVERAGE

3.6.2 Classification of programming languages

3.6.3 Types of translator

LEARNING OBJECTIVES

In this chapter you will learn:

- how programming can be performed in machine code, assembly language and high-level languages
- that there are different methods of programming known as paradigms
- how interpreters and compilers are used to turn programming code into machine code (0s and 1s)
- what bytecode is.

INTRODUCTION

Most programming is now done using high-level programming languages. In this chapter we will look at how programming languages have developed over time, in particular looking at the main types of programming languages and the different methods (or paradigms) of programming. Linked with this is the way in which the code that you write as a programmer is converted into 0s and 1s so that it can be understood and processed by the processor.

Types of programming languages

Although computers as we know them have only been around since the mid-1940s they have changed beyond all recognition in that short time span. Programming the early computers was a very tedious business as programs were entered as a sequence of switch settings. Each switch was either on or off so by combining enough switches the programmer could represent different things.

Thankfully, programming computers has become a much simpler process, though the ever-increasing complexity of the computers themselves brings its own problems.

As computers have developed then so have the programming languages that can be used on them. There are three main types of languages:

- Machine code } These two are known as low-level languages
- Assembly language }
- High-level language

Machine code

KEYWORD

Machine code: the lowest level of code made up of 0s and 1s.

The processor of a computer can only work with binary digits (bits). This means that all the instructions given to a computer must, ultimately, be a pattern of 0s and 1s. Programs written in this format are said to be written in **machine code**. Having to write everything down as a sequence of 0s and 1s means that the programs are going to be very long-winded, and because everything is entered as bits there is a high risk of making a mistake. One way to make these bits easier to understand is to show a sequence of bits as either a decimal or hexadecimal number.

To compound the problem further, it is also going to be very difficult to track down any errors or bugs that exist in the code. One final problem is that because machine code is written for a specific processor, it is unlikely to be very portable. This means that it will only work on a computer with the same type of processor as it was written for.

Nevertheless because you are writing instructions that can be used directly by the processor, they will be executed very quickly and the processor will do exactly what you tell it to do.

Assembly language

KEYWORDS

Assembly language: a way of programming that involves writing mnemonics.

Mnemonics: short codes that are used as instructions when programming, e.g. LDR, ADD.

The need to make programs more programmer-friendly led to the development of **assembly languages**. Rather than using sequences of 0s and 1s, an assembly language allows programmers to write code using words. There is a very strong connection between machine code and assembly language. This is because assembly language is basically machine code with words. The number of words that can be used in an assembly language is generally small. Each of these commands translates directly into one command in machine code. This is called a one-to-one relationship.

Most, but not all, assembly codes use a system of abbreviated words called **mnemonics**. In the small section of assembly code below, the mnemonic `LDR` stands for `Load Register` and `STR` means `Store Register`. `ADD` and `SUB` should be rather more obvious. In this simplified example all the numbers following each mnemonic refer to memory addresses – the place in memory where numbers are written to or read from.

```
LDR   20
ADD   43
STR   20
SUB   41
STR   45
```

This code does the following:

- loads the accumulator with the contents of address 20
- adds the value held in memory location 43 to the value held in the accumulator
- stores the contents of the accumulator to address 20
- subtracts the value held in memory location 41 from the value held in the accumulator
- stores the contents of the accumulator to address 45.

Although assembly language uses words, some of which you might recognise, the code is not particularly easy to understand.

Before any code written in assembly language can be executed it has to be converted into machine code. This is because the processor can only understand binary digits (0s and 1s) – the words are meaningless to the computer, so the words must be converted into these binary digits. This conversion process is carried out by an assembler. The assembler will be able to identify some, but not all, of the errors that are likely to be hidden somewhere in the code. The code that the programmer creates is known as the **source code**. The **assembler** takes this source code and translates it into a machine code version that is known as the **object code**. As with machine code, assembly languages are based on one processor so they are generally not very portable.

KEYWORDS

Source code: programming code that has not yet been compiled into an executable file.

Assembler: a program that translates a program written in assembly language into machine code.

Object code: compiled code that can be run as an executable on any computer.

Assembly language is still used in certain situations today as it has a number of advantages over high-level languages:

- Programs are executed quickly as a compiler does not optimise the machine code that it produces as effectively as a programmer who codes in an assembly language that maps directly to machine code.
- Program code is relatively compact for the same reason.
- Assembly language allows direct manipulation of the registers on the processor, giving high levels of control.

Its main uses are where direct manipulation of hardware is required, for example in embedded systems where a low level of interaction is required with hardware. The processors in these systems may be relatively slow and have limited memory in which case the efficiency of an assembly language is needed. This may mean that very few commands are needed to operate the device. Therefore, these instructions can be programmed directly to the processor using an assembly language.

Real-time applications may also use assembly language as they need to respond very quickly to inputs. This is particularly relevant where it is an embedded system.

Assembly language does have other uses where only low-level coding is required. For example, some device drivers are written in assembly language, particularly for customised hardware and software.

High-level languages

Machine code and assembly languages are known collectively as **low-level languages**. The increasingly complex demands made on computers meant that writing programs in assembly code was too slow and cumbersome. **High-level languages** were developed to overcome this problem.

> **KEYWORDS**
>
> **Low-level language**: machine code and assembly language.
>
> **High-level language**: a programming language that allows programs to be written using English keywords and that is platform independent.

Whereas low-level languages are machine-oriented, high-level languages are problem-oriented. This means that the commands and the way the program is structured are based on what the program will have to do rather than the components of the computer it will be used with. This means that a program written in a high-level language will be portable. It can be written on one type of computer and then executed on other types of computer.

There are many different high-level languages available, each being written to cope with the demands of a specific type of problem. For example, some high-level languages are specifically designed for scientific applications, others for manipulating databases, whilst others are used to create web pages. The language the programmer chooses will depend to a large extent on the nature of the problem they are trying to solve.

High-level languages are often classified into three groups based on the programming paradigm we saw in Chapter 5. A paradigm is a concept for the way something works:

- **Imperative languages**: Also known as procedural languages, these work by typing in lists of instructions (known as subroutines or procedures) that the computer has to follow. Every time the program is run, it follows the same set of instructions.
- **Object-oriented languages**: These work by creating objects where the instructions and data required to run the program are contained within a single object. Objects can be further grouped into classes.
- **Declarative languages**: These work by describing what the program should accomplish rather than how it should accomplish it. One type of declarative language is logic programming, which is used widely in the fields of artificial intelligence and works by programming in facts and rules, rather than instructions. The program then uses these facts and rules to interrogate the data and provide results. Another type of declarative language is a **functional language**, which works by treating procedures more like mathematical functions. The building blocks of the program therefore are functions rather than lists of instructions.

> **KEYWORDS**
>
> **Imperative language**: a language based on giving the computer commands or procedures to follow.
>
> **Object-oriented language**: a programming paradigm that encapsulates instructions and data together into objects.
>
> **Declarative languages**: languages that declare of specify what properties a result should have, e.g. results will be based on functions.
>
> **Functional language**: a programming paradigm that uses mathematical functions.

The main characteristics of a high-level language are:

- It is easier for a programmer to identify what a command does as the keywords are more like natural language.
- Like assembly languages, high-level languages need to be translated.
- Unlike assembly code, one command in a high-level language might be represented by a whole sequence of machine code instructions. This is called a one-to-many relationship.
- They are portable.
- They make use of a wide variety of program structures to make the process of program writing more straightforward. As a result they are also easier to maintain.

Translating high-level languages

One of the main features of high-level languages is that they are programmer-friendly. Unfortunately this means the computer will not understand any of the high-level language source code, so it will have to be translated in some way. This process is called translation and in order to carry it out a special piece of systems software called a **translator** is needed.

The assembler is the translator used for low-level languages and there are two types of translator for high-level languages: an interpreter and a compiler.

KEYWORDS

Translator: the general name for any program that translates code from one language to another, for example translating source code into machine code.

Interpreter: a program for translating a high-level language by reading each statement in the source code and immediately performing the action.

Interpreter

An **interpreter** works by reading a statement of the source code and immediately performing the required action. It may do this by interpreting the syntax of each statement, or by calling predefined routines. In some cases, the source code is translated into an intermediate format before it is executed. Interpreters may examine every statement, which can reduce efficiency. For example, in an iterative statement, the interpreter will read the same statement once for each iteration.

Typically an interpreter only converts what needs to be translated. For example when the line of code:

```
If Age<17 Then Output = "Cannot drive a car"
```

needs to be translated, only the first section of the code `If Age<17 Then` is translated and executed. The second part of the code is only translated if the condition is true. This saves time as only the code that is needed is converted. Some interpreters translate an entire line of code before they execute it. These are known as line interpreters.

Benefits of using an interpreter:

- You do not need to compile the whole program in order to run sections of code. You can execute the code one statement at a time.
- As the code is translated each time it is executed, program code can be run on processors with different instruction sets.
- Because of this, an interpreter is most likely to be used whilst a program is being developed.

Drawbacks of using an interpreter:

- No matter how many times a section of code is revisited in a program it will need translating every time. This means that the overall time needed to execute a program can be very long.
- The source code can only be translated and therefore executed on a computer that has the same interpreter installed.
- The source code must be distributed to users, whereas with a compiled program, only the executable code is needed.

KEYWORD

Compiler: a program that translates a high-level language into machine code by translating all of the code.

Compiler

A **compiler** converts the whole source code into object code before the program can be executed. The good thing about this is that once you have carried out this process you will have some object code that can be executed immediately every time, so the execution time will be quick. This is ideal once you have sorted out all the bugs in your program.

Benefits of using a compiler:

- Once the source code has been compiled you no longer need the compiler or the source code.
- If you want to pass your object code on to someone else to use they will find it difficult to work out what the original source code was. This process of working out what the source code was is known as reverse engineering.

Drawbacks of using a compiler:

- Because the whole program has to be converted from source code to object code every time you make even the slightest alteration to your code, it can take a long time to debug.
- The object code will only run on a computer that has the same platform.

Bytecode

KEYWORD

Bytecode: an instruction set used for programming that can be executed on any computer using a virtual machine.

Some programming languages use **bytecode**, which is an instruction set that can be executed using a virtual machine. The virtual machine can emulate the architecture of a computer, meaning that the source code written using bytecode can be translated into a format that can be executed on any platform.

For example, Java bytecode is compiled using either one or two bytes to define the instruction and then any number of bytes to pass the parameters. This code can then be executed on any computer that is running the Java Virtual Machine regardless of what type of processor or operating system is being used.

Microsoft Common Intermediate Language (CIL) works on a similar basis where, as the name suggests, rather than translating source code into machine code that is specific to a platform, the source code is translated into an intermediate code. This intermediate code can then be executed by the virtual machine.

Practice questions can be found at the end of the section on page 264.

KEY POINTS

- There are three main types of programming languages: machine code, assembly language and high-level languages.
- Machine code and assembly language are known as low-level languages.
- Machine code uses 0s and 1s.
- Assembly languages use mnemonics.
- High-level languages use natural language keywords.
- Assembly language needs to be converted to machine code using an assembler.
- High-level languages need to be converted to machine code using an interpreter or a compiler.
- There are three main programming paradigms: imperative (procedural), declarative and object-oriented.
- Bytecode is an instruction set that can be implemented using a virtual machine and is therefore platform-independent.

TASKS

1. Under what circumstances would you compile a high-level computer program?
2. Explain the benefits of using a high-level language compared to an assembly language.
3. Explain the differences between a compiler and an interpreter.
4. Why are machine code and assembly languages said to be machine-oriented?
5. Explain why there are so many different high-level languages.
6. Explain why some programmers still write programs in an assembly language.
7. Explain why assembly language and machine code are said to have a 'one-to-one' relationship.
8. Why must assembly language programs be assembled before they can be executed?

STUDY / RESEARCH TASKS

1. What is meant by the term artificial intelligence and what type of programming language might you use if working in this area?
2. What features would you expect to find in a programming language that was designed to work in a control environment?
3. There are dozens of programming languages available. Select one particular language, e.g. Java, C#, Python or Visual Basic, and explain why the language was originally developed.

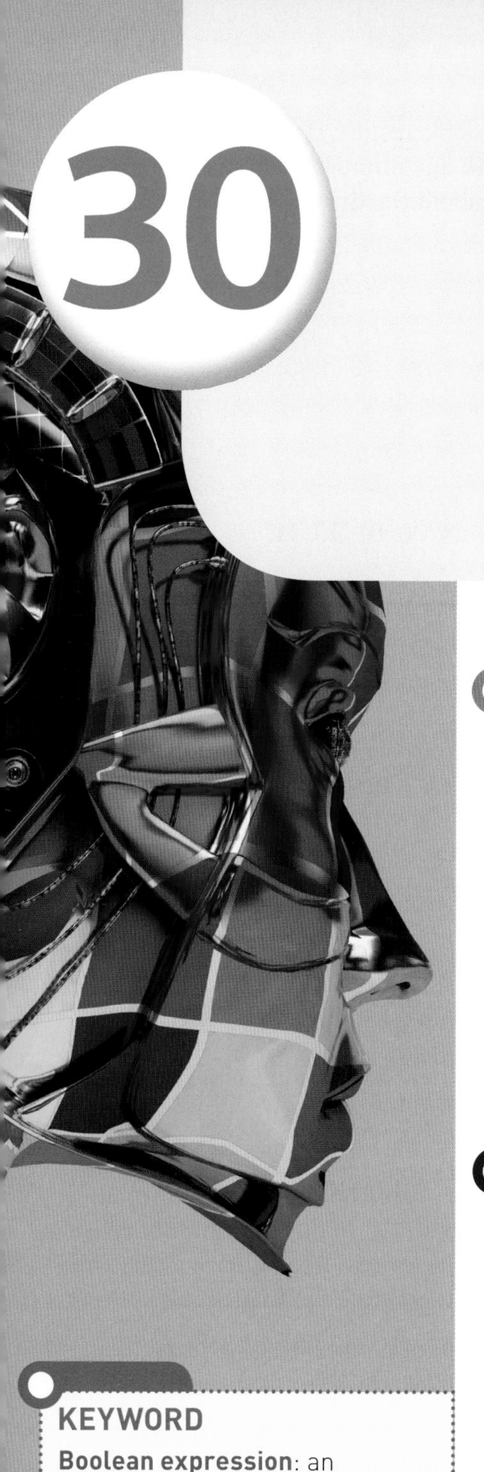

30 Boolean algebra

SPECIFICATION COVERAGE

3.6.4 Logic gates

3.6.5 Boolean algebra

LEARNING OBJECTIVES

In this chapter you will learn:

- that Boolean algebra produces a result that either equals TRUE or FALSE
- how truth tables are used to represent Boolean expressions
- how to use the AND, OR and NOT operators on their own or grouped together
- how to use NAND, NOR and XOR operators
- how to simplify Boolean expressions
- how De Morgan's Law allows Boolean expressions to be created using only NAND or NOR operators.

INTRODUCTION

Boolean algebra is a form of algebra named after George Boole who originally developed it in the mid-1800s. The study of Boolean algebra is closely linked to logic gates, so this chapter should be read in conjunction with Chapter 31.

The basic principle is that logical expressions can be evaluated that will result in one of two results/outcomes – either TRUE or FALSE.
For example, the following are examples of **Boolean expressions**:

```
The button has been pressed

5 < 10

Age > 17 and hold a driving licence
```

KEYWORD

Boolean expression: an equation made up of Boolean operations.

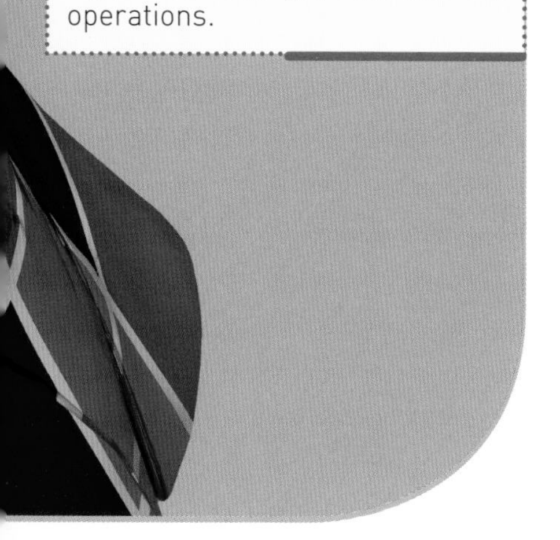

Although Boolean logic predates computers, it has become an important aspect of computing, as the result is one of two states, which equates to binary and the way the electronic circuitry of the processor works. For example, for the first expression: `The button has been pressed`, we could represent the input as A as follows:

- A = 1 where 1 means that the statement is TRUE
- A = 0 where 0 means the statement is FALSE.

As there is one input there are two possible results, 0 and 1. Boolean logic can be used to evaluate statements with any number of inputs to return a TRUE or FALSE value. The statement about the driving licence above has two inputs. To evaluate it:

`At least 17 = 1`

`Under 17 = 0`

and

`Hold a licence = 1`

`No licence = 0`

In this case there are four possible inputs: 00, 01, 10, 11.

Truth tables

KEYWORDS

Truth table: a method of representing/calculating the result of every possible combination of inputs in a Boolean expression.

AND gate: result is true if both inputs are true.

We can use a **truth table** to combine the permutations of 0s and 1s and work out which shows whether the answer is TRUE or FALSE. Table 30.1 shows all the possible inputs and the output of each combination as follows. In this example: A = `at least 17` and B = `Holds a licence.` Q shows the possible results.

Table 30.1 A truth table

Inputs		Output
A	B	Q
0	0	0
0	1	0
1	0	0
1	1	1

This particular truth table is an example of an **AND gate** as both inputs need to be 1 in order to generate an output of 1. In our example above you have to be at least 17 and hold a driving licence to legally drive a car. The two inputs need to be ANDed together to generate the final result.

In this case there is only one combination of A and B that will lead to a TRUE statement being returned, which is where A and B are both 1.

When creating Boolean statements it is possible to use the relational operators in Table 30.2.

Table 30.2 Relational operators

Operator	Name of operator
<	less than
<=	less than or equal to
==	equal to
!=	not equal to
>=	greater than or equal to
>	greater than

KEYWORD

Boolean operation: a single Boolean function.

Statements can be combined to form more complex expressions and this is done using six main **Boolean operations**: AND, OR, NOT, NAND, NOR and XOR. We will look at each of these in turn.

AND operation

As we saw in the first example, in an AND statement, all conditions (inputs) must be TRUE to generate a TRUE output. For example in an embedded system to control a lift, you might evaluate this statement:

```
Button has been pressed AND Door is closed
```

where:

A = Button pressed

B = Doors closed.

A = 1 means the button has been pressed, A = 0 means the button has not been pressed.

B = 1 means the door is closed, B = 0 means the door is not closed.

Q is whether or not the lift should move.

A AND B = Q, which can also be notated as A.B = Q

Table 30.3 Truth table for AND

Inputs		Output
A	**B**	**Q**
0	0	0
0	1	0
1	0	0
1	1	1

The expression therefore is only TRUE when A and B are both 1. You could also say that the expression is TRUE when Q = 1.

OR operation

KEYWORD

OR: Boolean operation that outputs true if either of its inputs are true.

An **OR** expression can return a TRUE result when any of the inputs are true. Consider the following expression that could be used to validate an employee's ID:

```
Proof of ID is Passport OR Driving Licence
```

where:

A = Passport

B = Driving Licence.

A = 1 means the employee has a passport, A = 0 means the employee does not have a passport.

B = 1 means the employee has a driving licence, B = 0 means the employee does not have a driving licence.

Q means that the employee has either a passport or a driving licence and therefore has at least one valid form of ID.

This is written as A+B = Q with the + representing the OR expression.

Table 30.4 Truth table for OR

Inputs		Output
A	B	Q
0	0	0
0	1	1
1	0	1
1	1	1

In this example there are three possible TRUE results. As long as the employee has one of the types of ID, then this results in a TRUE result.

NOT operation

KEYWORD

NOT: Boolean operation that inverts the result so true becomes false and false becomes true.

The **NOT** statement inverts the input so that TRUE becomes FALSE and FALSE becomes TRUE.

This is written as Q = NOT A or $Q = \overline{A}$. Notice the overbar above the A, which is standard notation for NOT.

Table 30.5 Truth table for NOT

A	Q
0	1
1	0

Notice that the results are inverted so FALSE becomes TRUE and TRUE becomes FALSE.

The NOT statement can be used in combination with other Boolean expressions to create more complex selections. For example, when searching the web, it is possible to use the minus key (–) to exclude results that are nothing to do with your topic. This effectively uses a NOT operation. Consider the following expression for a web search about Python programming:

`Python – snake` is the same as `Python NOT snake`

A = 1 if Python is found; A = 0 if Python is not found.

B = 1 if snake is found; B = 0 if snake is not found.

When we NOT the B input we get a TRUE if snake is not found (as we have inverted it).

This is then ANDed with the A input to generate the final outcome of TRUE or FALSE. This is because in a search engine there is an assumed AND between the Python and the NOT which would make the full statement read as `Python and NOT snake`. As you can see this is the logic we require for the search.

The overall search result can be represented as in Table 30.6.

Table 30.6 Extended truth table for A AND NOT B

Inputs		Intermediate step	Output (from A AND NOT B)
A	B	NOT B	Q
0	0	1	0
0	1	0	0
1	0	1	1
1	1	0	0

Combining AND and OR expressions

AND expressions can be combined with OR expressions to create more complex statements. For example, to input data from a barcode:

```
Barcode scanner on and barcode scanned
```

or

```
Barcode number input manually
```

A = Barcode scanner on: A = 1 means yes, A = 0 means no.

B = Barcode scanned: B = 1 means yes, B = 0 means no.

C = Barcode number input manually: C = 1 means yes, C = 0 means no.

Q indicates whether the data from the barcode has been read or not, either automatically or manually.

For data to be input from the barcode, the scanner must be on and barcode scanned, or the barcode must be input manually.

In this case,

Q = A AND B OR C

which can also be written as:

$Q = A.B+C$

In Boolean notation, + means OR and . means AND.

NAND operation

KEYWORDS

NAND: Boolean operation that outputs true if any of the inputs are false.

NAND gate: result is true if any of the inputs are false.

NAND is a combination of NOT and AND and produces a TRUE result if any of the inputs are false. It is commonly used to create **NAND gates** on integrated circuits, which can be used for example, to create solid state drives (see Chapter 35).

The truth table would be as shown in Table 30.7.

Table 30.7 Truth table for NAND

Inputs		Output
A	B	Q
0	0	1
0	1	1
1	0	1
1	1	0

This is written as $\overline{A.B}$, which means NOT A.B. This could be described as the inverted form of A.B.

NOR operation

KEYWORDS

NOR: Boolean operation that outputs true if all of its inputs are false.

NOR gate: result is true if both inputs are false.

The **NOR** or NOT OR expression results in a TRUE value only if all inputs are FALSE. It means that the answer is TRUE if it is neither A nor B. It is used to create **NOR gates** on integrated circuits, which can be used for example to make CMOS devices (see Chapter 35).

The truth table would be as shown in Table 30.8.

Table 30.8 Truth table for NOR

Inputs		Output
A	B	Q
0	0	1
0	1	0
1	0	0
1	1	0

This is written as $\overline{A+B}$, which means it is NOT A+B. This could be described as the inverted form of A+B.

XOR operation

KEYWORD

XOR: Boolean operation that is true if either input is true but not if both inputs are true.

The exclusive OR expression produces a TRUE result only when one of the inputs is TRUE and the other is FALSE. If they are both TRUE it returns a FALSE result. It can be used to carry out bitwise operations and to create an adder in logic circuits. There is an example of this in the next chapter.

The truth table would be as shown in Table 30.9.

Table 30.9 Truth table for XOR

Inputs		Output
A	B	Q
0	0	0
0	1	1
1	0	1
1	1	0

This is written as $Q = A \oplus B$, which means Q is true when either A or B are true, but not when both are true.

Simplifying Boolean expressions

When using Boolean expressions it is good practice to reduce the expression into its simplest form. As Boolean algebra is used to create logic gates, simplifying the expressions also simplifies the actual circuit that will be built, reducing the number of components needed, which in turn will make the circuit cheaper to make, more efficient in operation and more reliable as fewer gates are being used.

To help visualise the process for ensuring that Boolean expressions are in their simplest form you can run it through a truth table. For example, take the expression A.B+A. The values of A can be 0 or 1 and the values of B can be 0 or 1 leading to four possible inputs: 00, 01, 10, 11.

The first part of the statement is A.B so the result is true when A and B = 1. The truth table would look like Table 30.10.

Table 30.10 Truth table for A AND B

A	B	A.B
0	0	0
0	1	0
1	0	0
1	1	1

Next we look at the +A part of the expression, which means OR A in Boolean expressions. This means that (A.B)+A will be true when A and B is 1 or A is 1.

Table 30.11 Truth table for (A.B)+A

A	B	A.B	(A.B)+A
0	0	0	0
1	0	0	1
0	1	0	0
1	1	1	1

Looking at the final column of Table 3.11 you can see that (A.B)+A is only true when A is true. Therefore the expression can be reduced to A:

(A.B)+A = A

An expression may be made up of many variables, usually referenced as letters (A, B, C etc) each of which can produce a result of 0 or 1. This can lead to the creation of complex Boolean expressions. Therefore rules have been developed as a method of simplifying expressions. Table 30.12 shows the common rules associated with what are known as Boolean identities.

In this section we will look at how the rules can be used to simplify an expression. Note that De Morgan's Law is covered separately in the next section.

Table 30.12 Common rules associated with Boolean identities

Identity name	AND form	OR form
Identity	A.1 = A	A+0 = A
Null (or Dominance) Law	A.0 = 0	A+1 = 1
Idempotence Law	A.A = A	A+A = A
Inverse Law	$A.\overline{A} = 0$	$A+\overline{A} = 1$
Commutative Law	A.B = B.A	A+B = B+A
Associative Law	(A.B).C = A.(B.C)	(A+B)+C = A+(B+C)
Distributive Law	A+B.C = (A+B).(A+C)	A.(B+C) = A.B+A.C
Absorption Law	A.(A+B) = A	A+A.B = A
De Morgan's Law	$\overline{(A.B)} = \overline{A}+\overline{B}$	$\overline{(A+B)} = \overline{A}.\overline{B}$
Double Complement Law		$\overline{\overline{A}} = A$

Table 30.13 Explanations of the main identities and rules

A.B = B.A	The order in which two variables are ANDed makes no difference
A+B = B+A	The order in which two variables are ORed makes no difference
A.0 = 0	A variable ANDed with 0 equals 0
A+1 = 1	A variable ORed with 1 equals 1
A+0 = A	A variable ORed with 0 equals the variable
A.1 = A	A variable ANDed with 1 equals the variable
A.A = A	A variable ANDed with itself equals the variable
A+A = A	A variable ORed with itself equals the variable
$A.\overline{A} = 0$	A variable ANDed with its inverse equals 0
$A+\overline{A} = 1$	A variable ORed with its inverse equals 1
$\overline{\overline{A}} = A$	A variable that is double inversed equals the variable
(A.B).C = A.(B.C)	It makes no difference how the variables are grouped together when ANDed
(A+B)+C = A+(B+C)	It makes no difference how the variables are grouped together when ORed
A.(B+C) = A.B+A.C	The expression can be distributed or factored out, meaning that variables can be moved in and out of brackets either side of the expression. In English this expression would be A AND (B OR C) = (A AND B) OR (A AND C).

The rules can be used to simplify expressions.

A+A.B = A

This means A OR (A AND B) = A and can be proved by looking at the truth table (Table 30.11) above.

To use an example with an inverse:

$A+\overline{A}.B = A+B$

This means that A OR (NOT A AND B) = A OR B. It could be proved that this is true by drawing out a truth table for the two expressions. Alternatively this can be deduced by logical reasoning:

- Suppose that A is 1, then $A+\overline{A}.B$ will be 1.
- On the other hand, suppose that A is 0. Then $A+\overline{A}.B$ will only evaluate to 1 if $\overline{A}.B$ is 1, which will only be true if B is 1.
- So, $A+\overline{A}.B = 1$ when A = 1 or when B = 1, hence it is equivalent to A+B.

To use an example that uses distribution: (A+B).(A+C) = A+B.C

This means (A OR B) AND (A OR C) is the same as A OR (B AND C). This can be achieved by factoring out the A.

The more complex example below shows the stages that you might go through to simplify an expression:

$(A+B).C.\overline{C}+(A+\overline{A}).B$

Starting with the first part of the expression $(A+B).C.\overline{C}$:

(A+B).0 Any value ANDed with its inverse = 0

0 Any value ANDed with 0 = 0

Now taking the second part of the expression $(A+\overline{A}).B$:

$(A+\overline{A})$ Any value ORed with its inverse = 1

$1.B$ Any value ANDed with 1 is the variable = B

So putting both parts of the expression together we get 0+B. When you OR a variable with 0 you get the variable so the answer is B. Therefore we can say that the simplified expression of $(A+B).C.\overline{C}+(A+\overline{A}).B$ is B.

De Morgan's Law

KEYWORD

De Morgan's Law: a process for simplifying Boolean expressions

De Morgan's Law is another way of simplifying Boolean statements by inverting all the variables, changing ANDs to OR and ORs to ANDs and then inverting the whole expression. One application is to simplify statements so that only NAND or NOR gates are used. This makes it much simpler to create logic gates and circuits, which in turn makes it easier to design and build microprocessors. For example, solid state drives are made up of NAND gates.

In simple terms this means that ANDs can replace ORs and ORs can replace ANDs. This works as long as the rest of the expression is changed, or negated to take account of this.

The basic principles are:

- Rule 1: NOT (A AND B) is the same as (NOT A) OR (NOT B)
- Rule 2: NOT (A OR B) is the same as (NOT A) AND (NOT B)

In algebraic notation:

- Rule 1: $\overline{A.B}$ is the same as $\overline{A}+\overline{B}$
- Rule 2: $\overline{A+B}$ is the same as $\overline{A}.\overline{B}$

The Venn diagram in Figure 30.1 shows the concept. The area outside the Venn diagram is X. We can define X as being:

- NOT in A+B and
- NOT in A and also NOT in B.

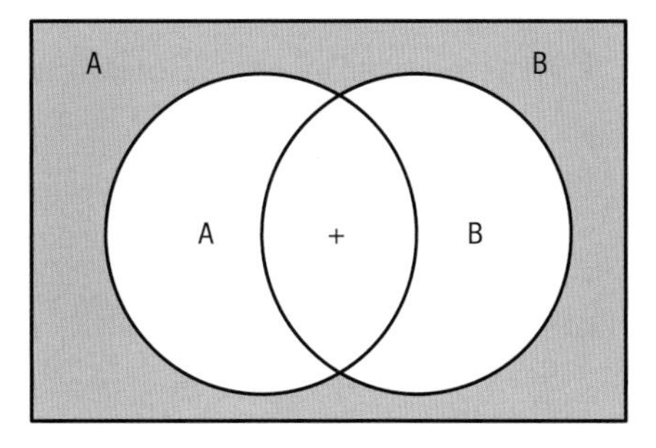

Figure 30.1 Venn diagram representing NOT (A and B)

This could be written as follows:

$X = \overline{A+B}$

$X = \overline{A}.\overline{B}$

$\overline{A.B} = \overline{A}+\overline{B}$

When using De Morgan's Law to write Boolean expressions, the following steps must be taken:

- You can only apply De Morgan's Law to one operator at a time.
- If the operator is an OR change it to an AND, and vice versa.
- Invert the terms on either side of the operator.
- Invert the entire expression.

For example, Figure 30.2 shows how to simplify the expression: $\overline{A+\overline{A}.B}$.

$$\overline{A + \overline{A} \cdot B}$$

Apply De Morgan's Law to this operator

$$= \overline{A + \overline{\overline{\overline{A}} + \overline{\overline{B}}}}$$

$$= \overline{A + \overline{A + B}}$$

Apply De Morgan's Law to this operator

$$= \overline{A} \cdot \overline{\overline{A + B}}$$

$$= \overline{A} \cdot (A + B)$$

$$= \overline{A} \cdot A + \overline{A} \cdot B$$

$$= 0 + \overline{A} \cdot B$$

$$= \overline{A} \cdot B$$

Figure 30.2 Applying De Morgan's Law

This can also be shown by a truth table:

A	B	$\overline{A}$	$\overline{B}$	$\overline{A}.\overline{B}$	$A+\overline{A}.\overline{B}$	$\overline{A+\overline{A}.\overline{B}}$
0	0	1	1	1	1	0
0	1	1	0	0	0	1
1	0	0	1	0	1	0
1	1	0	0	0	1	0

The final column of the truth table only has a 1 in the row where A = 0 and B = 1, therefore the result obtained using De Morgan's Law is confirmed by the truth table, i.e. that the expression is equivalent to $\overline{A}$.B.

Practice questions can be found at the end of the section on page 264.

TASKS

1. Write an example of a Boolean expression and draw the corresponding truth table for each of the following expressions.
   a) AND
   b) OR
   c) NOT
   d) NOR
   e) NAND
   f) XOR
2. Write an example of a Boolean expression for a real-life situation where you could use any combination of:
   a) AND
   b) OR
   c) NOT
3. Give an example of where you could use the following Boolean expressions.
   a) NAND
   b) NOR
   c) XOR

4 Simplify the following expressions.
   a) $(A+\overline{A}).B$
   b) $\overline{\overline{(A+B)}+B}$
   c) $A.(B+B)$
   d) $B.(\overline{A}+B)$
   e) $A.B.C+\overline{A}.B$

5 What are the principles of De Morgan's Law?

KEY POINTS

- Boolean algebra returns a values that is either TRUE or FALSE.
- Truth tables are a visual method of showing the results of a Boolean expression.
- You need to know how to construct AND, OR, NOT, NAND, NOR and XOR statements and combine them to create more complex expressions.
- You should always try to create Boolean expressions in their simplest form.
- De Morgan's Law is a method that can be used to simplify Boolean algebra expressions.

STUDY / RESEARCH TASKS

1 Use Venn diagrams to represent each of the six main expressions.

2 Research the relationship between the NAND expression and solid state drives.

3 Research the relationship between the NOR expression and CMOS.

4 How is Boolean logic applied to web searching?

5 Set yourself some Boolean expressions and then go through the process of simplifying them, using De Morgan's Law where necessary.

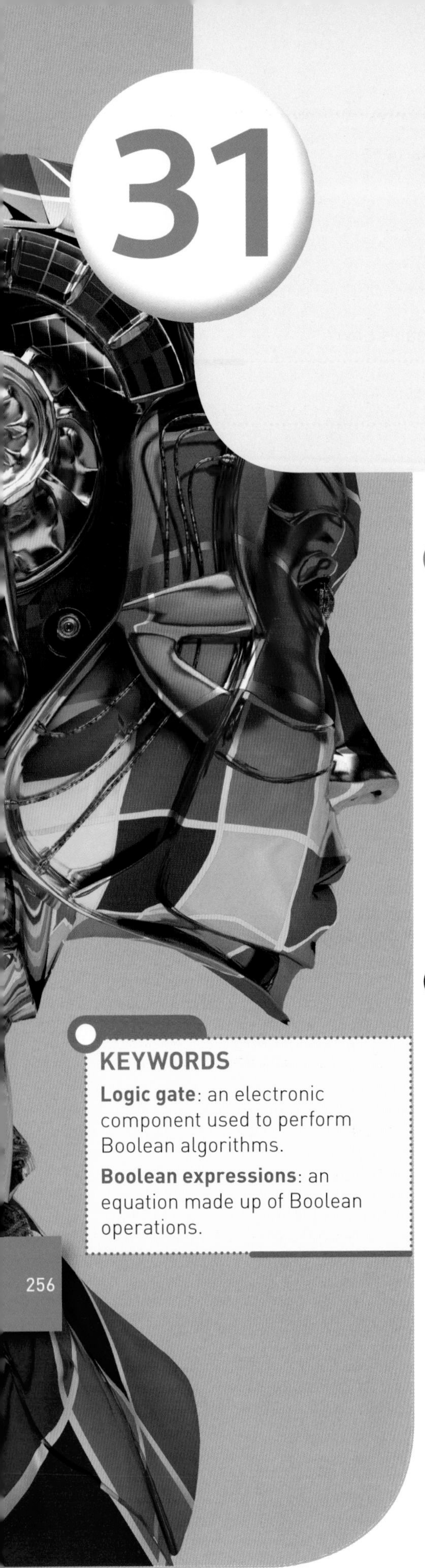

31 Logic gates

SPECIFICATION COVERAGE

3.6.4 Logic gates

3.6.5 Boolean algebra

LEARNING OBJECTIVES

In this chapter you will learn:

- that there is a one-to-one relationship between logic gates and Boolean expressions
- how logic gates are combined to build circuits within processors
- how logic gates are used to evaluate Boolean expressions to produce a result
- how logic gates can be combined to create full systems
- what full and half adders are.

A-level students will learn:

- what an edge-triggered D-type flip-flop is.

INTRODUCTION

Logic gates are electronic components used in registers, memory chips and processors to evaluate **Boolean expressions**. Therefore, before you read this chapter, you must understand the basic Boolean expressions described in the previous chapter. This is because there is a one-to-one relationship between Boolean expressions and logic gates.

Logic gates are represented as logic diagrams and also have a corresponding truth table. Consequently there is a logic diagram for each of the main Boolean expressions that we have looked at in the previous chapter.

KEYWORDS

Logic gate: an electronic component used to perform Boolean algorithms.

Boolean expressions: an equation made up of Boolean operations.

The logic gates for each of the six basic Boolean expressions are shown in Figure 31.1.

Gate	Symbol	Operator
and		$A \cdot B$
or		$A+B$
not		$\overline{A}$
nand		$\overline{A \cdot B}$
nor		$\overline{A+B}$
xor		$A \oplus B$

Figure 31.1 Logic gate symbols and corresponding Boolean expressions

KEYWORD

AND gate: result is true if both inputs are true.

Logic gates take inputs and produce a single output. In the electronic circuit, the inputs are voltages with a high voltage representing a 1 and a low voltage representing a 0. For example, the logic gate for an AND expression is shown below:

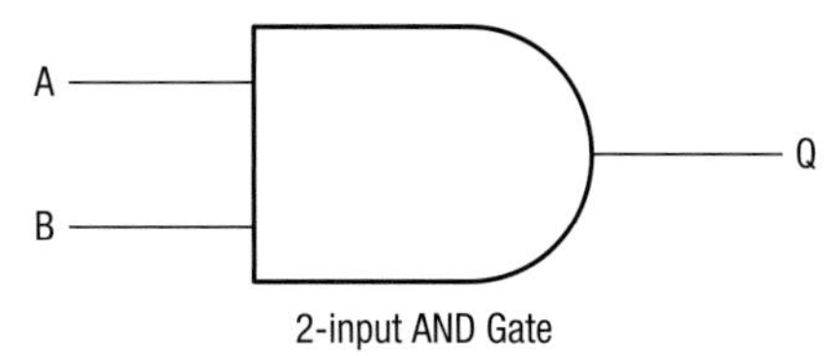

Figure 31.2 The AND gate

A and B are the inputs and Q is the output so this diagram is the equivalent of A.B = Q

The standard ANSI/IEEE standard 91-1984 diagram for each logic gate is shown below along with its truth table and Boolean notation for ease of reference.

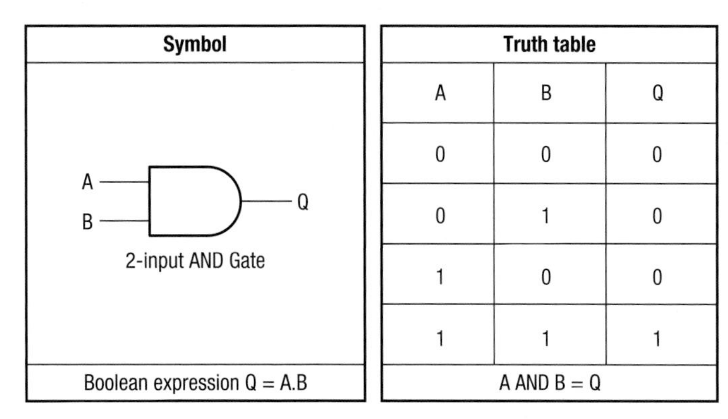

Truth table

A	B	Q
0	0	0
0	1	0
1	0	0
1	1	1

A AND B = Q

Figure 31.3 The AND gate with truth table

KEYWORD

OR gate: result is true if either input is true.

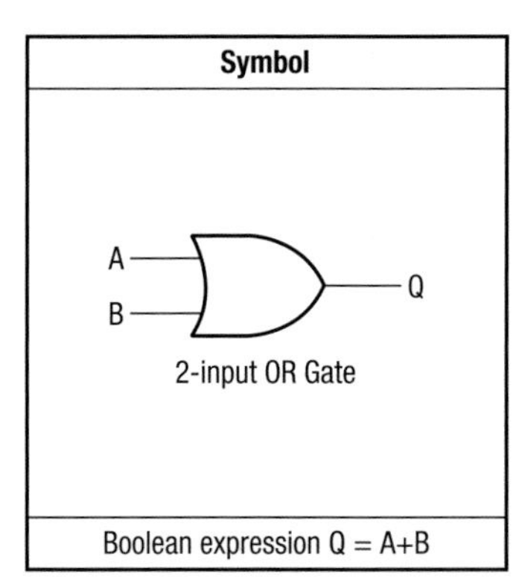

Truth table		
A	B	Q
0	0	0
0	1	1
1	0	1
1	1	1
A OR B = Q		

Figure 31.4 The OR gate with truth table

KEYWORD

NOT gate: inverts the result so true becomes false and false becomes true.

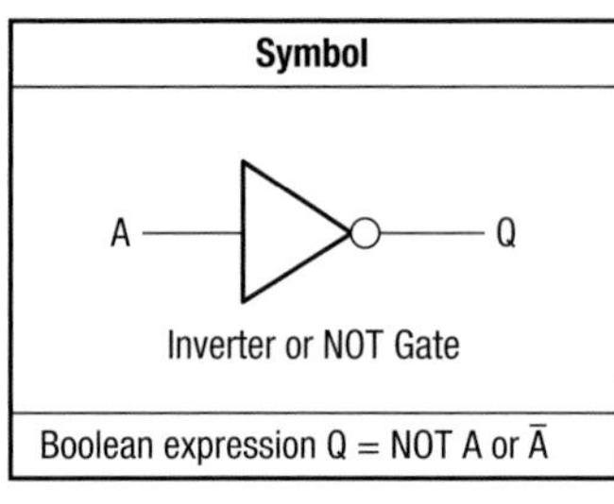

Truth table	
A	Q
0	1
1	0
The inversion of A = Q	

Figure 31.5 The NOT gate with truth table

KEYWORD

NAND gate: result is true if any of the inputs are false.

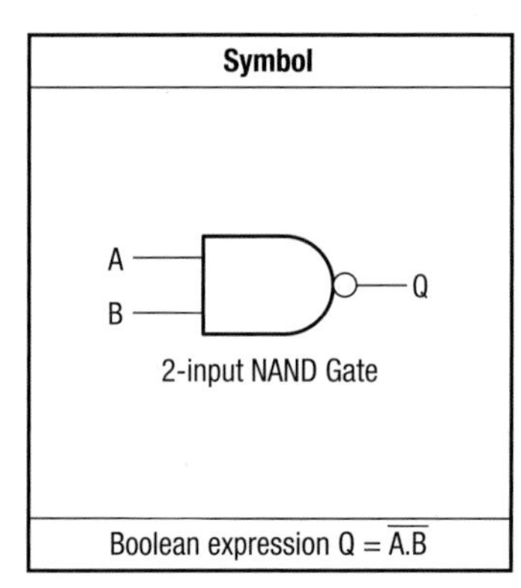

Truth table		
A	B	Q
0	0	1
0	1	1
1	0	1
1	1	0
A AND B = NOT Q		

Figure 31.6 The NAND gate with truth table

KEYWORD

NOR gate: result is true if both inputs are false.

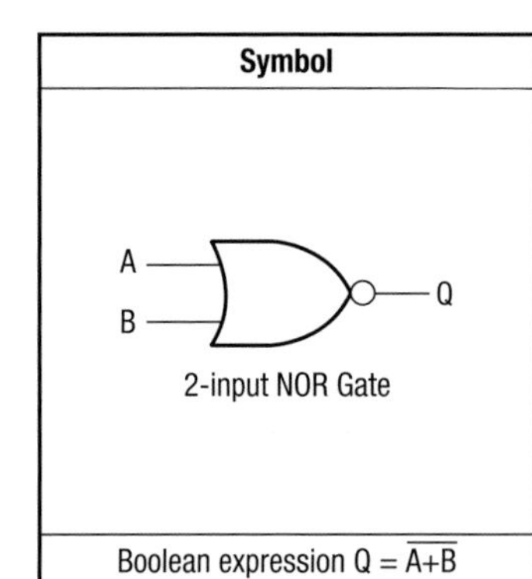

Truth table		
A	B	Q
0	0	1
0	1	0
1	0	0
1	1	0
A OR B = NOT Q		

Figure 31.7 The NOR gate with truth table

KEYWORD

XOR gate: result is true if either input is true but not if both inputs are true.

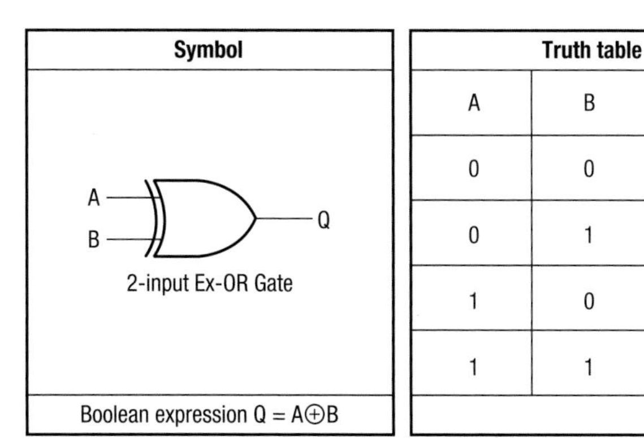

Truth table		
A	B	Q
0	0	0
0	1	1
1	0	1
1	1	0

Figure 31.8 The XOR gate with truth table

Combining logic gates

KEYWORD

Logic circuit: a combination of logic gates.

Logic circuits are made up of a series of logic gates to create full systems. These can get very complex as there may be thousands of gates connected together. The output from the first gate becomes the input for the second gate and so on. Therefore there will be various values generated until a final value of Q is arrived at.

For example, Figure 31.9 shows a simple alarm system where A and B are inputs from sensors and C is the manual override to turn the alarm off.

- A or B = 1 means that the sensors have picked up an intruder.
- A or B = 0 means that the sensors have not picked up an intruder.
- C = 1 means that the override button has been pressed to turn the alarm off.
- C = 0 means that the alarm will continue to sound.

Therefore the only scenario where the alarm would sound is if either A OR B = 1 AND C = 0, which could be written as $(A+B).\overline{C}$

This could be shown as $Q = (A+B).\overline{C}$ and represented in the logic diagram in Figure 31.9 with the corresponding truth table.

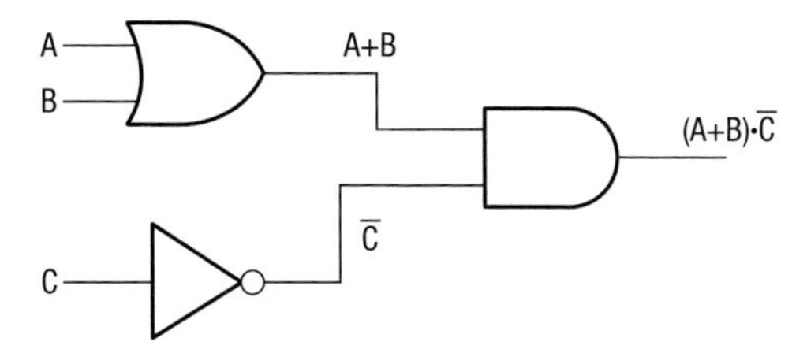

Inputs		Outputs	
A	B	C	Q
0	0	0	0
0	0	1	0
0	1	0	1
0	1	1	0
1	0	0	1
1	0	1	0
1	1	0	1
1	1	1	0

Figure 31.9 A logic circuit with truth table

To follow this through:

- Inputs from A and B pass through an OR gate.
- Input C passes through a NOT gate.
- These two results are passed through an AND gate to create the final result.

In the example above a logic circuit has been constructed from the Boolean expression. It is also possible to construct a Boolean expression from a logic circuit. For example:

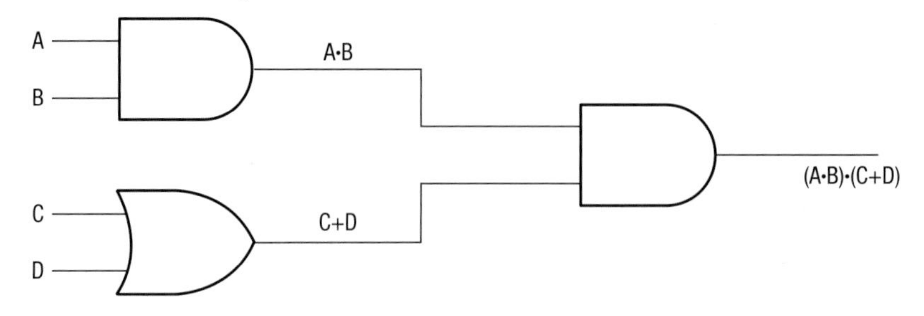

As the logic circuits get more complex, so the corresponding Boolean expression will get more complex:

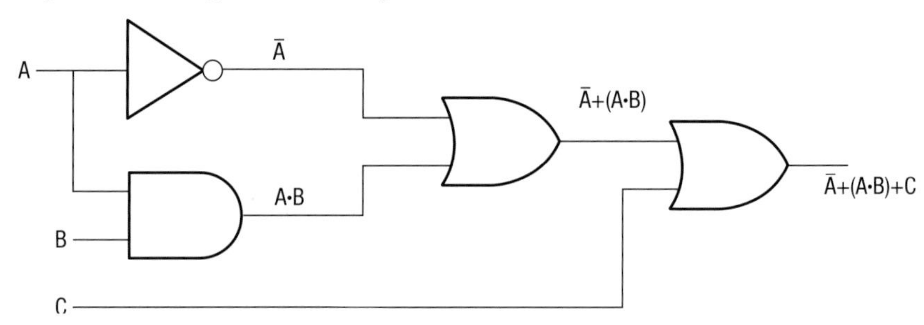

Full and half adder

To add together binary numbers, electronic circuits called adders are used. These are formed by combining logic gates together. Adders are commonly found in the **Arithmetic Logic Unit (ALU)** in the CPU. There are two main types, the **half adder** and the **full adder**. To understand this section, you need to make sure you have read Chapter 24 on adding binary numbers together.

Adders take in two bits A and B and add them to create a sum, S. So A + B = S. There is a further bit required called a **carry bit**, for when the result of the sum requires a further digit. For example, binary addition of 1 + 1 results in 10 (i.e. 0 carry 1), so the carry bit is needed to store this additional bit.

A half adder calculates the sum and stores the value of the carry bit C as well as the result S. The logic diagram to represent this is shown in Figure 31.10.

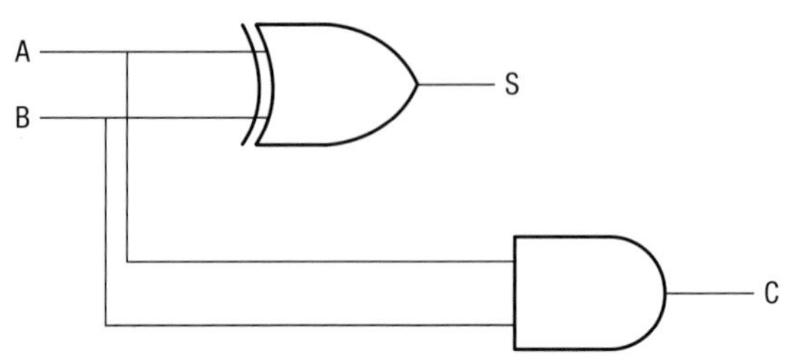

Figure 31.10 A logic circuit for a half adder

KEYWORDS

Arithmetic Logic Unit (ALU): part of the processor that processes and manipulates data.

Half adder: a circuit that performs addition using inputs from A and B only.

Full adder: a circuit that performs addition using inputs from A and B plus a carry bit.

Carry bit: used to store a 0 or 1 depending on the result of binary addition.

The XOR gate is used to look at inputs A and B:

- If A = 0 and B = 1 or if A = 1 and B = 0 then the value for S will be 1.
- Where A and B are both 0, the sum will be 0.
- Where A and B are both 1, then the answer is 10 in binary, which means that 0 is the result for S with the 1 put into the carry bit.

The truth table shows the values of C and S for every possible input of A and B.

Inputs		Outputs	
A	B	C	S
0	0	0	0
1	0	0	1
0	1	0	1
1	1	1	0

Half adders can be added/chained together to create a full adder. Full adders take three inputs, which are the two binary digits to be added, plus the carry bit from the previous addition. The output is the sum (S) plus the carry bit. Notice in the logic diagram below that C_{in} represents the value of the carry bit at the start of the addition and C_{out} is the value of the carry bit at the end. C_{out} from the first addition becomes C_{in} on the second addition and so on.

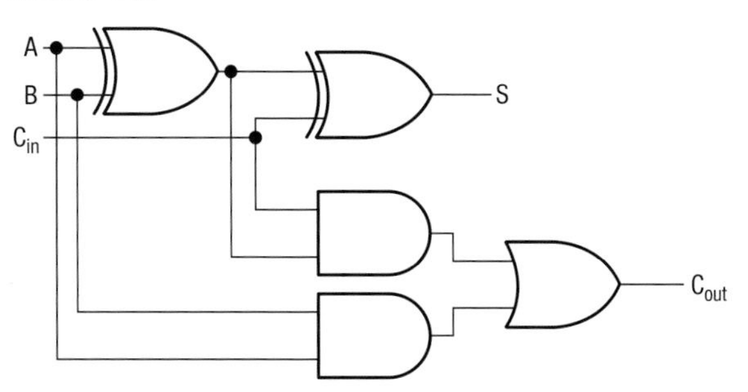

Figure 31.11 A logic circuit for a full adder

- The XOR gate works in the same way as a half adder to identify where the values of A and B are different. Where they are different the result is 1; where they are the same the result is 0.
- A second XOR gate is used to perform the same function on the result of A and B with the value of C_{in} to generate S.
- A and B are ANDed together.
- C_{in} is ANDed with the result A XOR B.
- These two are ORed together to calculate C_{out}.

The truth table shows the values of C_{in} and C_{out} and S for every possible input of A and B.

Inputs			Outputs	
A	B	C_{in}	C_{out}	S
0	0	0	0	0
1	0	0	0	1
0	1	0	0	1
1	1	0	1	0
0	0	1	0	1
1	0	1	1	0
0	1	1	1	0
1	1	1	1	1

KEYWORDS

Flip-flop: a memory unit that can store one bit.

Edge-triggered D-type flip-flop: a memory unit that changes state with each pulse of the clock.

Clock: a device that generates a signal used to synchronise the components of a computer.

Edge-triggered D-type flip-flop

A level only

Logic gates and logic circuits show how 0s and 1s can be manipulated to evaluate Boolean expressions. Data is passed around these gates and circuits at very high speed. However, as soon as the next set of inputs are fed in, the previous inputs are lost. As we have seen, often the next set of inputs come from the outputs of the previous calculation. Therefore some form of memory is needed and this is provided by a **flip-flop**, which is capable of storing one bit.

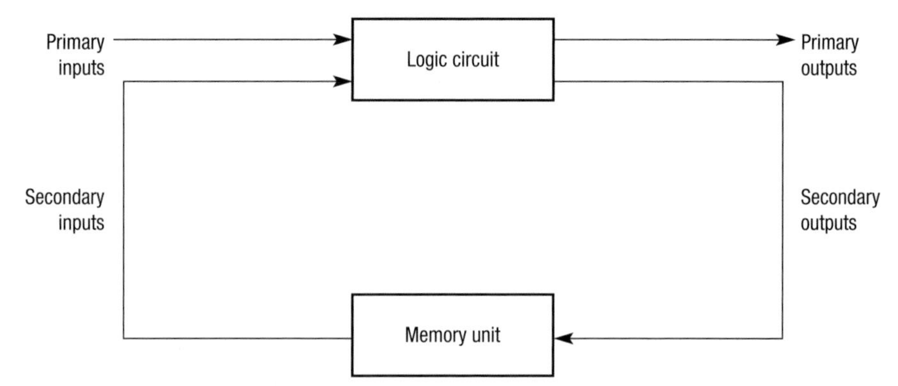

Figure 31.12 An edge-triggered D-type flip-flop

This means that the logic circuit is now receiving two sets of inputs, primary and secondary. It uses the system **clock** to synchronise these requests. Each pulse of the system clock has a rising edge and a falling edge, which represents each pulse of the clock.

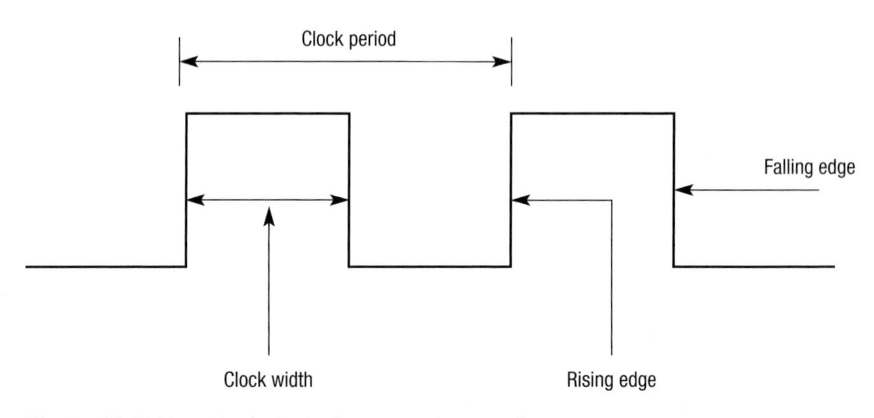

Figure 31.13 How the 'edge' triggers a change of state

In the case of an edge-triggered D-type flip-flop this means that on each pulse of the clock, the flip-flop will change state. This means that for each pulse of the clock, data coming from the input will be stored in the flip-flop and continue to be output until the next trigger pulse is received.

Practice questions can be found at the end of the section on page 264.

TASKS

1 What is the relationship between Boolean algebra and logic gates?

2 What is a logic circuit?

3 Draw the logic diagram and truth table for each of the six main logic gates:
   a) AND
   b) OR
   c) NOT
   d) NOR
   e) NAND
   f) XOR

4 Draw a logic diagram for the following Boolean expressions.
   a) (A+B)+(C.D)
   b) (A+B).$\overline{\text{C}}$
   c) $\overline{\text{(A.B)}}$.C.$\overline{\text{D}}$

5 Write the Boolean expression to build this circuit:

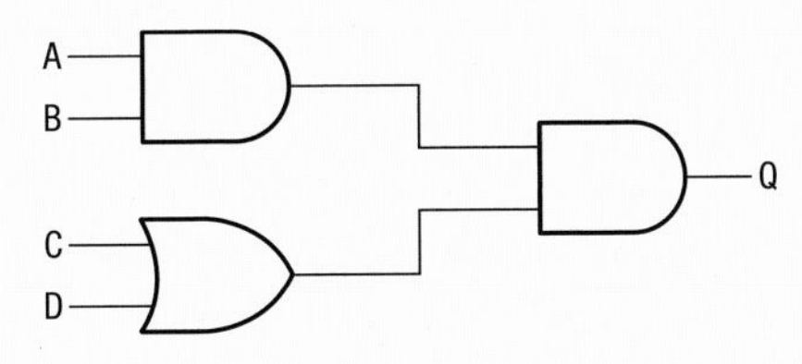

6 Draw a logic diagram and truth table for a half adder.

7 Draw a logic diagram using only NAND operations for A + B.

8 What is the purpose of a flip-flop and what is the specific purpose of the edge-triggered D-type flip-flop?

KEY POINTS

- Logic gates are a way of designing and describing the way in which electronic components within registers and processor are constructed.
- Logic gates evaluate Boolean expressions.
- There are six main symbols that relate directly to the six main Boolean expressions.
- Logic gates can be combined to represent more complex systems.
- Logic gates effectively show how 0s and 1s are manipulated and binary addition takes place within them using either a half or full adder.
- An edge-triggered D-type flip-flop is a memory unit that temporarily stores the result of an operation.

STUDY / RESEARCH TASKS

1 Create a full specification for an embedded system, e.g. operation of a lift or an alarm system, implemented in the form of logic diagrams.

2 Research SR, T and JK flip-flops. Find out how they work.

3 Explore how NAND gates can be used to create memory chips.

Section Six: Practice questions

1 Three categories of programming languages are machine code, assembly language and high level language.

a) Describe what is meant by machine code.

b) A programmer writes a program using assembly language. What has to be done to this program before it can be executed?

c) Some high-level languages are classified as imperative. What is meant by imperative?

d) Give an example of an imperative high-level language.

e) What is the relationship between an imperative high-level language statement and its machine code equivalent?

f) Give two disadvantages of programming in machine code or assembly language, compared with programming in imperative high-level languages.

2 In addition to the procedural paradigm, two other common paradigms are object-oriented and functional.

a) Describe one of the main features of object-oriented programming and one situation where it could be used.

b) Describe one of the main features of functional programming and one situation where it could be used.

3 An operating system creates a virtual machine.

a) What is meant by the term virtual machine?

b) Explain how the operating system is able to have two different applications running at the same time.

c) Give two examples of utility programs.

4 Complex systems can be represented using logic gates and truth tables.

a) Draw the symbol and truth table for the OR gate.

b) Give an example of how logic gates can be combined to represent a logic circuit.

c) Logic gates have corresponding Boolean expressions. Write Boolean expressions to demonstrate:

i) the NAND gate

ii) the NOR gate.

d) Simplify the following expressions:

i) $A.B.(A+B)$

ii) $A+\overline{A}.B$

iii) $A+B+\overline{A.B}$

Section Seven: Fundamentals of computer organisation and architecture

32 Internal hardware of a computer

SPECIFICATION COVERAGE

3.7.1.1 Internal hardware components of a computer

LEARNING OBJECTIVES

In this chapter you will learn:

- how the processor works
- about the different types of memory and what memory is used for
- how buses are used to pass data and instructions
- how external devices are handled
- the difference between the von Neumann and Harvard architectures.

INTRODUCTION

A computer is any machine or device that processes data. The word computer also implies that the machine is electronic or digital. In simple terms this means that it will contain one or more microprocessors that can be programmed to control the device. Microprocessors are made up of microscopic electronic circuits and belong to a group of devices commonly referred to as chips.

This definition is deliberately broad. It is important as A-level students that you realise that your PC (Personal Computer) is only one type of computer and that there are other types in common use. For example, you could look at any number of devices and describe them as computers – a burglar alarm, a microwave oven and a mobile phone all fit this definition. Computers used in this context are referred to as embedded systems as the chip is embedded within another device.

KEYWORD

Processor: a device that carries out computation on data by following instructions, in order to produce an output.

Processor

The **processor** is a device that carries out computation on data by following instructions. It handles all of the instructions that it receives from the user and from the hardware and software. For example, you may press 'A' on the keyboard in which case an electrical signal is sent either wirelessly or through the cable into the USB port at the back of the computer. This signal is routed through the processor which recognises the signal as an 'A'. It then sends a signal to the monitor which displays the letter.

Figure 32.1 A modern processor

Obviously, it is more complicated than this and the process of displaying a letter on a monitor requires the processor to carry out a number of processes which are invisible to the user. The processor will also be receiving instructions from other programs and other devices at the same time. A further complexity is added by the fact that many computers now contain processors with multiple cores and the actions of each will need coordinating.

Physically, the processor is made up of a thin slice of silicon approximately 2 cm square. Using microscopic manufacturing techniques, the silicon is implanted with millions of transistors. Microscopic wires called buses connect groups of transistors together. The transistors are used to control the flow of electrical pulses that are timed via the computer's clock. The pulses of electricity represent different parts of the instruction that the processor is carrying out. Each of these pulses is routed around the circuitry of these transistors at very high speeds. In theory a 3 GHz processor could process 3000 million instructions per second.

Generally speaking, the higher the clock speed of the processor, the faster it will carry out instructions and the faster your computer will work. There are other factors that affect performance, which are covered later. Manufacturers of processors such as Intel and AMD bring out newer, faster chips each year.

Main memory

KEYWORDS

Main memory: stores data and instructions that will be used by the processor.

Fetch-execute cycle: the continuous process carried out by the processor when running programs.

Random Access Memory (RAM): stores data and can be read to and written from.

Chip: an electronic component contained within a thin slice of silicon.

Memory is used to store data and instructions. It is connected to the processor which will fetch the data and instructions it needs from memory, decode the instructions and then execute them. This is commonly known as the **fetch-execute cycle** and is a key principle in modern computing. There is more on this in the next chapter. Any new data created will be stored back into memory. Put simply, memory is a medium of storage. There are two main types: RAM and ROM.

RAM – Random Access Memory

RAM is temporary storage space that can be accessed very quickly. This means that applications such as word processors and spreadsheets will run at high speed. Speed of operation becomes more apparent when you use your computer for games or videos as these require more memory in order to refresh the graphics. Physically, RAM is a chip or series of **chips** on which the data is stored electronically. It is made up of millions of cells, each of which has its own unique address. Each cell can contain either an instruction or some data.

The cells can be accessed as they are needed by the processor, by referencing the address. That is they can be accessed randomly, hence the name. Because they are electronic they are able to be accessed quickly. However, RAM is volatile, which means that when you turn your computer off, all of the contents of RAM are lost.

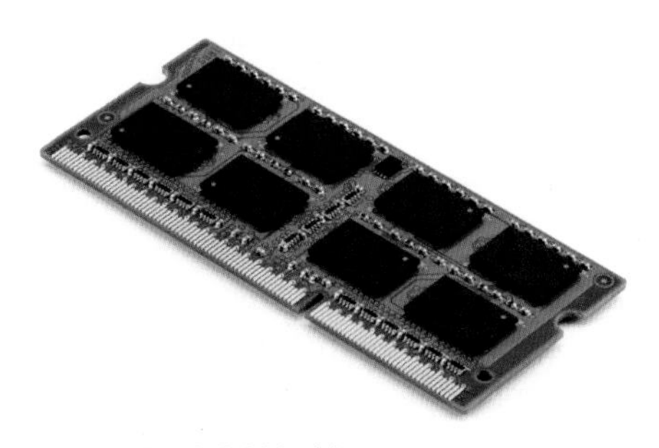
Figure 32.2 RAM chips

Whenever a program is run on your computer, the entire program or parts of it are loaded into RAM. The more memory you have, the more applications you can have loaded at any one time. For example, if you load a spreadsheet file, both the file and the spreadsheet application are stored in RAM. When you are creating formulae, these are stored temporarily in

RAM. If you turn the computer off without saving it will close down the spreadsheet and your work will be lost.

As for processors, manufacturers are always bringing out more powerful memory chips and the price of memory has actually fallen over the years. In 1997, 32 MB was fairly standard. By 2003, 512 MB was standard. By 2014, 4–8 GB became the standard range for laptops and desktops. Software manufacturers are also bringing out new products all the time that take advantage of the larger memory available. You may have noticed this yourself as certain programs will not run on machines that do not have enough memory.

KEYWORD

Read Only Memory (ROM): stores data and can be read from, but not written to (unless programmable ROM).

ROM – Read Only Memory

ROM is also a method of storing data and instructions. However, it is not volatile which means that the contents of ROM are not lost when you switch off. Unlike RAM, the user cannot alter the contents of ROM as it is read-only. It is important to note that it is possible to have programmable ROM, which is used in memory sticks and other devices. The definition here is of traditional ROM used within a PC.

In this case, ROM is used to store a limited number of instructions relating to the set up of the computer. These settings are stored in the BIOS which stands for Basic Input/Output System.

When you switch on your computer it carries out a number of instructions. For example, it checks the hardware devices are plugged in and it loads parts of the operating system. All of these instructions are stored in ROM. The instructions are programmed into ROM by the manufacturer of the PC.

Addressable memory

Memory is made up of millions of addressable cells and the various instructions and data that make up a program will be stored across a number of these cells. Each address can be uniquely identified. It is the job of the processor to retrieve each instruction and data item and to carry out instructions in a sequential manner.

Memory is organised in a systematic way. Using the addresses, different programs can be stored in different parts of memory. For example, a block of memory addresses will be allocated for the operating system, another block for the application software and so on. This way, the processor is able to find the data and instructions it needs much more quickly than if the programs were stored completely randomly.

A memory map can be produced which shows which programs are stored at which addresses. You will see that memory addresses are normally shown in hexadecimal format rather than binary as the hex version is shorter.

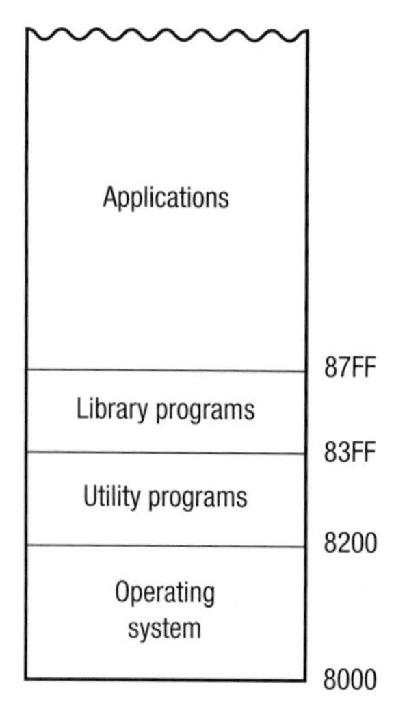

Figure 32.3 A basic memory map

Buses

Buses are groups of parallel microscopic wires that connect the processor to the various input and output controllers being used by the computer. They are also used to connect the internal components of a microprocessor, known as registers (more on these in Chapter 33), and

> **KEYWORDS**
>
> **Bus**: microscopic parallel wires that transmit data between internal components.
>
> **Data bus**: transfers data between the processor and memory.
>
> **Input/Output (I/O) controller**: controls the flow of information between the processor and the input and output devices.

to connect the microprocessor to memory. There are three types of **bus**: data, address and control.

Data bus

The instructions and data that comprise a computer program pass back and forth between the processor and memory as the program is run. The **data bus** carries the data both to and from memory and to and from the **I/O controllers**, that is, they are bi-directional or two-way. The instructions and data held in memory will vary in size. Each memory cell will have a width measured in bits. For example, it may have a width of 32 bits.

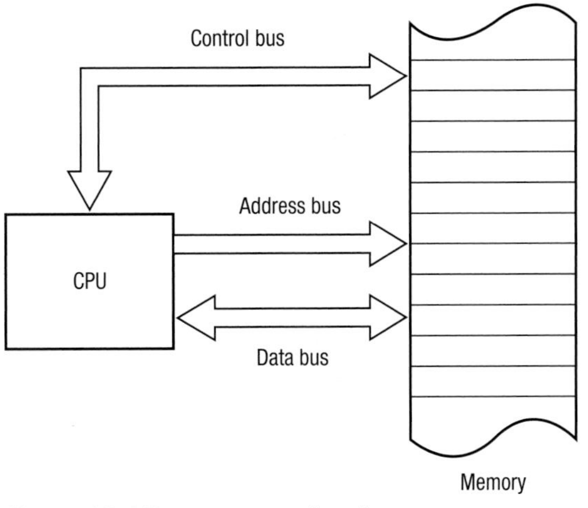

Figure 32.4 Buses connecting the processor to memory

The data bus connects the registers to each other and to memory. The amount of data that can be passed along the bus depends on how many wires are in the bus. An 8-bit data bus has eight wires. There are only two things that can pass down each wire, that is a 0 or a 1. Therefore, by using eight wires on the data bus, we can transmit any item of data that can be represented using 2^8 combinations which is 256. As we saw in Section Five, these patterns can be used to represent text, numbers, sound and graphics or instructions.

Therefore, when large data items are transmitted, the data will have to be split into smaller parts which are sent one after the other. The greater the width of the data bus, in terms of wires, the more data can be transmitted in one pulse of the clock. Consequently, the size of the data bus is a key factor in determining the overall speed and performance of the computer. 32-bit and 64-bit buses are the norm at the time of writing. The data bus width is usually the same as the **word length** of the processor and the same as the memory word length.

> **KEYWORDS**
>
> **Word length**: the number of bits that can be addressed, transferred or manipulated as one unit.
>
> **Address bus**: used to specify a physical address in memory so that the data bus can access it.

Address bus

The **address bus** only goes in one direction – from the processor into memory. All the instructions and data that a processor needs to carry out a task are stored in memory. Every memory location has an address. The processor carries out the instructions one after the other. The address bus is used by the processor and carries the memory address of the next instruction or data item. The address bus therefore is used to access anything that is stored in memory, not just instructions.

KEYWORDS

Addressable memory: the concept that data and instructions are stored in memory using discrete addresses.

Control bus: controls the flow of data between the processor and other parts of the computer.

The size of the address bus is also measured in bits and represents the amount of memory that is addressable. An 8-bit bus would only give 256 directly **addressable memory** cells. This means that a program could only consist of a maximum of 256 separate instructions and/or data items. If we assume that each memory address can store 8 bits (one byte) of data then we would have 256 bytes of memory available. This would be useless on modern computers.

You may have realised from Section Five that each additional wire will double the capacity. Consequently 24 lines on the address bus would give 2^{24} combinations, which means it can access 16 MB of memory. A 32-bit address bus, which is common for most PCs, would provide 4 GB of addressable memory. A 64-bit address bus would provide, in theory, addressable memory of 16.8 million terabytes.

Control bus

The **control bus** is a bi-directional bus which sends control signals to the registers, the data and address buses. There is a lot of data flowing around the processor, between the processor and memory, and between the processor and the input and output controllers. Data buses are sending data to and from memory while address buses send only to memory.

The job of the control bus therefore is to ensure that the correct data is travelling to the right place at the right time. This involves the synchronisation of signals and the control of access to the data and address buses which are being shared by a number of devices.

For example, a signal on the control bus would dictate the direction of data transmission through the data bus; it would also indicate whether it was reading to or writing from an I/O port. The control bus will also be transmitting the pulses being delivered by the system's clock.

Input/Output (I/O) controllers

In addition to the direct link between the processor and main memory, the processor will also receive and send instructions and data to the various input and output devices connected to the computer. Basic I/O devices would be the keyboard, monitor, mouse and printer, though modern computer systems would typically include several other devices.

Physically, these I/O devices are connected via the I/O ports (usually USB ports) on your computer as shown in Figure 32.5. The ports are physical connections that allow I/O devices to be plugged in. For example, the printer will be plugged into one of the USB ports. Signals will be passed in both directions through the printer cable, via the port and through the processor to send and receive the instructions.

Inside the computer, the data buses carry the signals to and from the processor. In order to do this the processor is working in the same way as if it were sending data to or from memory. The difference, however, is that the processor does not communicate directly with the I/O devices. Instead, there is an interface called an I/O controller.

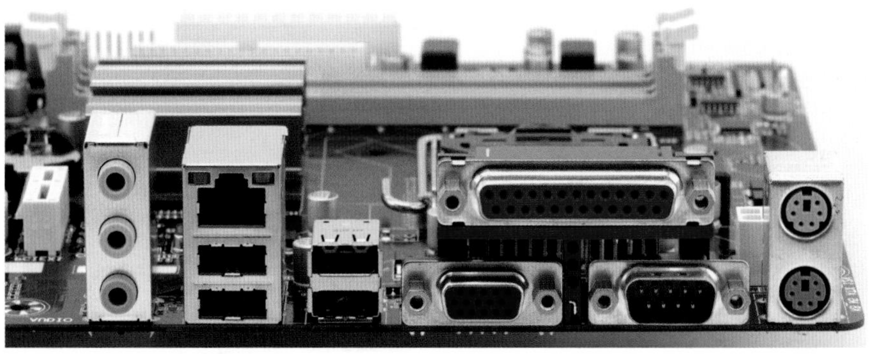

Figure 32.5 Physical ports on a standard PC

Controllers consist of their own circuitry that handle the data flowing between the processor and the device. Every device will have its own controller which allows new devices to be connected to the processor at any time. As a minimum, therefore, a typical computer will have a monitor controller, a mouse controller, a keyboard controller and a hard disk controller.

A key feature of an I/O controller is that it will translate signals from the device into the format required by the processor. There are many different devices and many different types of processor and it is the I/O controller that provides the flexibility to add new devices without having to redesign the processor.

Another important feature is that the I/O devices themselves respond relatively slowly compared to the speed at which a processor can work. Therefore the I/O controller is used to buffer data being sent between the processor and the device, so that the processor does not have to wait for the individual device to respond.

Von Neumann and Harvard architectures

In Section One we identified that a program is a series of instructions that the processor will carry out. Programs also require the data on which these instructions will be carried out. As we have seen, a program is loaded into main memory when it is run. In simple terms it means that the instructions and data that comprise a program are both stored in main memory and must both pass through the same bus (the data bus) in and out of memory.

The early computers that used this concept were known as 'von Neumann' machines after the man who first invented the technique in the 1940s. Most modern PCs use this technique and so are also von Neumann machines.

The word 'architecture' is widely used in computing and usually refers to the way that something is built. For example, a microprocessor has an architecture that refers to the way that the chip is built. The von Neumann method of building computers therefore is often referred to as **von Neumann architecture**.

An alternative method of building chips is the **Harvard architecture**. The key difference between this and von Neumann is that separate buses are used for data and instructions, both of which address different parts of memory. So rather than storing data and instructions in the same memory and then passing them through the same bus, there are two separate buses addressing two different memories.

KEYWORDS

Von Neumann architecture: a technique for building a processor where data and instructions are stored in the same memory and accessed via buses.

Harvard architecture: a technique for building a processor that uses separate buses and memory for data and instructions.

Von Neumann architecture

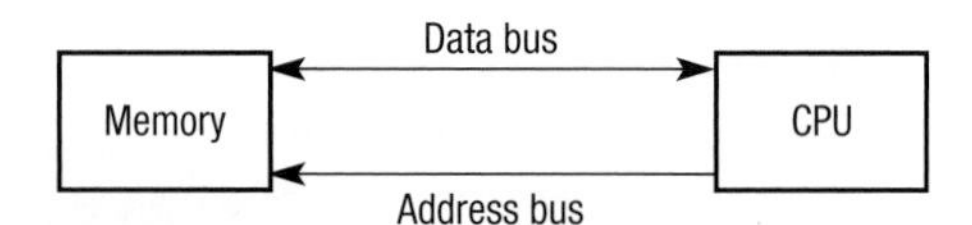

Harvard architecture

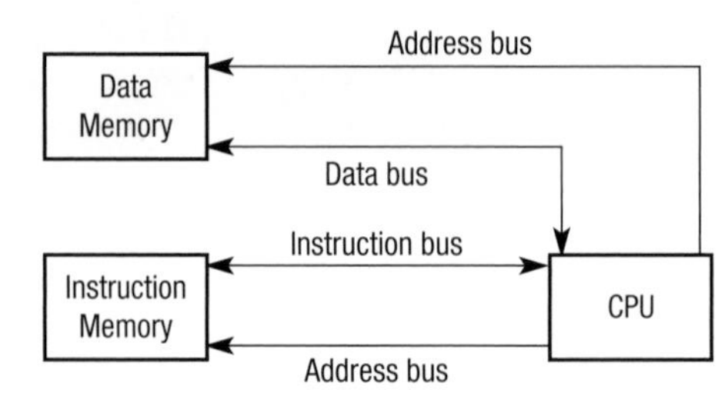

Figure 32.6 The von Neumann and Harvard architectures

The advantage of this is that the instructions and data are handled more quickly as they do not have to share the same bus. Therefore a program running on Harvard architecture can be executed faster and more efficiently. Harvard architecture is widely used on embedded computer systems such as mobile phones, burglar alarms etc. where there is a specific use, rather than being used within general purpose PCs.

Many such devices use a technique called Digital Signal Processing (DSP). The idea of DSP is to take continuous real world data such as audio or video data and then to compress it to enable faster processing. Chips that are optimised for DSP tend to have much lower power requirements, making them ideal for applications such as mobile phones where power consumption is critical.

Practice questions can be found at the end of the section on page 298.

TASKS

1. Identify the three other terms that are commonly used to mean main memory.
2. Identify one advantage of increasing the amount of RAM in a PC.
3. Describe the purpose of the following components:
   a) processor
   b) random access memory
   c) read only memory.
4. Explain why it is necessary to have ROM, RAM and a hard disk within a computer system.
5. Identify one advantage of increasing the capacity of the hard disk.
6. On a games console, explain why it can take over a minute to load a game.
7. What is the maximum amount of data that can pass down a 16-bit data bus in one stage of the fetch–execute cycle?
8. What is the largest amount of addressable memory available with a 16-bit address bus?
9. Name and describe the function of the three different buses.
10. Identify three functions carried out by the control bus.
11. Give two reasons why I/O devices are handled by controllers rather than being connected directly to the processor.

KEY POINTS

- The processor handles and processes instructions from the hardware and software.
- Processors can handle millions of instructions every second.
- Memory is made up of millions of addressable cells.
- Data and instructions are fetched, decoded and executed.
- There are two main types of main memory, RAM and ROM.
- The processor is connected to main memory and data and instructions are passed around circuitry known as buses.
- In von Neumann architecture, instructions and data are stored together in memory.
- In Harvard architecture, separate memory is used for data and instructions.

STUDY / RESEARCH TASKS

1. Recommend a suitable specification for a computer system for online gaming. How would the specification of this system vary from one that was designed to handle a large database?
2. Moore's Law broadly states that processor performance will double every two years. He made this prediction in 1965. Is it true and do you think it will continue to be true?
3. Some people believe that the next big advance in microprocessor technology is when we move on from the silicon chip. What are the limitations of the silicon chip and what might replace it in the future?

33 The stored program concept and processor components

SPECIFICATION COVERAGE

3.7.2.1 The meaning of the stored program concept

3.7.3.1 The processor and its components

3.7.3.2 The fetch–execute cycle and the role of registers within it

3.7.3.6 Factors affecting processor performance

3.7.3.6 Interrupts (A Level only)

LEARNING OBJECTIVES

In this chapter you will learn:

- how instructions are handles using the stored program concept
- what happens at each stage of the fetch–execute cycle
- that a processor is made up of components including registers and units
- how the registers and units are used to handle instructions
- what factors affect processor performance.

A-level students will also learn:

- what an interrupt is and how a processor handles it.

INTRODUCTION

In this chapter we look in more detail at what happens during the fetch–execute cycle, looking specifically at the physical components of the processor and how data and instructions are handled internally. The processor is made up of microscopic registers and units and each instruction will be passed around these components, manipulating 0s and 1s in order to create a result. This chapter explains what happens at each stage of the cycle. In addition, we consider the factors that affect the overall performance of the processor.

A-level students also need to be aware of the interrupt and this is covered at the end of the chapter.

The stored program concept

KEYWORDS

Stored program concept: the idea that instructions and data are stored together in memory.

Fetch–execute cycle: the continuous process carried out by the processor when running programs.

As you saw in the previous chapter, the von Neumann concept was to store instructions and data in the same memory unit. Each instruction or data item is fetched from memory, decoded and then executed, with any new data created being placed back into memory. Every time a program is run, the processor runs through this **fetch–execute cycle**.

Therefore, all the processor is doing is running through this cycle over and over again, millions of times every second. The computer's clock times the electrical pulses into the processor.

- Fetch – the processor fetches the program's next instruction from memory. The instruction will be stored at a memory address and will contain the instruction in binary code.
- Decode – the processor works out what the binary code at that address means.
- Execute – the processor carries out the instruction which may involve reading an item of data from memory, performing a calculation or writing data back into memory.

It is worth pointing out that a simple instruction for a user, for example, adding two numbers together, would actually involve a number of cycles for the processor. There is also an unanswered question in terms of how the processor fetches, decodes and executes each instruction. To understand this fully, you need to understand the architecture of the processor. Processors are made up of a number of components including the clock, control unit, arithmetic logic unit and various registers.

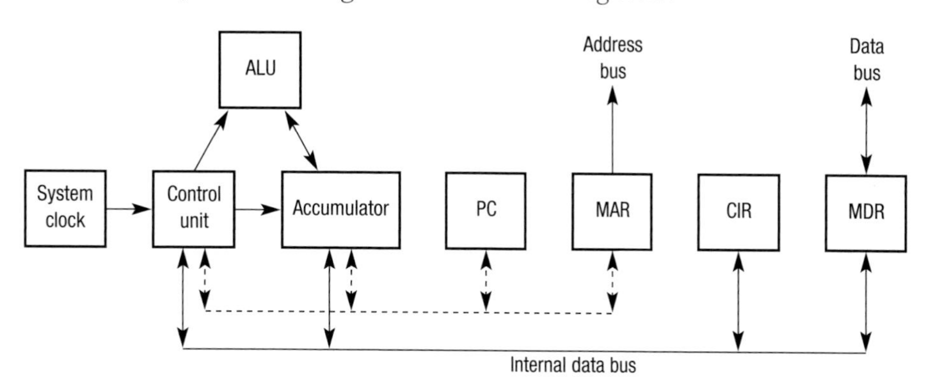

Figure 33.1 Machine architecture – the processor

KEYWORDS

Control unit: part of the processor that manages the execution of instructions.

Arithmetic Logic Unit (ALU): part of the processor that processes and manipulates data.

The control unit

The **control unit** is the part of the processor that supervises the fetch–execute cycle. The control unit also makes sure that all the data that is being processed is routed correctly – it is put in the correct register or section of memory.

The arithmetic logic unit (ALU)

The **ALU** carries out two types of operation – arithmetic and logic. The ALU can be used to carry out the normal mathematical functions such as add, subtract, multiply and divide, and some other less familiar processes such as shifting. This process is described in more detail later in the chapter.

The ALU is also used to compare two values and decide if one is less than, greater than or the same as another. Some comparisons will result in either TRUE or FALSE being recorded.

The ALU is sent an operation code (op-code) and the operands (the data to be processed). The ALU then uses logical operations such as OR, AND and NOT to carry out the appropriate process. In some computers, a separate arithmetic unit (AU) is used to cope with floating-point operations.

> **KEYWORD**
>
> **Clock**: a device that generates a signal used to synchronise the components of a computer.

The clock

All computers have an internal **clock**. The clock generates a signal that is used to synchronise the operation of the processor and the movement of data around the other components of the computer.

The speed of a clock is measured in either megahertz (MHz – millions of cycles per second) or gigahertz (GHz – 1000 million cycles per second). In 1990 a clock speed of between 4 and 5 MHz was the norm. In 2000, 1 GHz clock speeds were common. The typical clock speed at the time of writing is 2–3 GHz.

> **KEYWORDS**
>
> **Register**: a small section of temporary storage that is part of the processor. Stores data or control instructions during the fetch–decode–execute cycle.
>
> **Status register**: keeps track of the various functions of the computer such as if the result of the last calculation was positive or negative.
>
> **Interrupt register**: stores details of incoming interrupts.
>
> **Current Instruction Register (CIR)**: register that stores the instructions that the CPU is currently decoding/executing.
>
> **Program counter (PC)**: register that stores the address of the next instruction to be taken from main memory into the processor.
>
> **Memory Buffer Register (MBR)**: register that holds data that is either written to or copied from the CPU.
>
> **Memory Data Register (MDR)**: another name for the MBR.
>
> **Memory Address Register (MAR)**: register that stores the location of the address that data is either written to or copied from by the processor.

Registers

The control unit needs somewhere to store details of the operations being dealt with by the fetch–execute cycle and the ALU needs somewhere to put the results of any operations it carries out. There are a number of storage locations within the processor that are used to store this sort of data. They are called **registers** and although they have a very limited storage capacity they play a vital role in the operation of the computer.

A register must be large enough to hold an instruction – for example, in a 32-bit instruction computer, a register must be 32 bits in length. Some of these registers are general purpose but a number are used for a specific purpose:

- The **status register** keeps track of the status of various parts of the computer – for example, if an overflow error has occurred during an arithmetic operation.
- The **interrupt register** is a type of status register. It stores details of any signals that have been received by the processor from other components attached to it, for example, the I/O controller for the printer. This will receive input and output requests from processor and then send device-specific instructions to the printer. The I/O controller performs any necessary conversion of signals between the processor and a peripheral, ensuring that new peripherals can easily be connected. We will be looking at the role of interrupts later in the chapter.

There are four registers that are used by the processor as part of the fetch–execute cycle:

- The **Current Instruction Register (CIR)** stores the instruction that is currently being executed by the processor.
- The **Program Counter (PC)** stores the memory location of the next instruction that will be needed by the processor.
- The **Memory Buffer Register (MBR)**, also known as the **Memory Data Register (MDR)**, holds the data that has just been read from or is about to be written to main memory.
- The **Memory Address Register (MAR)** stores the memory location where data in the MBR is about to be written to or read from.

How the cycle works

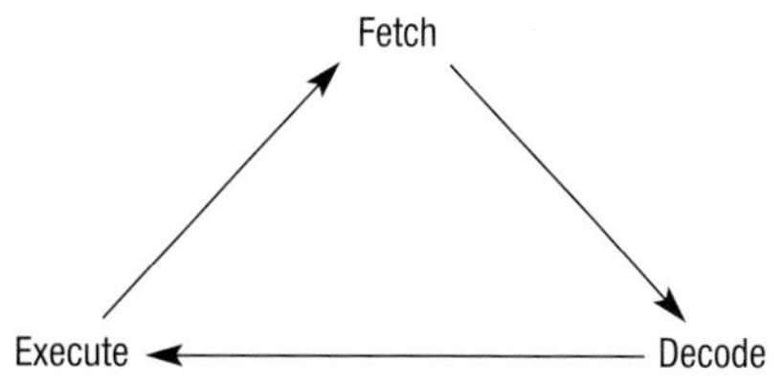

Figure 33.2 The fetch–execute cycle

- Fetch: The PC holds the address of the next instruction. The processor sends this address along the address bus to the main memory. The contents of the memory location at that address are sent via the data bus to the CIR and the PC is incremented. The details of addresses are initially loaded into the MAR and the data initially goes to the MBR. Some instructions need to load a number of bytes or words, so they may need to be fetched as successive parts of a single instruction.
- Execute: The processor then takes the instruction from the CIR and decides what to do with it. It does this by referring to the instruction set. These instruction sets are either classed as an RISC (reduced instruction set) or a CISC (complex instruction set). An instruction set is a library of all the things the processor can be asked to do. Each instruction in the instruction set is accompanied by details of what the processor should do when it receives that particular instruction. This might be to send the contents of the MBR to the ALU. There is a detailed example of how this works in the next chapter.
  Once the instruction that has just been taken from the memory has been decoded, the processor now carries out the instruction. It then goes back to the top of the cycle and fetches the next instruction. A simple instruction will require only a single clock cycle, whereas a complex instruction may need three or four. The results of any calculations are written either to a register or a memory location.

Factors affecting processor performance

There are a number of factors that affect processor performance. Often these factors have to be looked at in combination to understand how quickly a processor will work. For example, clock speed is seen as an important measure of performance, but increasing clock speed alone may not have a positive effect if other components within the processor are limited. There is no point fitting a faster clock to a computer if you do not change the components that are going to make use of that pulse as well.

- Clock speed: As we saw earlier, clock speed is one measure of the performance of the computer. It indicates how fast each instruction will be executed. In theory therefore, increasing the clock speed will increase the speed at which the processor executes instructions.
- Bus width: The processor needs to optimise the use of the clock pulse. One way of doing this is to increase the **bus width**. In the last chapter we looked at the data bus and address bus and saw how the width of the bus showed how many bits could be transferred in one pulse of the clock. Increasing the width of the data bus means that more bits and therefore more data can be passed down it with each pulse of the clock,

KEYWORD

Bus width: the number of bits that can be sent down a bus in one go.

which in turn means more data can be processed within a given time interval. Increasing the width of the address bus will increase the amount of memory that can be addressed and therefore allows more memory to be installed on the computer.

- Word length: Related to the data bus width is the **word length**. A word is a collection of bits that can be addressed and manipulated as a single unit. Computer systems may have a word length of 32 or 64 bits, indicating that 64 bits of data can be handled in one pulse of the clock. Word length and bus width are closely related in that a system with a 64-bit word length will need 64-bit buses.
- Multiple cores: Most computer systems have one processor. One way of increasing system performance is to use several processors. For convenience, multiple processors can be incorporated onto one single chip; this is known as a **multi-core** processor. A dual-core processor therefore has two processors on the one chip and will run much faster than a single-core system, which only has one processor. The term 'core' is used to define the components that enable instructions to be fetched and executed.
- Cache memory: Caching is a technique where instructions and data that are needed frequently, are placed into a temporary area of memory that is separate from main memory. The advantage of this is that the **cache** can be accessed much more quickly than main memory, so programs run faster. The key to this is ensuring that the most commonly used functions or data used in a program are placed into the cache.

KEYWORDS

Word length: the number of bits that can be addressed, transferred or manipulated as one unit.

Multi-core: a chip with more than one processor.

Cache: a high-speed temporary area of memory.

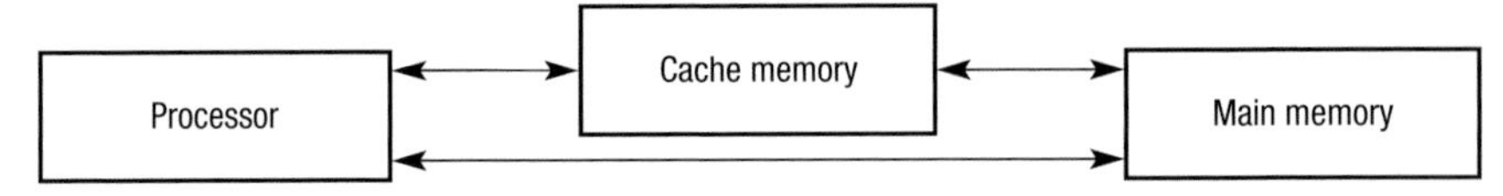

Figure 33.3 Cache memory

A level only

Interrupts

The processor in a computer is always working, irrespective of whether there is an application active or not. This is because the operating system, which is itself a large collection of programs, is always active. This means that the fetch–execute cycle is always in use. If an error occurs or a device wants the computer to start doing something else then we need some way to grab the processor's attention. The way to do this is to send an **interrupt**. An interrupt is a signal sent to the processor by a program or an external source such as a hard disk, printer or keyboard.

KEYWORD

Interrupt: a signal sent by a device or program to the processor requesting its attention.

There are a number of different sources of an interrupt. These are some typical examples:

- a printer sends a request for more data to be sent to it
- the user presses a key or clicks a mouse button
- an error occurs during the execution of a program, for example, if the program tries to divide by zero or tries to access a file that does not exist
- an item of hardware develops a fault
- the user sends a signal to the computer asking for a program to be terminated
- the power supply detects that the power is about to go off
- the operating system wants to pass control to another user in a multitasking environment.

How the interrupt works

What happens is that an additional step is added to the fetch–execute cycle. This extra step fits between the completion of one execution and the start of the next. After each execution the processor checks to see if an interrupt has been sent by looking at the contents of the interrupt register.

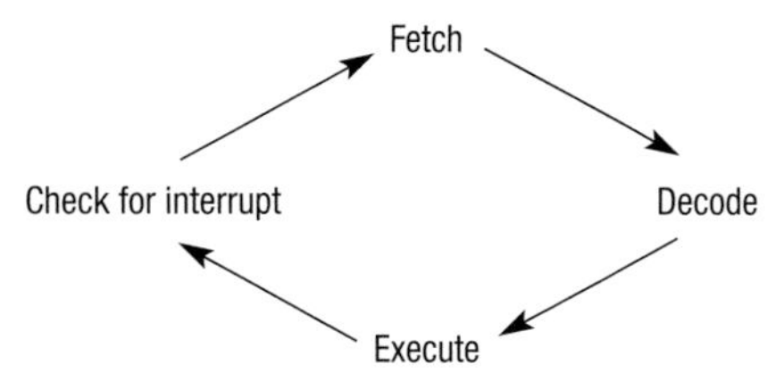

Figure 33.4 The fetch–execute cycle with interrupts

> **KEYWORDS**
>
> **Interrupt Service Routine**: calls the routine required to handle an interrupt.
>
> **Priorities**: a method for assigning importance to interrupts in order to process them in the right order.

If an interrupt has occurred the processor will stop whatever it is doing in order to service the interrupt. It does this using the **Interrupt Service Routine (ISR)** which calls the routine required to handle the interrupt. Most interrupts are only temporary so the processor needs to be able to put aside the current task before it can start on the interrupt. It does this by placing the contents of the registers, such as the PC and CIR on to the system stack. Once the interrupt has been processed the CPU will retrieve the values from the stack, put them back in the appropriate registers and carry on.

Priorities

Sometimes the program that has interrupted the running of the processor is itself stopped by another interrupt. In this case the processor will either place details of its current task on the stack or it will assess the priority of the interrupts and decide which one needs to be serviced first. Assigning different interrupts different priority levels means that the really important signals, such as a signal indicating that the power supply is about to be lost, get dealt with first.

Table 33.1 shows some of the processes that can generate an interrupt, and the priority level that is attached to that interrupt. Level 1 is the highest priority, 5 the lowest. Interrupts with the same priority level are dealt with on a first-come first-served basis.

Table 33.1 Priorities, interrupts and possible causes

Level	Type	Possible causes
1	Hardware failure	Power failure – this could have catastrophic consequences if it is not dealt with immediately so it is allocated the top priority.
2	Reset interrupt	Some computers have a reset button or routine that literally resets the computer to a start-up position.
3	Program error	The current application is about to crash so the OS will attempt to recover the situation. Possible errors could be variables called but not defined, division by zero, overflow, misuse of command word, etc.
4	Timer	Some computers run in a multitasking or multiprogramming environment. A timer interrupt is used as part of the time slicing process.
5	Input/Output	Request from printer for more data, incoming data from a keyboard to a mouse key press, etc.

Vectored interrupt mechanism

Once the values of the registers have been pushed to the stack, the processor is then free to handle the interrupt. This can be done using a technique called a **vectored interrupt mechanism**.

Each interrupt has an associated section of code that tells the processor how to deal with that particular interrupt. When the processor receives an interrupt it needs to know how to find that code. Every type of interrupt has an associated memory address known as a vector. This vector points to the starting address of the code associated with each interrupt.

So when an interrupt occurs, the processor identifies what kind of interrupt it is, then finds its associated interrupt vector. It then uses this to jump to the address specified by the vector, from where it runs the Interrupt Service Routine (ISR).

KEYWORD

Vectored interrupt mechanism: a method of handling interrupts by pointing to the first memory address of the instructions needed.

Practice questions can be found at the end of the section on page 298.

KEY POINTS

- The stored program concept is the idea of instructions and data being stored together in memory.
- The fetch–execute cycle explains how an instruction is fetched from memory, and executed to produce a result and place this back into memory.
- There are a number of key components of the processor including: the clock, the control unit, the arithmetic logic unit and various registers. You need to know how an instruction passes through all of these components.
- There are a combination of factors that affect the performance of a processor including clock speed, bus width, word length and caching.
- An interrupt is a signal (e.g. from a hardware device) that stops the processor from carrying out its current instruction in order to deal with another task.

TASKS

1. Describe what happens at each stage of the fetch–execute cycle.
2. What is the maximum amount of data that can pass down a 16-bit data bus in one stage of the fetch–execute cycle?
3. What is the ALU and what function does it perform?
4. Explain what the clock in a computer does.
5. The processor has access to many registers. What is a register?
6. The control unit uses four registers to control the execution of a program. They are the CIR, PC, MAR and MBR. Explain what each of these is and the part it has to play in the execution of a program.
7. Why might clock speed be an inaccurate way of measuring the performance of a computer system?
8. Why are the contents of the registers put on the stack before an interrupt is processed?
9. Why is it important that different types of interrupts have different priorities?
10. Explain how the vectored interrupt mechanism works.

STUDY / RESEARCH TASKS

1. Some of the major causes of interrupts are listed in the chapter. Find out about other causes of interrupts and try to decide what priority level you would give each.
2. What is over-clocking and what are the positive and negative effects of it on your computer?
3. What are the implications for the operating system of a multi-core processor?
4. Explain why having a dual-core system does not make your computer twice as fast.

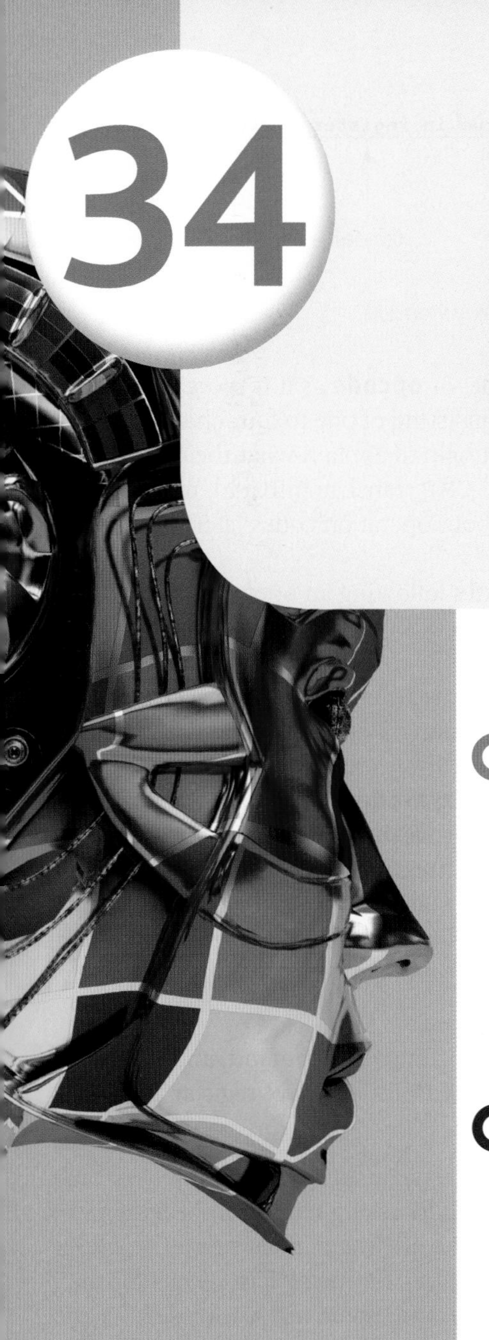

34 The processor instruction set and addressing modes

SPECIFICATION COVERAGE

3.7.3.3 The processor instruction set

3.7.3.4 Addressing modes

3.7.3.5 Machine-code/assembly language operations

LEARNING OBJECTIVES

In this chapter you will learn:

- how to use mnemonics to write code using assembly language
- the difference between immediate and direct addressing
- how to use different types of operation codes: data transfer, arithmetic operation, logical (bitwise) operations and branch operations.

INTRODUCTION

In Chapter 29, we looked at the different types of programming language and the distinction between low- and high-level languages. At machine code level, programming is carried out by directly manipulating 0s and 1s. The next level up is to use assembly language, where the code is made up of mnemonics. In this chapter we will look at how to write assembly language code.

Instruction set

KEYWORD

Instruction set: the patterns of 0s and 1s that a particular processor recognises as commands, along with their associated meanings.

In order to write this code you need to be familiar with the mnemonics that can be used and this will depend on what processor is being used. Each processor will have its own **instruction set**. These instruction sets are either classed as RISC (reduced instruction set) or CISC (complex instruction set). An instruction set is the patterns of 0s and 1s that a particular processor recognises as commands, along with their associated meanings.

A typical assembly language statement consists of four parts as shown in Figure 34.1.

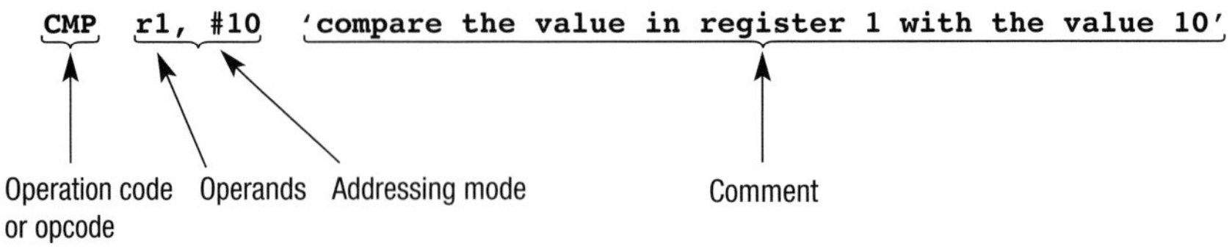

```
CMP  r1, #10  'compare the value in register 1 with the value 10'
```

Operation code or opcode | Operands | Addressing mode | Comment

Figure 34.1 A typical assembly language statement

> **KEYWORDS**
>
> **Opcode**: an operation code or instructions used in assembly language.
>
> **Operand**: a value or memory address that forms part of an assembly language instruction.
>
> **Addressing mode**: the way in which the operand is interpreted.
>
> **Assembly language**: a way of programming using mnemonics.
>
> **Mnemonics**: short codes that are used as instructions when programming, e.g. `LDR, ADD`.

- Operation code: The operation code, or **opcode** as it is more commonly called, is shown as a mnemonic consisting of one to four characters. The mnemonic usually uses letters that help to explain what the command does. For example, `ADD, MOV` and `CMP` translate into add, move and compare. There are more details about operation codes later in this chapter.
- Operands: The number of **operands** following an operation code and the way they are interpreted depends on the sort of code it is. For example, the command `CMP` must be followed by two operands – the first identifies the memory address or register that is to be accessed and the second the data that is to be compared with. Note that with the ARM6 architecture, the first operand always refers to a register.
- The use of # indicates the **addressing mode**. In this case the # refers to immediate addressing, which means that the value that follows it is the actual data item. There is more on addressing modes later in the chapter.
- Comments: The comment part of the statement is optional. **Assembly language** programs can be hard to follow. Assembly language programs tend to be very long, so being able to add comments makes them easier to understand.

Our example uses the **mnemonic** for a compare command, and this particular comparison entails comparing the contents of register 1 with the value 10.

In machine code, the instruction will operate using a fixed number of bits. Within that, the operator, operand and addressing mode will be assigned a certain number of bits. For example, a 32-bit system means that the whole instruction is 32-bits. It might assign 12 bits for the opcode, 4 bits for the addressing mode and 16 bits for the operand. Some instruction sets fix the number of bits that can be used for each part of the instruction, whereas others enable the bit allocations to vary. Increasing the number of bits assigned to each instruction will therefore increase the range of opcodes and operands that can be used within a particular instruction set.

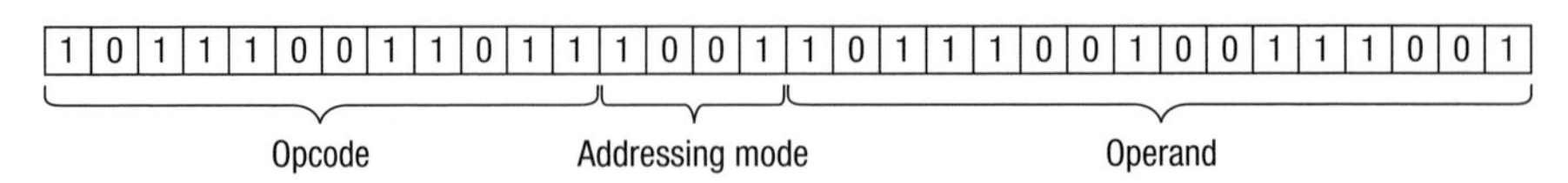

Figure 34.2 An example of how an instruction set might assign bits to instructions

A worked example

The following program shows how the assembly language instructions would be written for the following program code:

```
if y = 10 then
  x ← 9
else
  y ← y + 1
endif
```

The assembly language code based on the ARM 6 processor instruction set would be as follows:

```
CMP r1, #10       'compares the value of variable y
                  with the value 10'
BNE else          'branches if the value in variable y
                  is not equal to 10'
MOV r2, #9        'moves the value 9 into register 2,
                  which contain variable x'
B endif           'branch to the end of the statement
                  block
else
ADD r1, r1, #1    'adds 1 onto the value of variable y
                  already held in register 1
endif
```

Note that `r1` is being used to store the value of `y` and `r2` to store the value of `x`.

Immediate and direct addressing

In order to access anything that is held in memory you need to know its address. The address is a number that tells the computer where in memory to go to find a specific item of data. You might visualise the memory of a computer as a vast set of pigeon-holes, each with its own name or address. Data is put into or copied out of memory in different ways using addressing modes. Two of the main addressing modes are:

- Direct addressing: Using a **direct address** mode tells the CPU which address contains the data you want to access. So `LDR r1, 100` would copy the data held in memory location 100 into register 1.
- Immediate addressing: Rather than loading the contents of a memory address, the **immediate address** method loads the data directly. Therefore, the operand would have to be the actual number that you wish to use. A command such as `MOV r1, #10` would move the value 10 into register 1.

KEYWORDS

Direct address: the operand is the datum.

Immediate address: the operand is the memory address or register number.

Types of operation codes

The operation codes of an assembly language can be placed in one of four groups: data transfer operations, arithmetic operations, logical operations, and branch operations.

Data transfer operations

These include commands (such as the one detailed in the example above) that move data between the registers and main memory. Typical instructions include Move (**MOV**), Store (**STR**) and Load (**LDR**).

Arithmetic operations

Apart from the four normal arithmetic functions – add, subtract, multiply and divide – this section also includes the increment (increase by one), decrement (reduce by one), negate (reverse the sign), compare (two values) and shift instructions. The status register is used to record certain features of a calculation. These include whether the calculation has generated an overflow error or whether the result is zero or negative. **Shift instructions** are used to move the bits within a register. Shifts can move bits either left or right.

A logical shift can be used to extract the content of just one bit. This is achieved by repeatedly shifting until the bit you want is put in the carry bit. If this bit pattern:

1	0	1	1	1	0	0	1

is operated on with a shift right then it becomes:

0	1	0	1	1	1	0	0	→ 1 (Carry bit)

The least significant bit (the right-hand most) is placed in a carry bit; in this case it is a 1, and a 0 is placed in the most significant bit (left-hand most).

This type of shift is called a shift right. It is also possible to carry out a shift left. An alternative to this is rotate left and rotate right where the bit positions are shifted in a loop. A rotate right therefore would mean that the LSB became the MSB and all the other bits would shift one to the right.

> **KEYWORDS**
>
> **Data transfer operations**: operations within an instruction set that move data around between the registers and memory.
>
> **Arithmetic operations**: operations within an instruction set that perform basic maths, such as add and subtract.
>
> **Shift instructions**: operations within an instruction set that move bits within a register.

Logical operations

This includes the logical bitwise functions AND, OR, NOT and XOR. Bitwise means that each bit within a bit string is compared to a corresponding bit in another bit string of the same length. The results of these operations can be used to compare and calculate values. They can also be used to mask out or ignore the contents of some of the bits in a byte.

> **KEYWORD**
>
> **Logical operations**: operations within an instruction set that move the bits around within the operand.

```
    0011
AND 0010
=   0010
```

AND will compare each bit in the string. Where they are both 1, the answer is 1. Otherwise the result is 0. This effectively produces a mask allowing just parts of the bit string to be used. This would be useful for example, if identifying a parity bit.

```
    0011
OR  0010
=   0011
```

OR will compare each bit in the string. If either or both bits are 1 then the result is 1. Otherwise the result is 0. This could be used to set up a mask where if any of the bits are 1, the flag is TRUE.

```
NOT 0011
=   1100
```

NOT will negate each value so that a 0 becomes a 1 and 1 becomes a 0. The result is the equivalent of the two's complement value –1. In this case, +3 becomes –4.

```
    0011
XOR 0011
=   0000
```

XOR will compare each bit and return a 0 if both bits are 0 or both bits are 1. If one bit is 0 and the other is 1 then it will return a 1. This is commonly used to set a register to 0 as if you perform an XOR operation of a number on itself, it will always return a zero.

Branch operations

KEYWORD

Branch operations: operations within an instruction set that allow you to move from one part of the program to another.

Without the ability to branch or jump, all assembly programs and by extension all high-level languages, would have to be linear. There are a number of ways you can create a branch. A `B` command carries out an unconditional jump round a section of code. This means that there are no conditions attached so the jump will take place regardless of any prevailing conditions.

Conditional branches take the form:

- `BNE` – branch if not equal
- `BEQ` – branch if equal
- `BGT` – branch if greater than
- `BLT` – branch if less than.

The result of the last comparison will determine whether the jump is executed or not. This is where the labels come in.

All the complex structures that are taken for granted in high-level languages such as arrays and iterative routines can be constructed from these basic operation codes. Using a high-level language hides a lot of the complex nature of the assembly language.

Table 34.1 The ARM processor set codes

Instruction	Description
`LDR Rd, <memory ref>`	Load the value stored in the memory location specified by **<memory ref>** into register **d.**
`STR Rd, <memory ref>`	Store the value that is in register **d** into the memory location specified by **<memory ref>**.
`ADD Rd, Rn, <operand2>`	Add the value specified in **<operand2>** to the value in register **n** and store the result in register **d**.
`SUB Rd, Rn, <operand2>`	Subtract the value specified by **<operand2>** from the value in register **n** and store the result in register **d**.
`MOV Rd, <operand2>`	Copy the value specified by **<operand2>** into register **d**.
`CMP Rn, <operand2>`	Compare the value stored in register **n** with the value specified by **<operand2>**.
`B <label>`	Always branch to the instruction at position **<label>** in the program.
`B<condition> <label>`	Conditionally branch to the instruction at position **<label>** in the program if the last comparison met the criteria specified by the **<condition>**. Possible values for **<condition>** and their meaning are: • EQ: equal to • NE: not equal to • GT: greater than • LT: less than.
`AND Rd, Rn, <operand2>`	Perform a bitwise logical AND operation between the value in register **n** and the value specified by **<operand2>** and store the result in register **d**.
`ORR Rd, Rn, <operand2>`	Perform a bitwise logical OR operation between the value in register **n** and the value specified by **<operand2>** and store the result in register **d**.
`EOR Rd, Rn, <operand2>`	Perform a bitwise logical exclusive or (XOR) operation between the value in register **n** and the value specified by **<operand2>** and store the result in register **d**.

`MVN Rd, <operand2>`	Perform a bitwise logical NOT operation on the value specified by `<operand2>` and store the result in register **d**.
`LSL Rd, Rn, <operand2>`	Logically shift left the value stored in register **n** by the number of bits specified by `<operand2>` and store the result in register **d**.
`LSR Rd, Rn, <operand2>`	Logically shift right the value stored in register **n** by the number of bits specified by `<operand2>` and store the result in register **d**.
`HALT`	Stops the execution of the program.

`<operand2>` can be interpreted in two different ways, depending upon whether the first symbol is a # or an R:

- # – Use the decimal value specified after the #, e.g. **#25** means use the decimal value 25.
- **Rm** – Use the value stored in register **m**, e.g. **R6** means use the value stored in register 6.

Practice questions can be found at the end of the section on page 298.

TASKS

1 Explain the difference between direct and immediate addressing.

2 What is an operation code?

3 Some, but not all, operation codes are followed by one or more operands.
   a) What is an operand?
   b) Why does the number of operands vary from one operation code to the next?

4 Give an example to illustrate each of the following types of operation code:
   a) data transfer
   b) arithmetic operation
   c) logical operation
   d) branch operation.

5 Write assembly language instructions to create a counter that counts from 0 to 10 and then halts.

6 Write assembly language for the following pseudo-code. X should be stored as a variable in memory.

```
if x > 0
then x = x -1
Halt
```

KEY POINTS

- In order to write assembly code you need to be familiar with the mnemonics that can be used and this will depend on what processor is being used.
- Instructions are made up of opcode and operand.
- Direct addressing tells the CPU which memory or register address contains the data you want to access.
- Immediate addressing loads the data directly from the operand.
- Logical operations include the logical bitwise functions AND, OR, NOT and XOR.
- Transfer operations move data between the registers and main memory.
- Arithmetic operations include add, subtract, multiply, divide, increment, decrement, negate, compare and shift instructions.

STUDY / RESEARCH TASKS

1 Describe the key differences between a CISC and a RISC and give examples of where each different type of instruction set might be used.

2 Download and experiment with a CPU or instruction set simulator.

3 Research other addressing modes including: indirect addressing, displacement addressing, indexed addressing and base register addressing.

35 External hardware devices

SPECIFICATION COVERAGE

3.7.4 External hardware devices

LEARNING OBJECTIVES

In this chapter you will learn:

- how a digital camera works
- how a barcode reader works
- how RFID works
- how a laser printer works
- how magnetic, solid state and optical disks work
- the applications of each of these devices.

INTRODUCTION

In Chapter 32 we looked in detail at the main internal hardware components. This chapter considers external hardware including a range of input, output and storage devices. The definition of external hardware includes the hard disk as it is a form of secondary storage that is external to the processor.

There is an enormous range of devices that are available for computer systems and as AS- and A-level students you do not need to know how all of them work. In this chapter we concentrate on those devices that are stipulated in the specification as you must understand the main characteristics, purposes, suitability and principles of operation of these: the digital camera, barcode reader, laser printer and RFID.

Digital camera

KEYWORD

Digital camera: a device for creating digital images of photographs, which can be printed or transferred onto a computer to be manipulated and stored.

A **digital camera** is a device for recording still and moving images in digital form that can then be processed further using specialised software. In common with other devices, the camera is taking analogue data, in this case light waves, and converting them into binary (0s and 1s). It does this in the following way:

- When a photograph is taken the shutter opens and lets light in through the lens.

KEYWORDS

Charge coupled device (CCD): in digital cameras it is a sensor that records the amount of light received and convert it into a digital value.

Complementary metal oxide semiconductor (CMOS): is an alternative technology that performs the same functions as a CCD.

- The light is focused onto a sensor, which is usually either a **charge coupled device (CCD)** or a **complementary metal oxide semiconductor (CMOS)**.
- The sensors are made up of millions of transistors, each of which stores the data for one or more pixels. (A pixel is a picture element or individual dot, and the whole image will be made up of millions of pixels.)
- As the light hits the sensor, it is converted into electrons and the amount of charge is recorded for each pixel in digital form.
- With light, all colours can be created from red, green and blue (RGB). Therefore to record colour, the camera will either have three different sensors, or use three different filters – one for red, one for green and one for blue.
- The data are typically stored on removable storage devices, usually referred to as flash memory, which uses programmable ROM (see solid state disks later in this chapter).
- Data are usually stored in compressed files, for example, TIFF, JPG or PNG.
- RAW files can also be generated, which are uncompressed and therefore contain all of the data from the original photograph.
- This digital data can now be decoded and manipulated using specialised software.

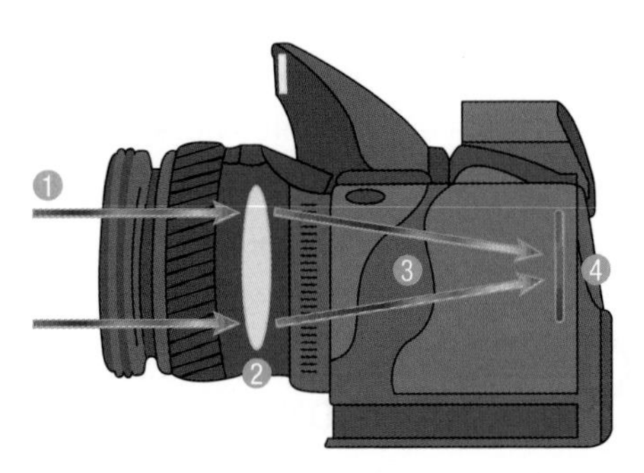

Figure 35.1 The workings of a digital camera

KEYWORD

RGB filter: red, green and blue filters that light passes through in order to create all other colours.

Light is let in through the shutter (1) and focused by the lens (2). It is directed through **RGB filters** (3) before being focused onto the CCD or CMOS sensor (4).

Figure 35.2 shows how the light is passed through the RGB filters to enable all possible colours to be created.

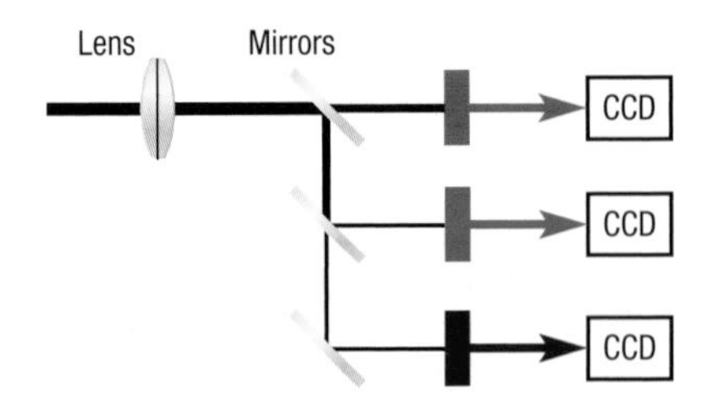

Figure 35.2 Red, green and blue (RGB) filters

The capability of digital cameras is often quantified in terms of how many megapixels it uses to record images. For example, a 12 megapixel camera will create an image made up of 12 million separate picture elements. This means that the sensor is breaking the image down into very small units and taking separate readings for each unit. The consequence is that the image can be recreated very accurately without blurring or pixellation. This creates high resolution images and is useful if the image is going to be printed and enlarged.

Figure 35.3 shows an enlarged area to demonstrate the effect of pixellation.

Figure 35.3 The effect of pixellation

However, for many users, a lower resolution is sufficient as it is more likely that the image will be taken using a smart phone and then transmitted over a mobile network. Lower resolution images will not be as accurate a reflection of the real image but they do have much smaller file sizes, making it more suitable for this application. Software can be used to alter the resolution of the image to make it suitable for the way in which the image is being used. It does this by compressing the image and many of the common file types used in digital photography such as JPG and TIFF are examples of compressed files. All of the original data collected by the camera's sensors are still available in the original file, which is said to be in raw format.

KEYWORDS

Compression: the process of reducing the size of a file.

Resolution: the number of pixels used to create an image.

Compression and **resolution** were covered in detail in Chapter 26.

Barcode reader

Barcode readers are one of a series of input devices that use scanner technology. These work in the following way:

KEYWORD

Barcode reader: a device that uses lasers or LEDs to read the black and white lines of a barcode.

- A light, usually an LED or laser is passed over an image.
- Some form of light sensor is used to measure the intensity of light being reflected back. This is converted into a current effectively generating a waveform. This could be achieved using a photodiode or a CCD sensor in the same way as a digital camera.
- White areas reflect most light and black areas the least, making it possible to use the waveform to distinguish the patterns of black and white bars.
- The waveform is analogue and therefore needs to be converted into digital form using an analogue to digital convertor.
- The encoding will convert the black and white into binary codes, for example, black = 0 and white = 1.
- The signal is decoded into a form that can then be interpreted by software.

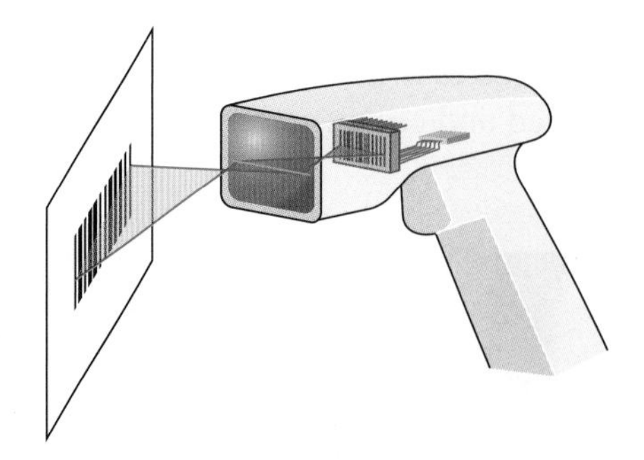

Figure 35.4 A barcode reader and barcode

Figure 35.5 A barcode

Figure 35.6 A QR code

There are many different types of barcode. Perhaps the most common is the UPC (universal product code) barcode, which uses a series of black and white lines of different widths with a printed number underneath. The lines are one of four widths and are encoded to represent the values 1 to 4. They are designed to be read reliably by a machine. The numbers are there as a manual over-ride and include a check digit.

Barcodes are used primarily for inputting product details at the point of sale. Typical uses include food, electrical products and books. Barcodes on food products are passed over a scanner built in to the checkout. Products that are physically bigger than food items, such as those sold in DIY stores, are more difficult to scan so hand-held scanners tend to be used. There are different classifications of barcodes, a common iteration being the European Article Number (EAN) which is standard for food products sold in the UK. The encoded data in the barcode is linked to a point-of-sale (POS) database system that matches the code to a particular item or product, where price details are also stored.

More recently, the same technology has been applied to codes that are made up of blocks of black and white symbols rather than lines. One example is the QR code, which has been widely adopted as it can be read with a scanner embedded within a smart phone and can contain a wider range of information than a barcode.

RFID

KEYWORD

Radio frequency identification (RFID): a microscopic device that stores data and transmits it using radio waves – usually used in tags to track items.

Radio frequency identification (RFID) is a technique where small wireless tracking devices or tags are embedded onto or into other items. The tags, which are typically about the size of a grain of rice, can be attached to almost anything and will contain data about the item being tracked. Typical uses include tagging pets or livestock or tracking products through a production line. Increasingly, RFID is being used in retail environments to tag products or to enable customer payments.

RFID works in the following way:

- The tag, which can be microscopically small, contains a chip, which contains the data about the item and a modem to modulate and demodulate the radio signals.
- The tag also contains an antennae to send and receive signals.
- Tags can be either active, which means they have their own power source in the form of a small battery, or passive, which means that they will pick up electromagnetic power when they are in range of a RFID reader.

- Signals and therefore data can be transmitted in both directions using radio frequencies. This may be over a short or long distance depending on what the tags are being used for and how they are powered. The typical range of RFID tag is between 1 and 100 metres.
- Tags may be used simply to track the physical location of the tagged item or the item may transmit data back.

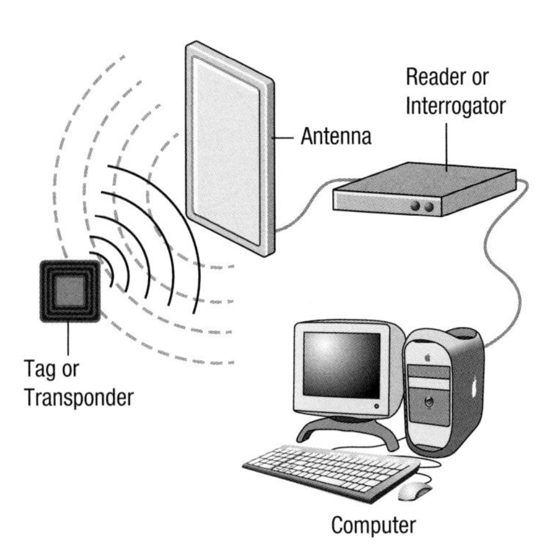

Figure 35.7 An RFID tag system

RFID is a relatively new technology and it is being put to various uses, some of which are controversial. Applications include:

- tracking individuals, particularly vulnerable adults such as Alzheimer's patients
- use in electronic passports to keep track of where people travel
- tags have been added to credit and debit cards to allow users to make contactless payment via RFID in a shop
- transport and distribution companies can use RFID to track shipments and deliveries
- tags have been added to high value items, for example artworks in museums or equipment in hospitals.

Laser printer

A **laser printer** works in the same way as a photocopier to produce high quality black and white and/or colour images. In fact, many laser printers are now 'all-in-one' combining the functions of scanning, copying and printing. They work in the following way:

- A rotating drum inside the printer is coated in a chemical which holds an electrical charge.
- The laser beam is reflected onto the drum and where the light hits the drum the charge is discharged, effectively creating the image on the drum.
- As the drum rotates it picks up toner which is attracted to the charged part of the drum.
- Paper is passed over the drum and by charging the paper with the opposite charge to the toner, the toner is attracted to the paper and away from the drum.
- The paper is heat treated to fuse the toner onto the paper.

KEYWORD

Laser printer: a device that uses lasers and toner to create mono and colour prints.

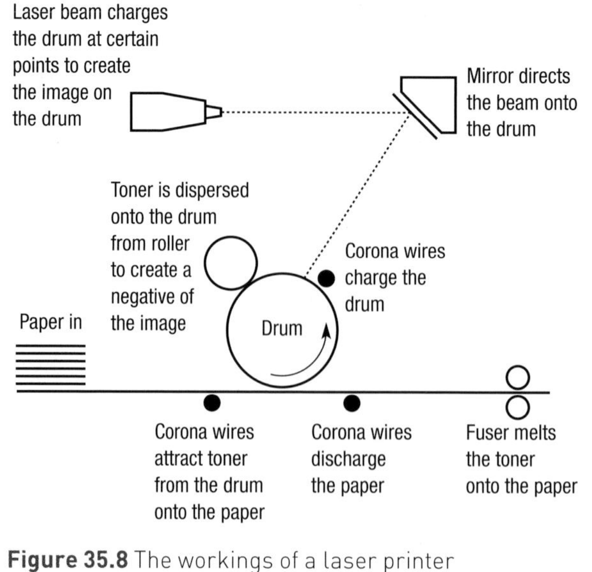

Figure 35.8 The workings of a laser printer

To achieve colour printing, four different coloured toners are used, and the process of transferring the toner to the drum is repeated for each colour. In some printers, a transfer belt is used to hold the four-colour image and therefore transfer it just once from the belt onto the paper, rather than having to pass the paper round the drum four times. Using light on a screen, creating all possible colours needs three primary colours: red, green and blue (RGB). When printing, four colours are needed: cyan, magenta, yellow and black (CMYK).

The cost of laser printers and toner cartridges has been reducing over recent years making them a common choice for personal and business users. Laser printers vary in size from small home printers costing around £100 through to large commercial machines that can cost tens of thousands of pounds.

One of the main advantages of laser printing is the speed, with home printers typically printing around 20 pages per minute with an output of a few hundred pages a month. A typical commercial laser printer can produce 200 pages per minute and is designed to print millions of copies a month.

Magnetic hard disk

Within a computer system, main memory or the Immediate Access Store (IAS) is used to store programs and data. It gives the user high speed access to applications and data, but is only temporary so the contents will be lost when the computer is switched off. To get around this there needs to be some form of permanent storage. There are a number of devices, known as secondary storage devices, that will permanently store data.

Hard disks are constructed of hard metallic material and are hermetically sealed. This is to protect them from being corrupted by dust or other debris. Most hard disks are in fact made up of a number of disks arranged in a stack. The disks are coated with a thin film of magnetic material. Changes in the direction of magnetism represent zeros and ones.

KEYWORD

Hard disk (HDD): a secondary storage device made up of metallic disks that stores data magnetically.

Hard disks spin at speeds between 3600 and 12 500 rpm as a series of heads read from and write to the disks. The heads do not actually touch the surface of the disk but float slightly above it by virtue of the speed at which the disk spins. There is an actuator arm which moves the head across surface of the disk as it spins. The combination of the rotating of the disks with the lateral movement of the arm means that the heads can access every part of the disk surface.

The surface of the disk is organised into concentric tracks and each track is split into sectors each of which can be individually addressed by the operating system. Because the head assembly can read any one of several disks, a cylinder reference is also used to identify which of the disks in the stack is being addressed.

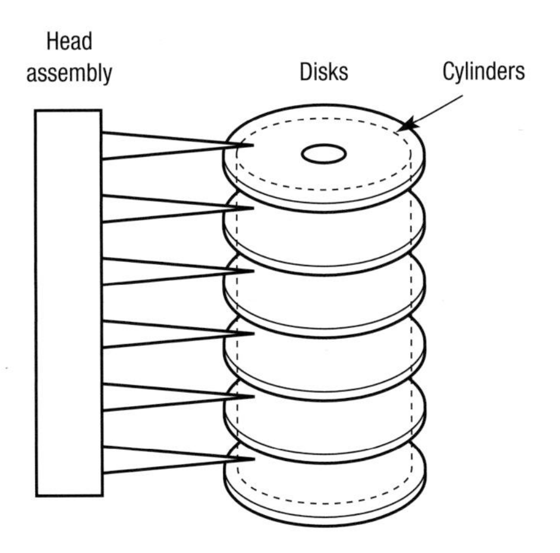

Figure 35.9 A magnetic hard disk array

Each sector has the same capacity and a large file will be stored over a number of sectors. The operating system groups sectors together into clusters to make storage easier to manage. There will be many occasions when a whole cluster is not needed. For example, a file may require five whole clusters and only part of a sixth. In this case, the whole cluster is allocated to the file even though it is not needed. This means that the disk is likely to have redundant space on it.

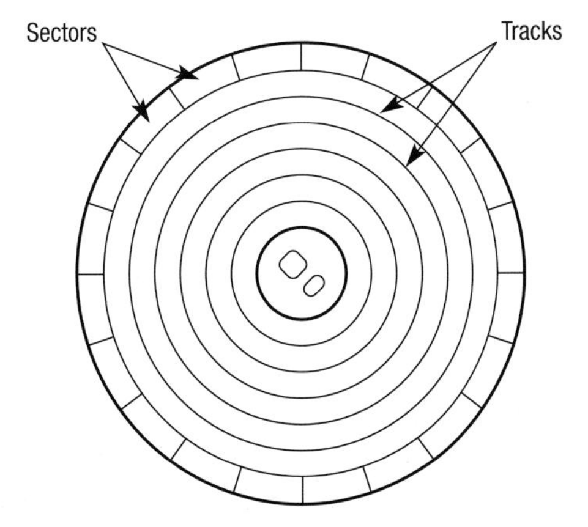

Figure 35.10 Tracks and sectors on a magnetic hard disk

The typical capacity of a hard disk at the time of writing is 1 TB. As hard disks are created with larger capacities there is an issue with the speed at which data can be retrieved from the disk. This is a feature of the physical speed at which it can spin along with the rate at which data can be transferred. In terms of relative speed, magnetic hard disks enable faster access than optical disks, but slower access than solid state disks.

Optical disk

Optical disk is a generic term for all variations of CD, DVD and Blu-Ray that use laser technology to read and write data. An optical disk is made up of one single spiral track that starts in the middle and works its way to the edge of the CD. The laser will read the data that are contained within this track by reading the pits and lands in combination with a sensor that measures how much light is reflected.

(a) Side view of a CD showing pits and lands

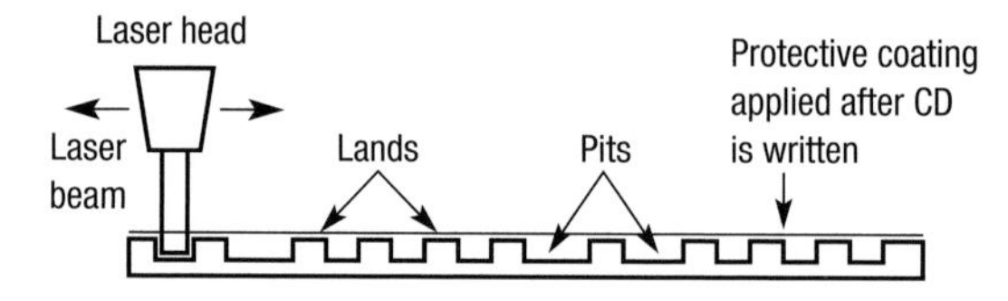

(b) Top view of a CD showing single spiral

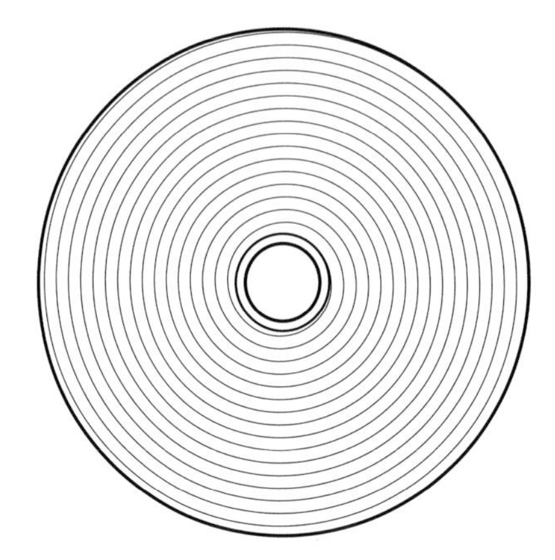

Figure 35.11 The workings of an optical disk

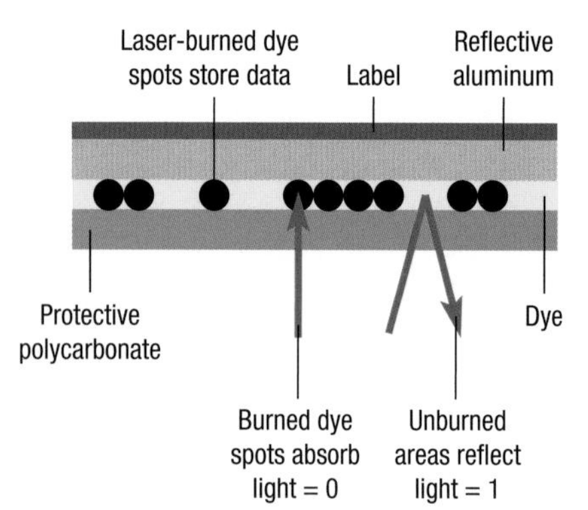

Figure 35.12 The workings of a writeable optical disk

For read-only optical disks, when data is written it is encoded as of a series of bumps, or pits and lands within the track on the disk. A protective layer is then put over the surface to prevent any corruption of the data. The pattern of pits and lands are used to represent data. When the CD is read, the pits and lands are read by the laser which then interprets each as different electrical signals. In turn the electrical signals can be converted into binary codes.

For writeable optical disks, rather than using pits and lands the disk is coated with a photosensitive dye, which is translucent. When writing to the disk, the laser will alter the state of a dye spot that is coated onto the surface making it opaque. The dye reflects a certain amount of light. A write laser alters the density of the dye and a read laser interprets the different densities to create binary patterns which in turn can represent data. Write lasers are higher powered than read lasers.

Solid state disk (SSD)

High-speed access to memory is achieved using memory cards, made up of semiconductors. As these are entirely electronic and have no mechanical parts, high-speed data transfer can be achieved. However, the problem is that they are volatile, which means that as soon as the power is lost, they lose their data. This is why hard disks are used as they use magnetic memory, which is non-volatile so is not lost when the power is switched off. However, access times for magnetic hard disks are relatively slow as the disk has to spin and an arm has to move across the surface of the disk until the data is found. This is known as latency.

Figure 35.13 Inside a solid state disk

A relatively new development is the solid state disk, also known as a solid state drive, which is made up of semiconductors, but is also non-volatile, meaning that data is not lost when there is no power. A common implementation of this is the flash drive or memory stick, but this technology is also used to replace hard disks in computer systems.

The term solid state disk is misleading is there is no actual disk, instead they use programmable ROM chips, similar to memory cards, hence the alternative description as a drive rather than a disk. However these are stored inside a unit that looks like a hard disk and commonly uses a type of memory called NAND memory. This organises data into blocks in a similar way to a traditional hard disk as described earlier, with a **controller** being used to manage the blocks of data.

KEYWORDS

Controller: in SSDs a controller is needed to organise data into blocks for storage purposes.

Block: in data storage it is the concept of storing data into set groups of bits and bytes of a fixed length.

Floating gate transistor: in SSDs it is a type of non-volatile transistor that stores data even without a power source.

Blocks of a set physical size will be made up of binary data. When reading and writing, data can only be accessed in blocks. On a traditional hard disk, blocks will be allocated to different clusters on the disk. With SSD, blocks are allocated to particular semiconductors. The advantage of this is that data can be added and deleted in blocks to different areas of the drive, so that only small parts of the drive have to be erased and written to. This enables very fast access times.

The semiconductors are able to retain their data due to the type of transistor used. It uses what is a called a **floating gate transistor**, which is able to trap and store charge. A floating gate transistor contains two gates: a floating gate and a control gate. A thin layer of oxide is placed between the two gates, effectively trapping the charge inside the floating gate even when the power is turned off.

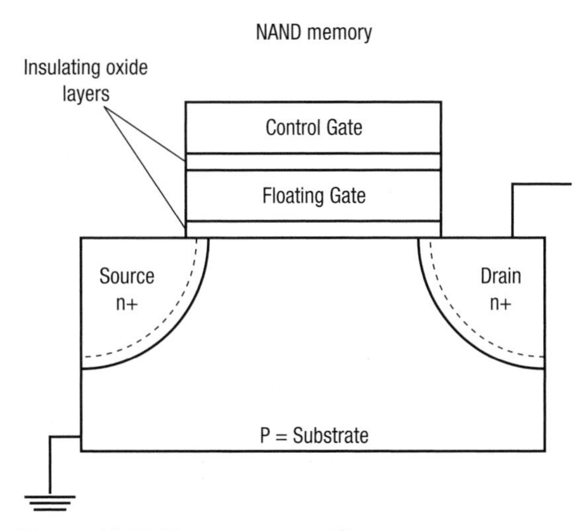

Figure 35.14 The workings of 'flash' memory

As there are no moving parts, SSDs are considered to be more reliable than HDDs as there is less chance of mechanical failure. Also, as there are no physical sectors, an SDD never needs to be defragmented, which means that its performance will not degrade over time as the disk gets full and it won't take longer to find the blocks of data.

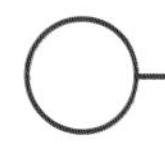

Storage devices compared

In this chapter we have looked at three main types of secondary storage. In addition to knowing how they work, you should also have a good understanding of where you would use one type rather than the other and be able to make comparisons between the capabilities of each. Table 35.1 shows a relative comparison.

Table 35.1 Comparison of HDD, SSD, CD/DVD and Blu-ray

	Hard disk (HDD)	Solid state disk (SDD)	Optical disk (CD/DVD)	Optical disk (Blu-ray)
Typical capacity	High (1 TB)	Medium (500 GB)	Low (900 MB to 1.7 GB)	Low to medium (25–50 GB)
Relative cost	Medium	High	Low	Low
Easily portable	External disks are available	External disks are available	Yes	Yes
Relative power consumption	High	Low	High	High
Relative speed of access	Medium	High	Low	Low
Latency	High	Low	Very High	High
Fragmentation		None		
Reliability	Good	Very good	Fair	Fair
Relative physical size	Large	Small	Large	Large

The differing characteristics of each device lend themselves to particular applications. HDD and SSD are broadly comparable in terms of what you would use them for, which is as the main secondary storage device in a PC or laptop.

SSD enable access times up to 100 times faster than HDD and are physically smaller, lending themselves perfectly for use in laptops. However, the relative cost per GB compared to HDD is higher and the largest capacity is limited to about 1 TB (at the time of writing), although this will inevitably increase over time. This has implications for business and organisational users who may require large amounts of storage capacity.

Large-scale users may consider a hybrid implementation where some aspects use SSD and others use HDD. For example, the operating system and applications could be run from computers using SSD and the data storage could be handled using HDD.

Optical disks are a cheap form of portable storage and lend themselves to creating inexpensive back-ups of important data and programs. Original copies of software are also generally stored on optical disks as they are easily transportable and can be stored safely off-site. However, their very limited storage capacity means that they are only generally suitable for small-scale back-ups, normally of home systems. Access time is also very slow on an optical drive, again limiting their use to back-up purposes.

Practice questions can be found at the end of the section on page 298.

KEY POINTS

- Digital cameras work by directing light into sensors made up of millions of cells and then converting this data into digital form.
- Digital cameras are often compared in terms of the number of pixels that they use to create an image.
- Three colours (red, green, blue) are needed to create all possible colours.
- Laser printers work by transferring toner off a drum and onto paper with electrical charge.
- Four print colours (cyan, magenta, yellow and black) are required to create all possible colours.
- RFID tags are tiny devices that can be attached to anything and transmit a signal containing data that can be picked up by a reader.
- Optical disks such as CDs and DVD use lasers to read pits and lands on the surface of the disk that are encoded to represent data.
- Magnetic hard disks are made up of an array of metallic disks that are read by a reading head that floats across the surface.
- Solid state disks used programmable ROM chips.

TASKS

1 What are the key differences between the way an HDD works and the way an SSD works?

2 What is the purpose of an RGB filter?

3 Identify the most appropriate storage device in the following scenarios. Justify your choice.
  a) Creating a back-up of a school network each night.
  b) Transferring a document from home to school.
  c) Creating a back-up of all the work on a stand-alone computer.
  d) Storing a feature-length movie.
  e) Storing a number of audio files.

4 Explain how a digital camera turns analogue data into digital data.

5 Explain the common formats that are possible for digital images. Why are the different formats available?

6 How does RFID work and what are the possible applications of it?

7 How do laser printers work?

STUDY / RESEARCH TASKS

1 Explain the difference between DVD-ROM, DVD-R and DVD-RW.

2 How do USB 'sticks' work?

3 Research the latest specification for SSDs. What is the limiting factor on the storage capacity of this type of drive?

4 Provide a detailed comparison of inkjet and laser printers in terms of speed, quality, initial cost and ongoing costs.

5 Will increasing the number of megapixels on a digital camera always lead to a better quality image?

Section Seven: Practice questions

1 Processors use the fetch–execute cycle.

a) Describe in full sentences how the fetch–execute cycle works in relation to the main registers: Current Instruction Register (CIR), Memory Address Register (MAR), Memory Buffer Register (MBR), Program Counter (PC) and Status Register (SR).

b) Explain the three types of buses used in a computer and what they do during the fetch–execute cycle.

c) An interrupt will suspend execution of the current program by the processor. What is an interrupt?

2 Explain the difference between the Von Neumann and Harvard architectures.

3 Write assembly language instructions that would perform the same task as the pseudo-code below. Use the registers r1 and r2 to store the variables A and B.

```
If A = 1 THEN
  B ← 2
ELSE
  A ← A + 1
ENDIF
```

4 Two alternatives for data storage are hard disk drives (HDD) and solid state disks (SSD).

a) Describe the key principles of how each work.

b) State two advantages of HDD over SSD.

c) State two advantages of SSD over HDD.

5 Explain how the following can lead to faster execution of programs:

a) increasing the clock speed

b) modifying the width of the data bus

c) utilising cache memory.

6 'Police officers would be able to respond faster to emergency calls and spend more time on the street if they made better use of modern technology.' Why might some people object to police officers using new technology?

7 Name the most suitable storage medium for each of the following.

a) Backing up a 30 Kb email message

b) Backing up 2 Gb of data

c) Distributing a software package requiring 500 Mb of storage space.

Section Eight: Consequences of uses of computing

36 Moral, ethical, legal and cultural issues

SPECIFICATION COVERAGE

3.8 Consequences of uses of computing

LEARNING OBJECTIVES

In this chapter you will learn:

- how technological innovation leads to moral, ethical, legal and cultural issues
- how our own moral and ethical beliefs can be applied to computer science
- what legal issues are related to the world of computing
- how computers and modern technology effect our culture.

INTRODUCTION

We are living through a technological revolution. There have been massive developments in computer science over recent years that have fundamentally changed the way that people use technology, the way in which information is used and even the way in which we live our lives. All of these changes have brought about a large number of moral, social, legal and cultural issues, which have implications for us as individual members of society but also for us as computer scientists.

Technological change

Consider the major innovations over the last 30 years:

- 1985 – Network file systems developed
- 1989 – Tim-Berners Lee invents the World Wide Web
- 1991 – Linux is created by Linus Torvalds
- 1993 – The widespread adoption of email
- 1997 – Broadband is available to home users
- 1998 – Google is developed
- 1999 – WiFi becomes a recognised standard
- 2004 – Facebook is founded
- 2005 – YouTube is founded
- 2006 – Twitter is created
- 2010 – Apple unveils the iPad.

At the same time, look how quickly hardware technology has advanced over recent years. In 1985, when the first network file system was created, most computers were found in the offices of businesses and large organisations. During the 1990s, people started to buy PCs for the home. The typical home user will now have a number of wireless devices that connect seamlessly to their own home network and to the Internet. Recent telecommunications developments such as 4G mean that we now live in an 'always connected' society.

All of these changes bring massive opportunities for individuals and organisations working in computing and related industries. For example:

- Five of the worlds' top 25 richest companies are IT businesses.
- The owners of Google and Facebook are personally worth billions of pounds even though the businesses only started in 1998 and 2004 respectively.
- Lesser known businesses like Dropcam and Skybox Imaging have been sold to Google for over 500 million US dollars, even though they only started in 2009.
- IT managers and directors are in the top ten list of best paid jobs in the UK.
- Programmers can earn over £60 000 a year.

Moral and ethical issues

A moral issue is one that concerns our own individual behaviour and our own personal concept of right and wrong. We learn our moral values from other people such as our parents, teachers and peers, and we learn them for ourselves from experience. Ethics vary slightly from morals in that they are a way of trying to define a set of moral values or principles that people within society live by. **Ethical issues** are sometimes referred to therefore as social issues.

> **KEYWORD**
> **Ethical issues**: factors that define the set of moral values by which society functions.

There are said to be 'no rights and wrongs' with morals and ethics as all issues are a matter of personal opinion. Some people argue that even actions that are illegal might still be ethical. As A-level students you need to be aware of some of the issues, and understand the implications. One of the main issues is the widespread collection and use or misuse of personal data.

The use and misuse of personal data

The collection, use and misuse of personal data has become one of the most topical issues in computer science. Most organisations collect data on an ongoing basis and much of these data are personal. At a basic level this might be name and address information, but may also include data about individuals' finances, health, relationship status, family, employment history and even their personal views.

This presents a number of issues. For example:

- Personal privacy: A lot of data are collected as a matter of routine and as individuals we may not have explicitly consented to them being collected and used.
- Data security: Much of the data are stored online or on networks connected to the Internet. How can we be sure they are safe from unauthorised users?

- Misuse of data: Data collected for one purpose are used for a different purpose. In many cases, data are sold on to other organisations.
- 'Big Brother': Many people have concerns that personal data are being used by government to monitor individuals and that this is a breach of our basic human rights.
- Online profile: Every time you do anything online such as use social media or contribute to a forum, that data may stay online for years and contribute to what people know about you.
- Profiling: Large organisations often accumulate data in order to build up a profile of individuals. This could have a negative impact on an individual.

In many cases, personal data are being collected, stored and used in an ethical way and in a way that benefits the individual. The case study below shows how technology has massively improved the way we do our banking.

CASE STUDY 1: BANKING – THE BENEFITS OF TECHNOLOGY

Around 30 years ago, if you wanted to carry out any banking transaction you had to do it between the hours of 9 a.m. and 3 p.m. on a weekday as this was when banks used to open. The invention of cash machines in the 1980s was a technological revolution giving customers access to their money 24 hours a day. The invention of online banking in the 1990s meant that almost all transactions could be done at any time on any day of the week, including paying bills, setting up direct debits and moving money from one account to another. Some estimates suggest that as many as half of all web users now do their banking online.

However, there are risks and there is a trade-off between the value of the system for organisations and individuals and the possible threats.

CASE STUDY 2: BANKING – THE THREAT OF TECHNOLOGY

According to some sources there are as many as 250 000 phishing attacks every year. This is where fraudsters attempt to get bank account details by sending emails that appear to be from your bank. An estimated £20m worth of online frauds are carried out on an annual basis using this and other methods. Some customers have lost hundreds of thousands of pounds in individual attacks that use sophisticated software to emulate the online banking websites.

Other moral and social issues

There are a number of other moral and ethical issues relating to computer science:

- **Unauthorised access**: Hackers gain access to systems for different reasons. Hacking for the purposes of committing fraud is considered to be wrong by many people. However, there are groups of people called ethical hackers who claim that they hack in order to expose weaknesses in system security. They claim that their actions therefore are for the good of society.

KEYWORD

Unauthorised access: where computer systems or data are used by people who are not the intended users.

- Unauthorised use of software: Some people believe that software companies and programmers spend hours developing programs and should therefore be rewarded for their work. Some people believe that software is too expensive, requires too many updates and that software companies are exploitative. Therefore, downloading or copying software is morally defendable.
- Inappropriate behaviour: There is evidence that people's behaviour changes when they are online. In the worst cases this can lead to online bullying, trolling and other forms of abuse that may then spread into the real world.
- Inappropriate content: A lot of content on the Internet is what most people would consider to be inappropriate. This includes pornography, violence or sites promoting religious or ethnic hatred. These sites may not be illegal but there is concern about what effect they have on the society, particularly younger people.
- Freedom of speech: Some people believe that you should be able to say whatever you like, even if that offends other people. The Internet gives almost everyone the ability to do that. It therefore raises the issue of whether there should be some code that all Internet users should adhere to when expressing their views.
- Unemployment: A broader social issue relates to the impact that new technology has on people's working lives. For example, many businesses such as retail and banking may no longer need to employ as many people in their stores and branches. On the other hand, they may create more jobs in IT for employees working in their online businesses.
- Access to the Internet: It is difficult to know how many people have access to the Internet. Some estimates are that there are 2.5 billion Internet users. There are 7 billion people in the world, so that means only around 35% of the world's population have access to it. An estimated 15% of the UK population do not have Internet access. Are they disadvantaged by this?

KEYWORDS

Moral issues: factors that define how an individual acts and behaves.

Code of conduct: a voluntary set of rules that define the way in which individuals and organisations will behave.

As is clear from the topics discussed above, ethical and **moral issues** become a matter of debate. When you are using your own computer at home, you make your own moral decisions about these issues. When you are using a computer in a school, college or any other organisation, you normally have to agree to a **code of conduct**.

The British Computer Society (BCS) has produced a code of conduct and a code of ethics that guide individuals and organisations on the ethical use of computer systems in general, including Internet usage. Observing the code is a condition of membership to the society and although it is not legally enforceable, any breaches of the code could lead to dismissal of an employee, or a student being asked to leave a college. The codes applies to users of computer systems and also to programmers and developers who create computer systems.

The main principles of the code of conduct are that members should:

- always operate in the public interest
- have a duty to the organisation that they work for, or the college they attend
- have a duty to the profession
- maintain professional competence and integrity.

Legal issues

As AS- and A-level students you are not expected to learn the details of all of the Acts listed in this section. However, you may be asked to consider issues in a legal context, so it is important to have some understanding of the current legal framework in terms of how it relates to computing.

KEYWORD

Legal issues: factors that have been made into laws by the government.

Legal issues relate to those issues where a law has been passed by the government. There are very few Acts of Parliament that are specific to the world of computing. The two main ones are the Data Protection Act and the Computer Misuse Act. In addition, the Freedom of Information Act, the Regulation of Investigatory Powers Act and the Copyright, Designs and Patents Act are of particular relevance to computing.

Also, using a computer does not exempt you from all the other laws of the land. For example, someone who carries out an act of fraud on the Internet can be prosecuted under the Fraud Act. Someone who steals computer data can be prosecuted under the Theft Act. Someone who makes false allegations about someone else in an email can be prosecuted for libel.

Legislators and those who enforce the law have two main issues:

- Geographical limitations: Most UK laws only apply in the UK. With the global nature of the Internet it can be difficult to prove where a particular offence took place. Also, if the perpetrator breaks a UK law but they are based in another country, it can be difficult to prosecute them. Different countries have different laws and therefore there is no universal way of regulating the computer industry or the Internet.
- Constant change: Many acts are introduced in response to current events. As technology develops so rapidly, laws often become out of date quite quickly. The Computer Misuse Act is a good example of this as it was introduced before the widespread adoption of the Internet.

Data Protection Act

The Data Protection Act was first introduced in 1984 as a result of public concerns about the increasing use of computers to store personal information. It has since been updated to reflect the enormous changes in the use of information during the 1990s. It places controls on organisations and individuals that store personal data electronically. The definition of personal data is any data on an individual where the person (known as the data subject) is alive and can be individually identified.

The Act states that with a few exemptions, any person or organisation storing personal data must register with the Information Commissioner. 'Information Commissioner' is a confusing term in that it relates to an actual person and the organisation that they run. The organisation itself is independent but was set up by the government to oversee Data Protection and Freedom of Information. The commissioner's mission is:

'We shall develop respect for the private lives of individuals and encourage the openness and accountability of public authorities

- by promoting good information handling practice and enforcing data protection and freedom of information legislation; and
- by seeking to influence national and international thinking on privacy and information access issues.'

There are eight main principles behind the Data Protection Act. Anyone processing personal data must comply with the eight enforceable principles of good practice. They say that data must be:

- fairly and lawfully processed
- processed for limited purposes
- adequate, relevant and not excessive
- accurate
- not kept longer than necessary
- processed in accordance with the data subject's rights
- secure
- not transferred to countries without adequate data protection.

Another feature of the Act is that data subjects have the right to know what data are stored about them by any particular individual or organisation. These are known as subject access rights. If this information is incorrect then the data subject has the right to have it corrected. The organisation must be given notice and may charge a small fee to the data subject.

Freedom of Information Act

The Freedom of Information Act extends the subject access rights of the Data Protection Act and gives general rights of access to information held by public authorities such as hospitals, doctors, dentists, the police, schools and colleges. Both Acts are overseen by the Information Commissioner.

The Act gives individuals access to both personal and non-personal data held by public authorities. The idea behind the Act was to provide more openness between the public and government agencies. Therefore, the agencies are obliged to give the public access to information and to respond to individual requests for information. Much of this is done through websites and email communications.

Computer Misuse Act

KEYWORD

Data misuse: using data for purposes other than for which it was collected.

The Computer Misuse Act was introduced primarily to prevent hacking (**data misuse**) and contains three specific offences relating to computer usage:

- Unauthorised access to computer programs or data: This includes some forms of hacking including breaking through password protection and firewalls, decrypting files and stealing another user's identity.
- Unauthorised access with further criminal intent: An extension of the first offence where there is a clear intention to carry out a further criminal act such as an act of fraud or a copyright breach.
- Unauthorised modification of computer material: This includes falsifying bank details or exam grades, spreading viruses designed to corrupt data and programs and interfering with system files.

The Act was introduced before the widespread use of the Internet, which has led to problems with enforcement. Prior to the Internet, hacking did take place, but not on the scale that it does today. There are now millions of computers and networks connected to the Internet and the opportunities for hackers have increased enormously.

There have been some amendments to the Act and there is pressure on the government from the computer industry and other businesses operating on the Internet, to introduce new laws to reflect the current activities of the cyber-criminal.

Regulation of Investigatory Powers (RIP) Act

The RIP Act was introduced to clarify the powers that government agencies have when investigating crime or suspected crime. It is not specific to the world of computing but was introduced partly to take account of changes in communication technology and the widespread use of the Internet.

There are five main parts to the Act. The most relevant to computing are Part 1 which relates to the interception of communications, including electronic data, and Part 3 which covers the investigation of electronic data protected by encryption. In simple terms, it gives the police and other law enforcement agencies the right to intercept communications where there is suspicion of criminal activity. They also have the right to decipher these data if they are encrypted even if this means that the user must tell the police how to decrypt the data.

It also allows employers to monitor the computer activity of their employees, for example, by monitoring their email traffic or tracking which websites they visit during work time. This raises a number of issues relating to civil liberties.

Copyright, Designs and Patents Act

This Act gives rights to the creators of certain kinds of material allowing them control over the way in which the material is used. The law covers the copying, adapting and renting of materials.

The law covers all types of materials but of particular relevance to computing are:

- original works including instruction manuals, computer programs and some types of databases
- web content
- original musical works
- sound recordings
- films and videos.

KEYWORD

Copyright: the legal ownership that applies to software, music, films and other content.

Copyright applies to all works regardless of the format. Consequently, work produced on the Internet is also covered by copyright. It is illegal to produce pirate copies of software or run more versions on a network than have been paid for. It is an offence to adapt existing versions of software without permission. It is also an offence to download music or films without the permission of the copyright holder.

In computing, two techniques are used to protect copyright:

- Digital Rights Management (DRM): This uses access control software to limit the way in which users can control, use, copy, print or edit digital content that they have bought.
- Licensing: Normally used for software, this provides users with a paper-based or digital proof that they have purchased software legally and details what they are allowed to do with the software.

Other acts relevant to computing

Other acts that are particularly relevant to computing are:

- The Official Secrets Act prevents the disclosure of government data relating to national security.
- The Defamation Act prevents people from making untrue statements about others which will lead to their reputation being damaged.
- The Obscene Publications Act and the Protection of Children Act prevent people from disseminating pornographic or violent images.
- The Health and Safety (Display Screen Equipment) Regulations provides regulation on the correct use screens and is a specific addition to the Health and Safety at Work Act, which contains more general regulation on keeping employees safe.
- The Equality Act makes it illegal to discriminate against anyone of the grounds of sex, sexual orientation, ethnicity, religion, disability or age. This includes the dissemination of derogatory material.

Cultural issues

KEYWORD

Cultural issues: factors that have an impact on the ways in which we function as a society.

Cultural issues are all of the factors that influence the beliefs, attitudes and actions of people within society. Common cultural influences are family, the media, politics, economics and religion. There are cultural differences between different groups of people. For example, people from different countries often have a different culture.

There are elements of computer use that have a cultural impact in that they can change our attitudes, beliefs and actions:

- Over-use of data: There are fears that we are becoming completely dependent on data. Data are being collected about us by every single organisation we deal with including government agencies and businesses. Many decisions about the way in which the country is run are based on data analysis.
- Invasive technologies: A lot of data are collected without our consent. Satellite images and Google StreetView enable anyone to look at your house. Zoopla and other websites tell everyone how much you paid for it.
- Over-reliance on computers: What happens when computer systems fail? At a simple level you might lose some data on your computer. At a more serious level people may be in physical danger or even die as a result of computer failure.
- Over-reliance on technology companies: According to some sources, two-thirds of all Internet searches are done through Google. That is around 115 billion searches a month. Wikipedia often appears on the front page of search results. This gives these two organisations a massive influence over the information we access.
- 'Big brother' culture: The original meaning of 'big brother' is that the government is watching everything we do and that we have to modify our behaviour to meet expected behaviours. With the increasing use of CCTV, the desire for national identity cards and the monitoring of emails and mobile phone calls, some people believe that we heading in that direction.
- Globalisation: As we become more connected to other cultures, we are more likely to be influenced by them. For example, many individuals and organisations use technology to try and influence the debate on religion and politics.

TASKS

1 Define the following terms from the Data Protection Act:
   a) data subject
   b) personal data
   c) subject access rights.
2 A large bank uses data collected from customers with personal loans to provide a car manufacturer with a list of people who have bought new cars using their loans. Which of the eight data protection principles have been breached?
3 Describe the three main principles of the Computer Misuse Act.
4 Apart from the laws above, identify three other pieces of legislation that apply when using computers.
5 Why might some people be worried about the amount of personal data being stored about them?
6 Has the Internet had a positive or negative effect on society as a whole?
7 What moral responsibilities do you think programmers have?
8 Is it morally acceptable to download software without paying for it?

KEY POINTS

- We are living through a technological revolution and as computer scientists we must consider the consequences of computing on individuals and society as a whole.
- Computing can bring about massive benefits but can also have a negative effect on individuals and society.
- There are a number of laws relating specifically to computing and other common laws also apply to actions that are undertaken on a computer.
- The Internet and World Wide Web have had a massive influence on our culture and will continue to do so.

STUDY / RESEARCH TASKS

1 Explain why the Computer Misuse Act is inadequate as a measure for preventing hacking.
2 There is currently a lot of pressure on the government to update the laws relating to computer misuse. Why is the current legislation considered by some to be ineffective and what could be done to improve it?
3 Do you consider the RIP Act to be an infringement of civil liberties?
4 Should the Internet be censored?
5 Some countries do censor the Internet. Find a country that does impose censorship and describe how they do it.

Section Nine: Fundamentals of communication and networking

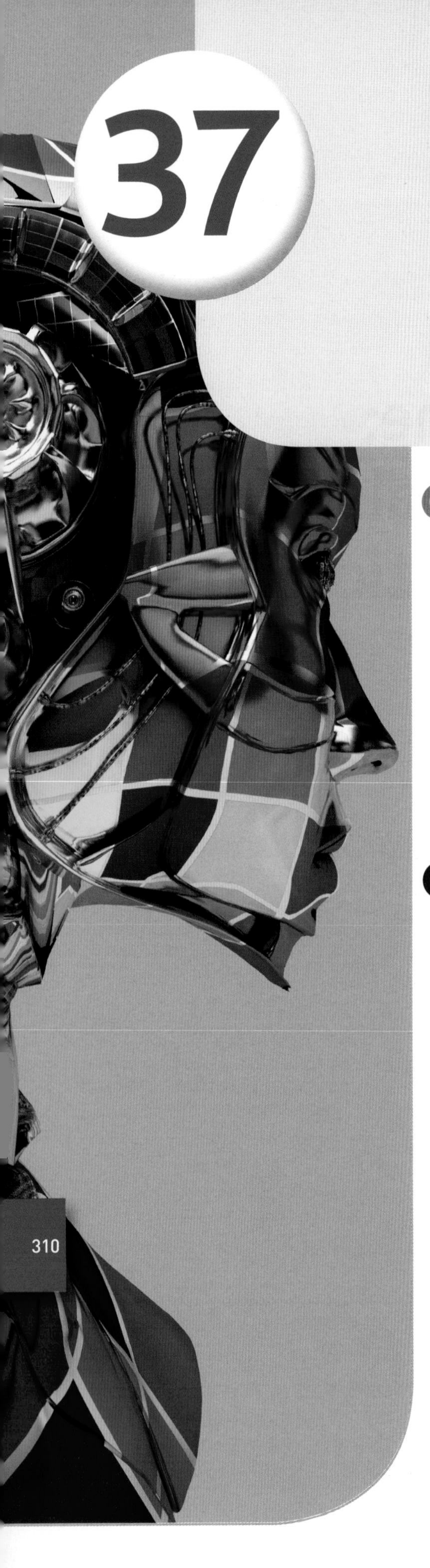

37 Communication basics

SPECIFICATION COVERAGE

3.9.1 Communication methods

LEARNING OBJECTIVES

In this chapter you will learn:

- what serial and parallel data transmission are
- what bandwidth, bit rate and baud rate are and the differences between them
- how latency affects the actual speed of transmission
- what synchronous and asynchronous data transmission are
- how devices establish communication through handshaking
- which protocols are used to establish rules and standards by which transmissions can take place.

INTRODUCTION

One of the key aspects of computing is communication. For example, input and output devices need to communicate with the processor, the hard disk needs to communicate with memory and so on. Communication in this sense takes place through the transmission of data and instructions. We have already looked at many examples of data transmission inside the computer. In this section, we are more concerned with communication between computers and peripheral devices and also between one computer and another across local and global networks. This section will also include a detailed analysis of the infrastructure that makes up the Internet.

Computer data can be transmitted using a variety of media. For example, there are a number of different cable types that can be selected, or microwave links can be used where wireless applications are needed. As we saw in Chapter 26, data is transmitted either in digital form, or is modulated into analogue. In either case, you should view the transmission of data as a series of signals being sent that represent different binary codes which in turn can represent text, numbers, sound or graphics. It is important to understand the difference between analogue and digital as most communication methods involve converting one to the other. Therefore, it is recommended that you read Chapter 26 again before reading this chapter.

Serial and parallel transmission

Serial transmission sends and receives data one bit at a time in sequence. Serial connections are used to connect most of the peripherals to the computer such as the mouse and keyboard, and it is serial cables that connect computers together to form a network.

The speed of the transmission will depend on the type of cabling used so it is not necessarily the case that serial transmission is slow. For example, the Universal Serial Bus or USB is a high-speed serial connection that allows peripheral devices to be connected to your computer. Serial network cables are capable of transmissions rates of 1 Gbps (1000 million bits per second).

KEYWORDS

Serial transmission: data is transmitted one bit at a time down a single wire.

Parallel transmission: data is transmitted several bits at a time using multiple wires.

Parallel data transmission uses a number of wires to send a number of bits simultaneously. The more wires there are, the more data can be sent at any one time. We have already come across **parallel transmission** in relation to the buses used inside the computer. A 32-bit parallel connection, for example, may connect the processor and memory together.

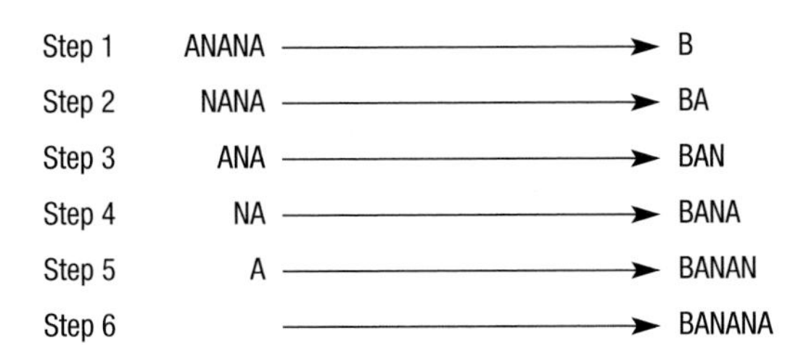

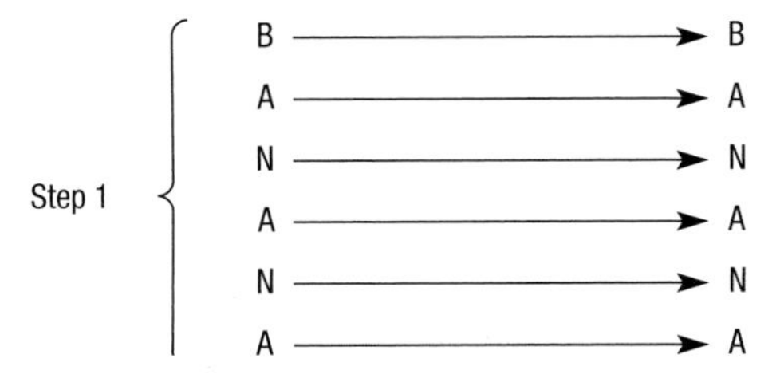

Figure 37.1 Serial and parallel communication

Parallel cables use more wires and are therefore more expensive to produce than serial cables. The signal will also degrade as distance or speed increases due to interference between the lines. Another problem is timing the signals so that the data sent down each wire arrive at the other end at the correct time and in sequence with data being transmitted through the other wires. This is known as synchronisation and this becomes more difficult as the number of wires increases.

Bandwidth

KEYWORD

Bandwidth: a measure of the capacity of the channel down which the data is being sent. Measured in hertz (Hz).

Bandwidth is the term used to describe the amount of data that can be transmitted along a communication channel. It relates to the range of frequencies that are available on the carrier wave that carries the data. The range in this case is the difference between the upper and lower frequencies. As the range of frequencies increases so does the amount of data that can be transmitted within the same time frame. We have already touched on the relative speed at which data can be transmitted. Speed is a vital factor in communications.

Bandwidth is measured in hertz (Hz) and megahertz (mHz). Network cabling has a bandwidth of up to 500 mHz meaning there are 500 million cycles per second. As the number of cycles increases, more data can be carried.

In common with other aspects of computing, bandwidth increases over time and therefore more data can be transmitted more quickly as each year passes.

Bit rate

KEYWORD

Bit rate: the rate at which data is actually being transmitted. Measured in bits per second.

Bit rate is the term used to describe the speed at which a particular transmission is taking place. It is closely linked to the bandwidth because the bit rate will be limited by how much bandwidth is available.

Bit rate is measured in bits per second (bps). Bandwidth represents the frequencies and therefore the capacity that is available and bit rate represents the actual speed of transfer. It is important to note that bandwidth and bit rate are not the same thing. Bandwidth is the range of frequencies that can be transmitted and bit rate is the number of bits that can be transmitted per unit of time. The bit rate that can be achieved is directly proportional to bandwidth.

Baud rate

Baud rate is another term used to describe the speed at which data can be transmitted. One baud represents one electronic state change per second. An electronic state change may be a change in frequency of the carrier wave, a change in voltage, a change in amplitude or a shift of a waveform. Traditionally, one bit is sent on each state change so one baud roughly equates to one bit per second. However, it is possible to send more than one bit per state change by using different voltage levels to represent the bits. In this case rather than sending bits, you are sending 'symbols', which may have any number of bits in them. The baud rate is determined by the transmission medium.

As we have seen data is transmitted on carrier waves. If 400 bits of data were transmitted at a bit-rate of 400 bps then it would take 1 second. If 4 bits were encoded into each symbol the data would be transferred in a quarter of the time; the baud rate would be 100 baud. Whether or not this could be achieved would depend upon the transmission medium. Figure 37.2 shows how 4 bits are encoded and transmitted at each electronic stage.

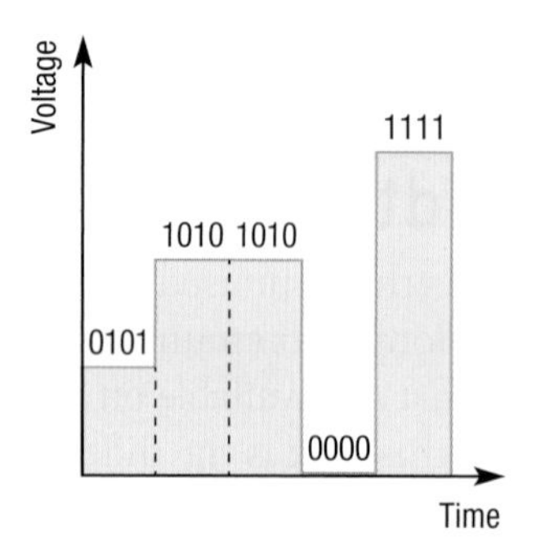

Figure 37.2 Bit rate and baud rate

This is slightly confusing but reflects the fact that the baud was originally developed in the late 1800s for use on telegraph machines and therefore predates the widespread use of computers and networks. It is more common now to find speeds quoted in terms of bits per second as this is an easier way of understanding the measurement of transmission rates.

Latency

KEYWORD

Latency: the time delay that occurs when transmitting data between devices.

Latency is the general term used to explain the time delay that occurs when any component within a computer system is responding to an instruction. This is because the instruction is being transmitted down cables, through buses and logic gates, all of which takes time. Therefore, latency can occur at any stage of the transmission process. These delays could be so short as to be unnoticeable.

For example, when you press a key on your keyboard, there is a latency of fractions of a microsecond as the instruction is transmitted down the cable, around the buses and registers in the processor and along another cable to the screen. When using the Internet, the latency may be more noticeable as the number of connections and components in the process is greater.

There are three general causes of latency when communicating data:

- Propagation latency: The amount of time it takes for a logic gate within a circuit to transmit the data.
- Transmission latency: The amount of time it takes to pass through a particular communication medium, for example, fibre optic would have a lower latency than copper cable.
- Processing latency: The amount of time it takes data to pass around a network depending on how many servers or devices it has to pass through.

You may be familiar with the concept of a ping test, which is often used to measure the speed of an Internet connection. It works by sending a data item to another point on the network and calculating how long it takes to come back again. Even if you have a notional speed of 8 Mbps, you may only be getting 4 Mbps and this may be due in part to latency.

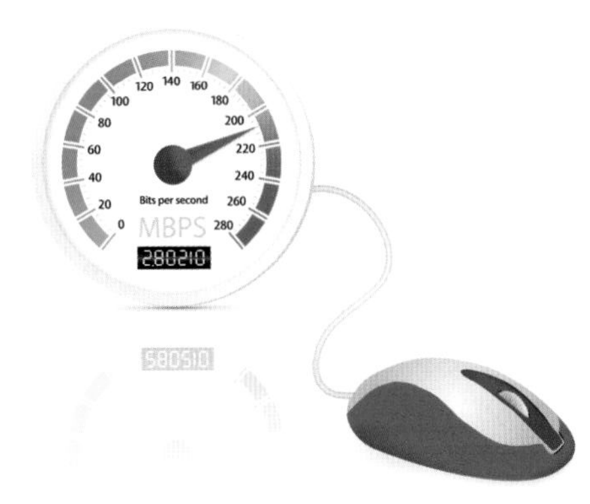

Figure 37.3 A ping test

Synchronous and asynchronous data transmission

Synchronous means 'occurring at the same time' or 'having the same speed'. In the context of transmissions this means that the two devices which are communicating will synchronise their transmission signals. Using the system clock, the computer sending the data will control the transmission rate to be in time with the device or computer receiving the signal. If the two devices are not synchronised then data could be lost during transmission. Once they are synchronised the two devices can send and receive data without need for any further information.

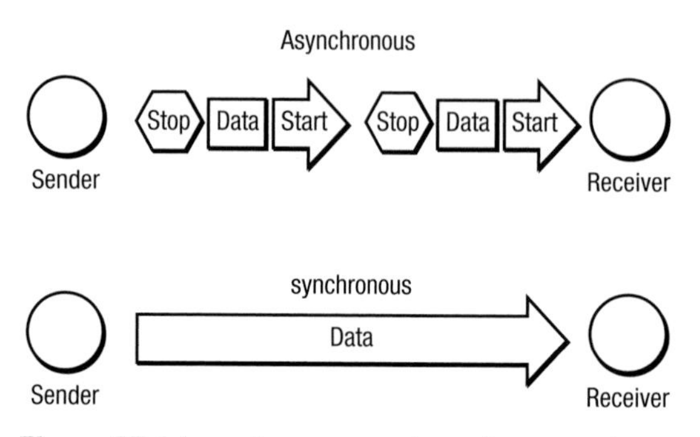

Figure 37.4 Asynchronous and synchronous data

Asynchronous transmission does not require the permanent synchronisation of the sender's and receiver's system clocks. Instead, it synchronises only for the duration of the transmission by sending additional bits of information called start and stop bits.

For example, to send a character may require an 8-bit code to be transmitted. In addition to the eight bits, **asynchronous data transmission** requires at least two other bits. At the start of the eight bits there is a **start bit** and at the end, a **stop bit**. The character may also include a **parity bit** as described in Chapter 30.

The process works as follows:

- The start bit causes the receiver to synchronise its clock to the same rate as the sender. This means that timing of the transmission and receipt are the same on both devices, that is, they are in step.
- Both devices must already have agreed on how many bits of data will follow (commonly 7 or 8 bits), whether a parity bit is being used, what type of parity it is, and how many stop bits there will be.
- The stop bit (or bits) indicate that the data has arrived so the processor on the receiver's device can now handle those bits, for example, by copying them into memory. The stop bit is also important as it allows the receiver to identify when the next start bit arrives, as the stop and start bits always have different values.
- If there is more data then another start bit will be sent and the cycle will continue.
- The sender's device sends data as soon as it is available rather than waiting for the clock pulse or a synchronisation signal from the receiving device.

KEYWORDS

Asynchronous data transmission: data is transmitted between two devices that do not share a common clock signal.

Start bit: a bit used to indicate the start of a unit of data in asynchronous data transmission

Stop bit: a bit used to indicate the end of a unit of data in asynchronous data transmission

Parity bit: a method of checking binary codes by counting the number of 0a and 1s in the code.

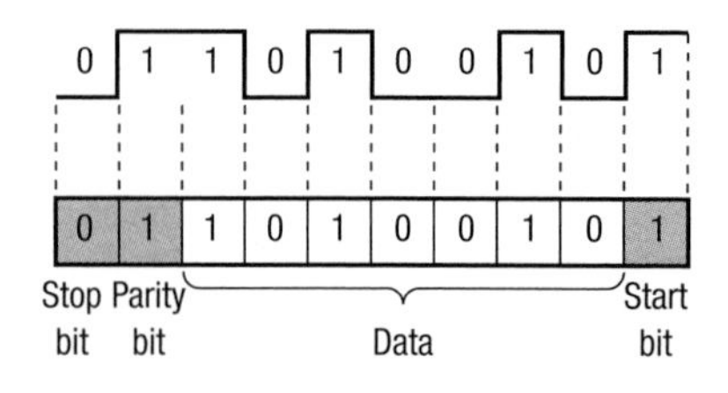

Figure 37.5 Start, stop and parity bits

> **KEYWORD**
>
> **Synchronous data transmission**: data is transmitted where the pulse of the clock of the sending and receiving device are in time with each other. The devices may share a common clock.

To send a character requires an 11-bit code of which only eight bits are the actual data. This is necessary so that the receiving device knows where each byte of data starts and stops. However, it does reduce the amount of actual data that can be transmitted in a given time frame. Transfer rates are measured in bits per second. In this case we are using three additional bits for every character we send. With **synchronous data transmission**, we could use all 11 bits for data.

Protocols

One of the biggest problems with computer communications is getting the various computers, networks, and peripheral devices to talk to each other. If we just take the example of accessing a website, there are a number of different transmissions that take place in order for this to happen: the mouse has to transmit data to the serial port, which in turn passes a signal to the processor; the processor transmits to the router which transmits over the telephone system, probably via a satellite to your ISP and so on.

There are so many different manufacturers of hardware and so many different ways of transmitting data that it is essential that there are accepted standards for data transmission. It might help to think of this in terms of the way that people from around the world communicate with each other. We all have our own languages and customs. When we deal with each other, we agree on what the rules of communication are going to be. For example, we may all agree to speak in the same language.

Protocols are a method for ensuring that different computers can communicate with each other.

> **KEYWORDS**
>
> **Protocols**: sets of rules.
>
> **TCP/IP**: a set of protocols (set of rules) for all TCP/IP network transmissions.
>
> **HTTP (Hypertext transfer protocol)**: the protocol (set of rules) to define the identification, request and transfer of multimedia content over the Internet.
>
> **FTP**: a protocol (set of rules) for handling file uploads and downloads.

A protocol is a set of rules. In the context of communications, there are a number of rules that have been established in relation to the transmission of data. **Protocols** cover aspects such as the format in which data should be transmitted and how items of data are identified.

From using the Internet, you will probably already have come across four common protocols:

- TCP/IP: Transmission Control Protocol/Internet Protocol is actually two protocols that are usually referred to as one and relate to the set of rules that govern the transmission of data around the Internet. Data sent around the Internet are split into packets. **TCP/IP** handles the routing and re-assembly of these data packets.
- HTTP: Hypertext Transfer Protocol. You may have seen http preceding Internet addresses, for example, http://www.aqa.org.uk, though you do not need to type it in yourself these days. **HTTP** is the set of rules governing the exchange of the different types of file that make up displayable web pages.
- FTP: File Transfer Protocol is similar to HTTP in that it provides the rules for the transfer of files on the Internet. **FTP** is commonly used when downloading program files or when you create a web page and upload to the ISP's server.

Practice questions can be found at the end of the section on pages 360 and 361.

KEY POINTS

- Serial transmission sends data one bit at a time.
- Parallel transmission sends data several bits at a time.
- Bandwidth is a measure of the physical capacity of a communication channel.
- Bit rate and baud rate are methods for quantifying how much data can actually be transmitted within a certain time frame.
- Latency is the delay that occurs when data is being transmitted.
- Synchronous data transmission takes place between devices that have synchronised their clocks.
- Asynchronous transmission does not require devices to be synchronised, instead it sends extra bits of data (start and stop bits).
- Protocols are the rules that define how transmission will take place.

TASKS

1. Identify two scenarios where a computer might use:
   a) serial connections
   b) parallel connections.
2. What are the advantages of USB connections over traditional serial connections?
3. Explain the difference between synchronous and asynchronous data transmission.
4. Explain the relationship between bit rate, baud rate and bandwidth.
5. Why are protocols needed in computing?
6. What is latency and what are the main causes of it?

STUDY / RESEARCH TASKS

1. Explain why you rarely achieve the connection speed that is theoretically possible from your Internet connection.
2. What are the highest bit rates that are theoretically possible between two devices using
   a) serial communications?
   b) parallel communications?
3. Which country has the highest average Internet speeds? Where does the UK come in the world rankings and why?

38 Networks

SPECIFICATION COVERAGE

3.9.2 Networking

LEARNING OBJECTIVES

In this chapter you will learn:

- what components are needed to construct a network, including servers, clients, routers, switches and network cards
- how networks can be constructed using different topologies
- the benefits of using different topologies
- how networks can be client–server or peer-to-peer
- how wireless networks work
- how data is transmitted in frames and packets
- how data being transmitted used protocols to prevent collisions of data and keep data secure.

INTRODUCTION

A network is any number of computers connected together for communication, sharing processing power, storage capacity and other resources. In its simplest form this could be two or three computers connected in someone's home or in a small office. At the other end of the scale, there are large global networks and, of course, the Internet, which is a global network of networks.

Connections between the computers are made using either various types of cables or wirelessly using radio signals as a means of connection. Each device or computer within a network is often referred to as a node.

Network basics

In order to connect to a **network**, a computer must have a network adapter, more commonly known as a **Network Interface Card (NIC)**. The NIC is a printed circuit board which is contained inside the computer like any other card (graphics and sound cards, for example). The NIC will be specifically designed to allow the computer to connect either via cable or wirelessly to the particular network topology being used. The type of card also dictates the speed of data transmissions that will be available between this device and the network.

KEYWORDS

Network: devices that are connected together to share data and resources.

Network adapter / Network Interface Card (NIC): a card that enables devices to connect to a network.

KEYWORDS

Network topology: the layout of a network usually in terms of its conceptual layout rather than physical layout.

Local Area Network (LAN): a network over a small geographical distance – usually on one site and typically used by one organisation.

Wide Area Network (WAN): a network spread over a large geographical distance.

Networks are usually described in terms of the geographical area that they cover and the way in which the connections are configured, known as **network topology**.

A **Local Area Network (LAN)** is a number of computers and peripherals connected over a small geographical distance, covering one building or site. LANs are common in businesses, educational establishments, hospitals and even the home. Most LANs are made up of one or more servers and clients. A server is a high specification computer with sufficient processing power and storage capacity to service a number of users. A client is any computer attached to the network.

A **Wide Area Network (WAN)** is a number of computers and peripherals connected together over a large geographical distance. This could mean any network that extends beyond a single site right up to global networks such as the Internet. WANs make use of a wider variety of communication media including telephone wires, microwave links, satellite connections and fibre optic cables.

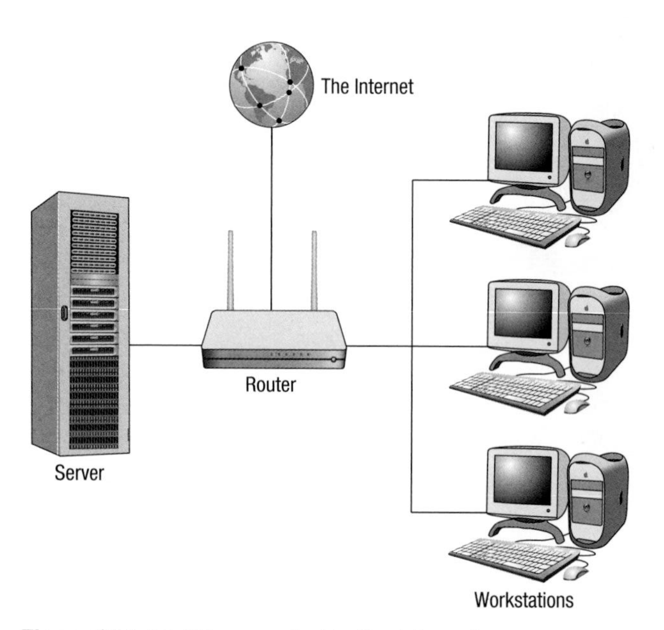

Figure 38.1 A LAN connected to the Internet

In addition to the server and client the other critical device within a network is a router. Modern routers are actually a number of devices merged together into a single device. The typical router for a home network:

- receives every packet of data being transmitted, reads the header of the packet and then forwards it to its destination
- acts as a firewall, preventing certain packets from being forwarded
- acts as an switch, creating a connection between two devices on a network
- provides a wireless access point transmitting a WiFi signal
- acts as a modem to convert digital signals to analogue so that they can be transmitted down standard telephone cables.

KEYWORD

Star topology: a way of connecting devices in a network where each workstation has a dedicated cable to a central computer or switch.

Star topology

There are two main topologies (layouts) for networks in use today – the star and the bus. A **star topology** takes its name from the simplified way in which it can be represented on paper as shown below. This shows the way that the devices are conceptually connected together.

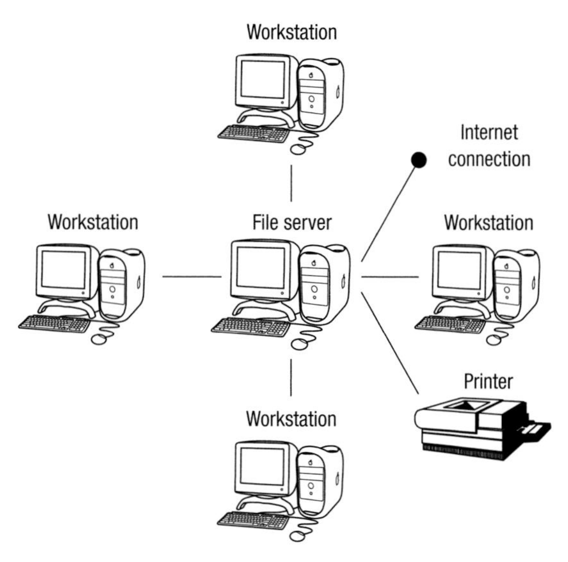

Figure 38.2 A star network

Figure 38.2 shows each client connected to a central server via an individual connection. The main feature is the dedicated connection between server and client. In reality it will be a switch in the centre with the server attached to one of the ports. The server will be a high specification machine with a large amount of processing power and storage capacity. The clients have access to the server through the cabling.

Software may be stored centrally on the server and can be installed, upgraded and maintained via the server. The server will also contain an operating system that controls the users' access to the system and also includes various administration functions such as managing the print queue.

The software can also be held locally at the client. If it is held locally the start-up time is low, but it is harder to maintain and upgrade. Holding the programs centrally means the administrator has much better control over the software and who has access to it.

A real star network may not look anything like the diagram as it is unlikely that the clients will just happen to make a star shape. Also, there will be additional hardware devices such as switches and routers between the server and the clients.

However, this is what topology is – it shows the *conceptual* or *logical* layout rather than the *actual* or *physical* layout.

Table 38.1 Advantages and disadvantages of the star topology

Advantages of star topology	Disadvantages of star topology
Fast connection speed as each client has a dedicated cable.	Expensive to set up due to increased cabling costs.
Will not slow down as much as other network topologies when many users are online.	If the cable fails then that client may not be able to receive data.
Fault-finding is simpler as individual faults are easier to trace.	Difficult to install as multiple cables are needed. The problem is exaggerated where the LAN is split across a number of buildings.
Relatively secure as the connection from client to server is unique.	The server can get congested as all communications must pass though it.
New clients can be added without affecting the other clients.	
If one cable or client fails, then only that client is affected.	

Bus topology

KEYWORD

Bus topology: a network layout that uses one main data cable as a backbone to transmit data.

The other main network topology is the **bus topology**, where all of the nodes within the network are connected via one main cable. If there is a main server, all of the clients connect to it down this main cable. This cable carries data between the server and the clients with each client branching off the main bus cable.

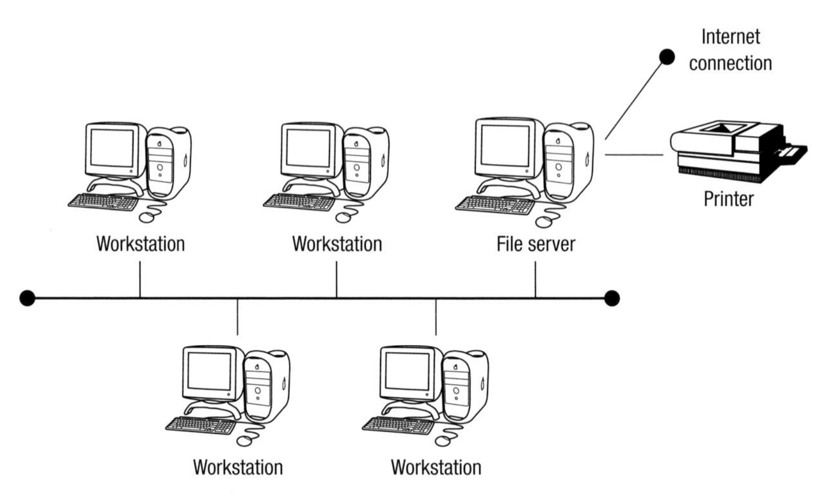

Figure 38.3 A bus network

The main cable or backbone must allow high-speed data transmission as all data must pass down this one channel. A common implementation of the bus system is an Ethernet network system, which has several variants. Ethernet is modelled on the bus network in that there is one central cable which carries all the data around the network. Ethernet cards are installed in each client to allow them to connect using the relevant protocols with a switch directing packets of data to the appropriate client using the address for the designated client.

Table 38.2 Advantages and disadvantages of a bus network

Advantages of bus topology	Disadvantages of bus topology
Cheaper to install than a star topology as only one main cable is required.	Less secure than a star network as all data are transmitted down one main cable.
Easier to install than a star topology.	Transmission times get slower when more users are on the network.
Easy to add new clients by branching them off the main cable.	If the main cable fails, then all clients are affected.
	Less reliable than a star network due to reliance on the main cable.
	More difficult to find faults.

KEYWORDS

Physical topology: the way in which devices in a network are physically connected.

Logical topology: the conceptual way in data is transmitted around a network (see Physical topology).

There is a distinction to be made between the **physical topology** and **logical topology** of a network. Physical topology refers to the actual connection of cables. However, it is possible for networks that are connected in a particular physical topology to act in a different way with the addition of more hardware and software. For example, some Ethernet networks were physically laid out as a star, but used hubs to repeat signals effectively creating a bus network.

Client–server networks

In the star and bus topologies, the diagram shows a main server. Although the clients have local resources in terms of processing power, and storage capacity, they are controlled by the server. This means that when new software is installed, for example, it can be installed on the server and then distributed to the clients.

There are a wide range of services that the client may request including:

- access to a printer
- providing a secure connection to the Internet
- access to email
- access to applications
- access to files.

This is the most common way of constructing a LAN with a large number of users. The server will be a high-end computer with a large amount of processing power and storage capacity. It needs to be big enough and fast enough to cope with the demands placed upon it by the clients.

KEYWORD

Client–server: a network methodology where one computer has the main processing power and storage and the other computers act as clients requesting services from the server.

The clients on the other hand do not need to be of such a high specification. The current trend is to have a thin client which refers to the fact that the client will not have a CD drive or expansion slots, thus reducing the cost of the client.

A-level students need to have a more detailed understanding of **client–server** networks and this is covered in Chapter 42.

Peer-to-peer networks

> **KEYWORD**
>
> **Peer-to-peer**: a network methodology where all devices in a network share resources between them rather than having a server.

In a **peer-to-peer** network, no one computer is in overall control of the network. Instead the resources of each computer or workstation are available to all the computers in the network. Each workstation therefore can act either as a client or as the server, depending on the current task.

This is more common among smaller networks or for certain applications such as file-sharing. Peer-to-peer networks can be created without the need for a special network operating system. With the growth in home computing, it is increasingly common to find peer-to-peer networks in private houses. These are often set up to allow every computer in the home to share a connection to the Internet or printer.

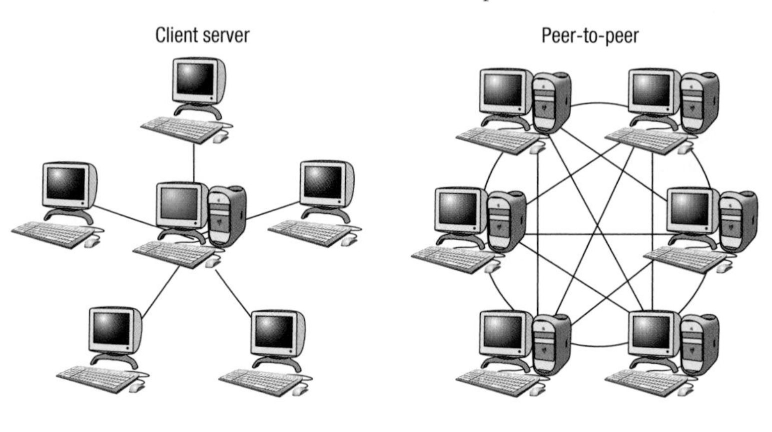

Figure 38.4 Client-server and peer-to-peer networks

Wireless networks

> **KEYWORDS**
>
> **Wireless Wide Area Network (WWAN)**: a WAN that does not use cables, but sends data using radio waves.
>
> **Media Access Control (MAC) address**: a unique code that identifies a particular device on a network.
>
> **WiFi**: a standard method for connecting devices wirelessly to a network and to the Internet.
>
> **Wireless Local Area Network (WLAN)**: a LAN that does not use cables but connects using radio waves.

A wireless network varies from a wired network in that it does not use cables to make the physical connections between devices. Instead, the data is sent using radio waves. Wireless networks can be implemented over small or large geographical distances so it is possible to have wireless LANs (WLAN) and **wireless WANs (WWAN)**. Many business and home networks are set up wirelessly doing away with the need for costly cabling and enabling easy access to the network from any device with a wireless network adapter (NIC).

All devices that are on a network have what is called a **Media Access Control (MAC)** address. This is a unique identifier encoded into the network interface card (NIC) in the format of six groups of two hex digits separated by colons, e.g. 02:32:45:77:89:ab. Any devices that connect to a network using WiFi will connect through a wireless access point and must have its own unique MAC address. Every NIC ever manufactured has a unique address meaning that they can be used to identify every device uniquely. The first half of the MAC address is the manufacturer code and the second half is the unique device code allocated by that manufacturer.

WiFi is the generic term for a **Wireless Local Area Network (WLAN)** where devices can connect wirelessly to each other and where a connection can be made to the Internet providing one of the devices in the network is online. WiFi operates to a generic standard called IEEE 802.11, ensuring that all devices are compliant and can connect and transmit data around the network.

Different communication devices are needed to create a wireless network, depending on the geographical coverage. For example, a WLAN in a home of office may have a wireless router that transmits a WiFi signal that is accessible within a few metres of the device. (802.11n is about 70 m indoors and 250 m outdoors. 802.11g is a little lower.) WiFi hotspots are set up by telecommunication companies and also use wireless routers to allow access over a larger distance of around 250 metres. These may be slower because the signal degrades with distance and there will be more users sharing the same signal and there may be interference.

WWANs usually make use of mobile phone networks, which in turn use satellites and transmitters and receivers located on towers around the country. These are capable of transmitting signals over long distances using set frequencies.

Figure 38.5 A wireless network connecting to the Internet

Protocols

> **KEYWORD**
> **Protocols**: sets of rules.

There are sets of standards or **protocols** for wireless communications and WiFi to ensure that all devices are able to connect with each other and transmit and receive data. A protocol called Carrier Sense Multiple Access with Collision Avoidance (CSMA/CA) was developed to enable the various devices to transmit data at high speeds without interfering with each other.

When data are sent around networks, they are sent in frames with all the frames being re-assembled at the receiving end. Any device on a wireless network may attempt to send frames. These data frames can be picked up by any nodes or devices within range. Before each frame is sent, the device uses the CSMA/CA protocol to see whether the transmission medium is idle or whether another device is using it. If the transmission media is idle, the data are sent. If it is busy, the device will wait and try again later. Each device will then wait a random amount of time before checking to see if the medium is free again so that it can send the data. This is known as a back-off mechanism and is random to reduce the chances of both devices trying to send simultaneously again.

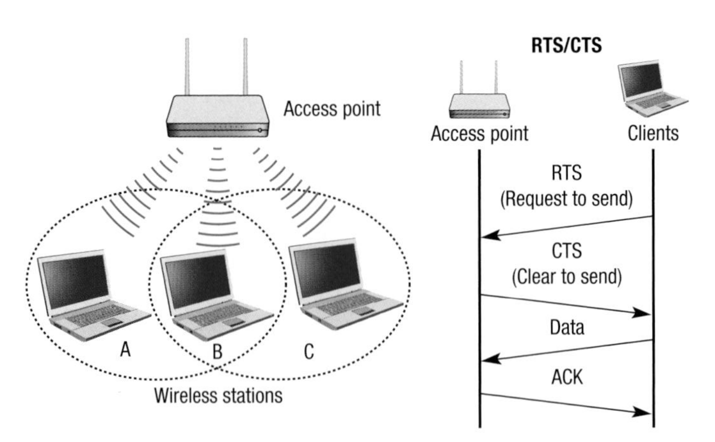

Figure 38.6 The CSMA/CA protocol using RTS/CTS

If the transmission medium is free then the data can be sent. On receipt of the data, an acknowledgment is sent back to the sending device to confirm that the data have been received and not corrupted. If this is not received, again it will wait a random amount of time before resending.

An optional extension to the protocol is a system called **Request to Send/ Clear to Send (RTS/CTS)**, which works between the nodes on a network. The RTS sends a message to the receiving node or access point and if a CTS message is received, it knows that the node is idle and that the data frame can be sent. If no CTS message is received, it will wait and send another RTS later.

> **KEYWORDS**
>
> **Request to Send / Clear to Send (RTS/CTS)**: a protocol to ensure data does not collide when being transmitted on wireless networks.
>
> **Service Set Identifier (SSID)**: a locally unique 32-character code that identifies a device on a wireless network.

SSID

One of the issues when using wireless networks is ensuring that the various devices are connecting to the correct WLAN. As all of the data are being sent through radio waves rather than cables, each device needs a way of ensuring that it is connecting to the correct network. The standard method of doing this is using a **Service Set Identifier (SSID)**, which is a 32-character code put into the header of each packet of data being sent.

Each code is locally unique to the particular WLAN that is being used and therefore acts as an identifier allowing that frame of data to be transmitted around the WLAN. The network interface card must also be programmed with the same 32-character code so that the device can connect to the WLAN in the first place.

Network security

Another issue with wireless technology is that it can be less secure than a wired system. This is because the signals travel through the air and are therefore easier to intercept than signals passing through wires. A potential hacker can tap into the radio signal being sent using receiving devices and read the data signals. There are a number of steps that can be taken to increase the level of security on a wireless network:

- Change the SSID from the default value and hide it from transmission.
- Ensure that all devices are **WiFi Protected Access (WPA/WPA2)** compliant.
- Use strong encryption (WPA/WPA2) (see Chapter 40).
- Create a 'white list' of MAC addresses from devices that you know to be trustworthy.

> **KEYWORD**
>
> **WiFi Protected Access (WPA/WPA2)**: a protocol for encrypting data and ensuring security on WiFi networks.

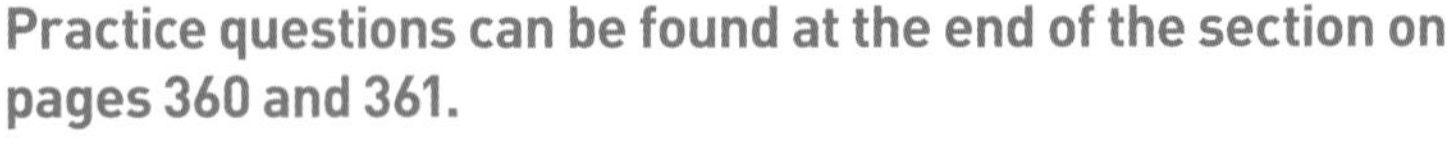

Practice questions can be found at the end of the section on pages 360 and 361.

KEY POINTS

- A network is a collection of connected devices.
- Networks are either Local Area (LAN) or Wide Area (WAN) and may be wired or wireless.
- Networks can be constructed in a star or bus topology.
- Client–server networks have a server providing resources and services to clients.
- Peer-to-peer networks share resources between a number of workstations.
- All devices need a network interface card (NIC) and media access control (MAC) address in order to connect to a network.
- Various protocols and standards exist for wireless networks including the 802.11 WiFi standard and CSMA/CA collision protocol.
- There are particular issues with wireless networks in ensuring that wireless devices connect securely to the correct network.

TASKS

1 Draw a diagram to show the following LAN topologies:
   a) bus
   b) star.

2 Give three reasons why an organisation may choose to install a star topology rather than a bus topology.

3 Give three reasons why an organisation may choose to install a bus topology rather than a star topology.

4 Define the following terms:
   a) local area network
   b) wide area network.

5 The Internet is often described as a 'network of networks'. It is also referred to as a Wide Area Network. Explain each of these terms and discuss which is the most accurate definition.

6 Explain how the CSMA/CA protocol is used to manage frames of data being transmitted around wireless networks.

7 Give one example of an application that requires a client–server network.

8 Give an example where a peer-to-peer network might be used.

9 Why are protocols so important in networks? Give examples.

10 What is the difference between a white list and a black list?

STUDY / RESEARCH TASKS

1 'No more wires.' How likely is it that wireless networks will completely replace networks with cables?

2 Identify three advantages and three disadvantages of wireless networking.

3 Networking is the connection of computers and other devices. Explain how a network may incorporate:
   a) mobile phones
   b) tablets.

4 What are the advantages of integrating these devices into a network?

5 How likely is it that the phone landline will become a thing of the past?

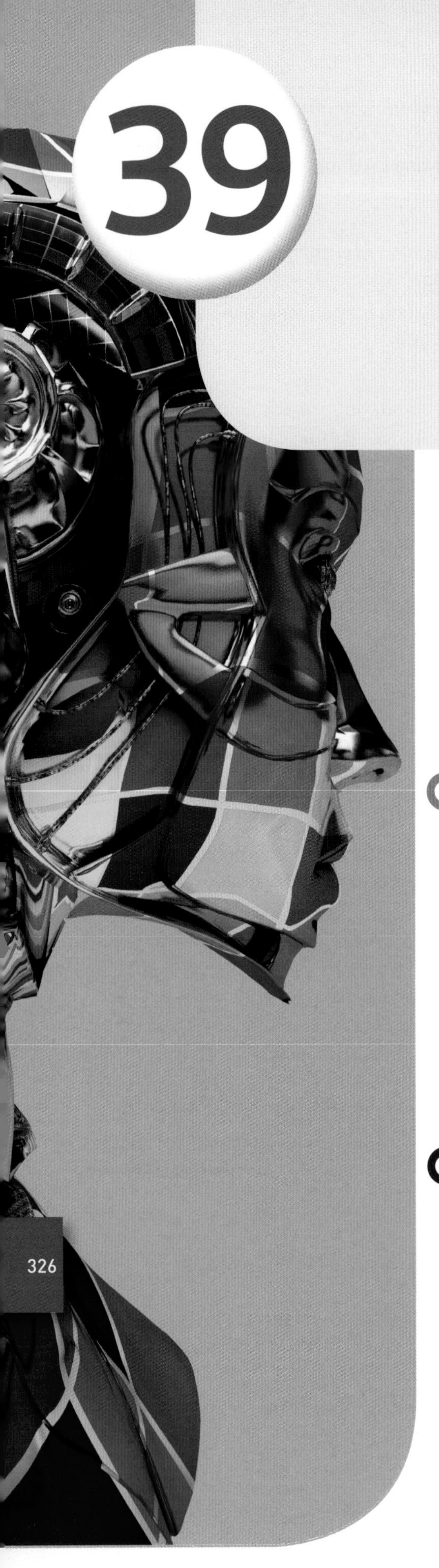

39 The Internet

A level only

SPECIFICATION COVERAGE

3.5.5.3 Error checking and correction

3.9.3.1 The Internet and how it works

3.9.4.3 IP address structure

3.9.4.4 Subnet masking

3.9.4.5 IP standards

3.9.4.6 Public and private IP addresses

3.9.4.8 Network address translation (NAT)

4.9.4.9 Port forwarding

LEARNING OBJECTIVES

In this chapter you will learn:

- how the Internet started and how it is different from the WWW
- how resources are registered, identified and located on the Internet
- how IP addresses and domain names are used
- that data is sent in packets and is routed around the network
- that different parts of a network and individual devices can be identified using IP addresses
- what ports and sockets are and how they are used.

INTRODUCTION

The Internet is described as a network of networks. It is a global interconnection of computers and networks. The Internet has had an enormous impact on society and is changing the way we communicate, work, socialise, shop and bank. It has grown exponentially over the last 10 years and now has an estimated 2.5 billion users worldwide, which is around a third of the world's population. This chapter examines the way that the Internet is structured and the basics of how data is transmitted around it.

The Internet and the World Wide Web

KEYWORD

Internet: a global network of networks.

The **Internet** started life as ARPANET in the late 1960s. ARPANET was a collection of connected computers set up by the American military as a secure way of transferring sensitive data during the Cold War with Russia. During the 1980s the network expanded and was used by a much wider community including universities and research centres.

The Internet as we know it now started to take shape in the mid-1980s when Tim Berners-Lee, a British scientist working in Switzerland, created the World Wide Web (WWW). Berners-Lee had been using the Internet to transmit and receive research documents but had found the interface very clumsy. As a result he developed the concept of an organised browser to allow people to navigate and search the Internet more easily. As a consequence, the WWW is perceived as being the same as the Internet. In fact, the WWW is a service provided on the Internet, albeit one used by millions of people. It is possible to use the Internet without using the WWW.

After this, many other organisations began to use the WWW to offer services to users. During the 1990s there was an explosion in the range of services on offer from Internet Service Providers, from search engines to email. This coincided with a massive increase in the number of people who were buying PCs for home use. Manufacturers of home PCs began to supply computers with pre-installed browsers and modems ready to be connected to the Internet.

Uniform Resource Locator (URL)

KEYWORD

Uniform resource locator (URL): a method for identifying the location of resources (e.g. websites) on the Internet.

A **URL** is the full address used to find files on the Internet. For example:

http://www.awebsite.co.uk/index.html

The contents of the file that a URL locates will vary depending on the Internet protocol being used. In this example, hypertext transfer protocol (HTTP) is being used. The file it points to is an html file called index.html which contains hyperlinks to further pages. HTTP indicates that the file can be accessed using a browser. Consequently, most URLs start with HTTP although it is not always necessary to type it in the address line.

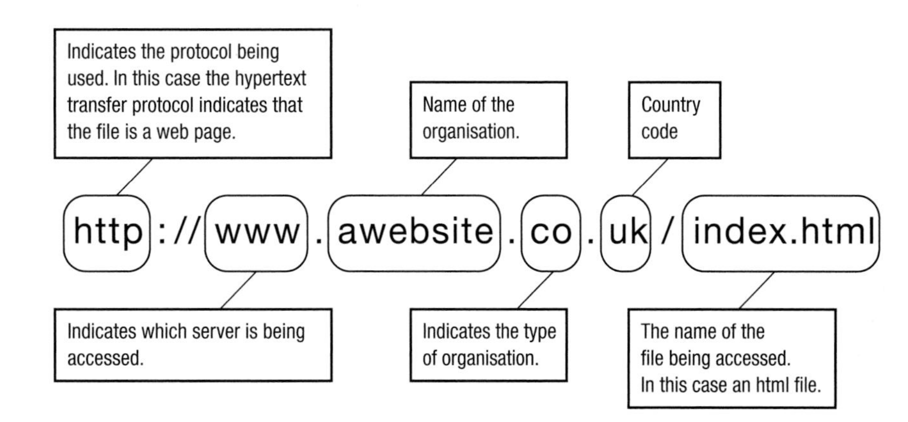

Figure 39.1 Uniform resource locator

KEYWORD

Domain name: the recognisable name of a domain on the Internet.

The address is made up of several parts:

- the protocol being used, which could be http, https or ftp
- the **domain name**, which is the location of the resource on the Internet
- the filename to locate the specific file needed. If the file is located within a subdirectory on the server then a pathname, including the directory name and filename, will be given instead of a simple filename.

Domain name

The domain name identifies organisations or groups on the Internet. For example:

bbc.co.uk

- bbc is the name of the organisation. Domain names have to be unique so organisations had to act quickly to secure a domain name that is the same as the name of their organisation. All domain names are registered with a central agency called ICANN to ensure uniqueness.
- .co indicates that it is a company. This part of the name is referred to as the top-level domain name. It indicates the type of organisation. There are some commonly used top-level domains which you should be aware of.
  - .com indicates that the organisation is commercial, that is, a business
  - .gov indicates that the organisation is part of the government
  - .ac indicates that the organisation is an academic institution, usually a college or university
  - .sch indicates that the organisation is a school
  - .org indicates an organisation other than a commercial business, for example, a charity or trade union
  - .net indicates a company providing Internet services.
- .uk indicates that the website is registered in the UK. There are a number of two-letter country codes and new ones are being added all the time. Country codes are abbreviated using their own language as in these examples:
  - .au is Australia
  - .de is Germany (Deutschland)
  - .it is Italy (Italia)
  - .es is Spain (Espania).
- Notice in the example above that it was not necessary to type in www before the domain name. www indicates the host server for the resource. Often the www does not need to be typed as most commonly used websites are accessed via www. Where the www is typed the domain name is known as a fully qualified domain name (FQDN) and is completely unambiguous as it can only relate to one host. The domain bbc.co.uk might also contain other hosts with different names, e.g. mail.bbc.co.uk or ftp.bbc.co.uk.

IP address

KEYWORD

Internet Protocol (IP) address: a unique number that identifies devices on a network.

An **Internet Protocol (IP) address** is a dotted quad number that identifies every computer that sends or receives data on a network and on the Internet. This was originally devised as a 32-bit or 4-byte code made up of four decimal numbers separated by dots as follows: 234.233.32.123. As one byte is allocated to each of the four sets of numbers, the range of each is between 0 and 255.

The numbers themselves make little sense to us as users, which is why we use the domain name. Domain names are designed to be easy to remember and relevant to the organisations.

However, the protocol used to transmit data (TCP/IP) can only work with numbers. Therefore, every domain name is mapped to a number. This number is the real Internet address and identifies the computer that is transmitting or receiving data. It is called an IP address because it uses the Internet Protocol.

KEYWORD

Domain name server (DNS): a server that contains domain names and associated IP addresses.

A domain name is sometimes described as a proxy for the IP address which means that the user types in a domain name which is transferred to a **domain name server (DNS)** which then translates the name into the IP address. An analogy could be the ability to store numbers on a mobile phone. The user selects a name from the list which is then looked at to find and dial the actual telephone number.

Some IP addresses are classed as private or non-routable addresses. Typically these are the IP addresses used by devices on a private network, perhaps in a home, school or business. The IP address is needed in order to route data around the network, however it does not need to be made public as that device is not directly connected to the Internet. It is hidden behind a router or firewall. Non-routable addresses only have to be unique within the LAN and therefore do not need to be allocated on a global basis.

When connection to the Internet is required, the device will be connected to a router or proxy server in order to connect. In this case the IP address of that router or server needs to be a public or routable address in which case it becomes a unique address that is registered under the domain name system (see below).

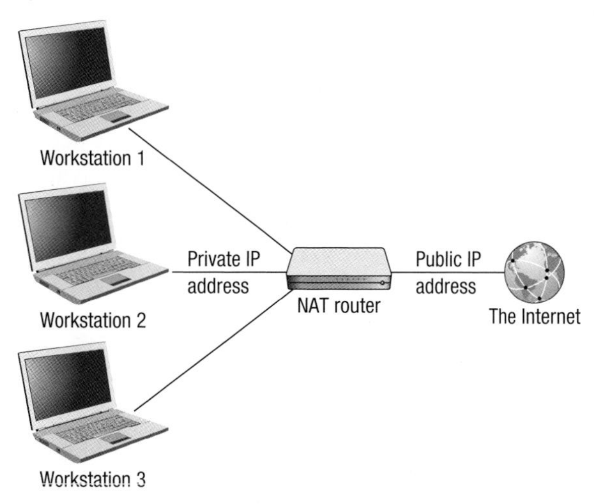

Figure 39.2 Public and private IP addresses

Ports

KEYWORDS

Port: used to identify a particular process or application on a network.

POP3: a protocol (set of rules) for receiving emails.

A **port** is used to identify a particular process or application on a network. The port address is a 16-bit number attached to the IP address. By addressing that port, a process or application will be accessed on the client.

Port addresses are often used to run processes for common networking tasks and many have been assigned port numbers that are in widespread use. For example, port 25 is used for the SMTP application that checks for incoming email on an email server. Port 110 is used for the **POP3** application that fetches email from the email server.

There are around 250 'well-known ports', which are used to launch various processes, many of which are applications related to other protocols, such as FTP, DHCP and SSH, all of which are covered later in the book. Table 39.1 shows some of the common well-known ports, which have been designated as such because they are some of the most widely used networking services.

Table 39.1 Well-known ports and services

Some well-known port numbers	Services
21	FTP
22	SSH
23	Telnet
25	SMTP
53	DNS
80	HTTP
110	POP3
143	IMAP

When a client sends a request to the server using a well-known port, the server needs to respond back to a client port and not the well-known port on the client side. For example, if a server receives a request on port 80, it does not send it back to port 80 on the client. Therefore as part of the client request, a source port must also be sent so that the server knows which port to send it back to.

KEYWORD

Secure shell (SSH) protocol: a protocol (set of rules) for remote access to computers.

The **Secure Shell Protocol (SSH)** can be run using port 22 on the well-known ports list. This protocol is explained in more detail in Chapter 41 and is used to provide remote access to computers. Routing through port 22 means that you have the advantage of the extra security when accessing files using HTTP, downloading or uploading files using FTP, or accessing mail using either SMTP or POP3.

Network Address Translation (NAT)

The system that is used to match up the private IP addresses with the public ones is called Network Address Translation (NAT). This has two main advantages. One is that a unique IP address is not needed for every single device on a network, only on the router or server that is physically connected. This means that only the public IP address needs to be registered with the **domain name server (DNS) system**. The second advantage is that there is an increased level of security as the private IP address is not being broadcast over the Internet, making that device more secure from unauthorised access.

KEYWORD

Domain name server (DNS) system: a system of connected domain name servers that provides the IP address of every website on the Internet.

The router will track connections and maintain a listing of the mappings between private IP addresses and port numbers and the corresponding public address. It does this by adding entries to a translation table which act as a look-up between the internal IP address and the external IP address.

The following is a common way that NAT can work when a workstation on the internal network wants to load data from a server on the Internet:

- The workstation on the internal network sends a packet to the server on the Internet to request some data, including its own internal IP address and port number in the packet so that the server knows where to return the data to.

- The router replaces the internal IP address and port number in the packet with its external IP address and a port number that it generates. This port number will be unique to this communication, within a certain time frame (port numbers are eventually reused or they would run out).
- The router stores the mapping information from the internal IP address and port number to external port number in the translation table.
- Data sent back from the server will be received by the router which will look up the port number in the translation table to identify which machine on the internal network sent the request to the server.
- The router's IP address in the packet will be replaced with the originating workstation's IP address and port number, as read from the translation table.
- The reply packet can then be sent on to the originating workstation.
- Where the translation table does not contain a match to a port number in a packet received from a computer on the Internet, the packet is dropped because it is not a packet sent in response to a request from within the network and may be a hacking attempt.
- The internal IP address and port are never made public on the external network.

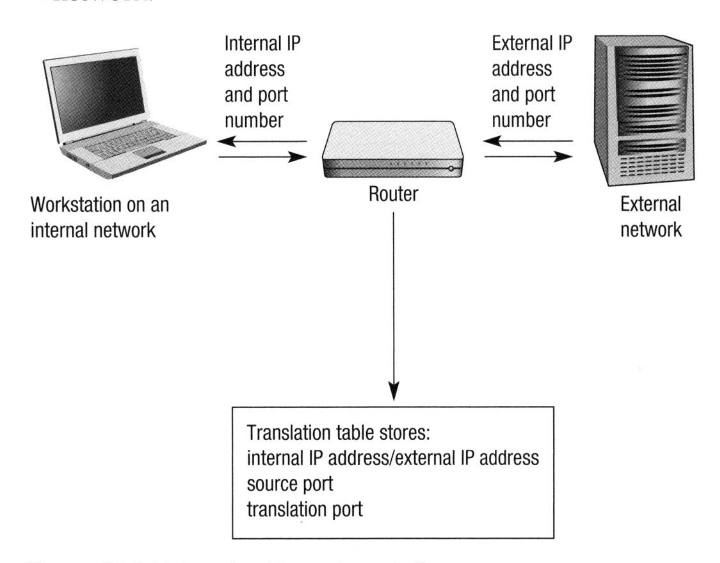

Figure 39.3 Network address translation

Port forwarding

KEYWORD

Port forwarding: a method of routing data through additional ports.

Port forwarding is commonly used when a server inside a private network, with a non-routable IP address, is to be used to provide services to clients on the Internet. As the server has a non-routable IP address, it cannot be accessed directly from the Internet. Therefore the client on the Internet must use the public IP address of the router that connects the private network with the server on it to initiate a connection. This router can be programmed so that requests sent to it on a particular port number are forwarded to a device with a specific IP address within the network. This is port forwarding. So, if the server was a web server with non-routable IP address 192.168.20.4 and the router had public routable IP address 103.12.94.56 then the router could be programmed to forward all requests made from the Internet to port 80 of its IP address to port 80 of the sever with IP address 192.168.20.4.

Sockets

> **KEYWORD**
> **Socket**: an endpoint of a communication flow across a computer network.

A network **socket** is an endpoint of a communication flow across a computer network. Sockets are created in software not hardware. A TCP/IP socket is made up from the combination of an IP address and a port number. When a computer needs to communicate with a server it will send a request to the server using the server's IP address and port number for that type of request (e.g. HTTP is usually port 80).

Sockets can be created at any time to enable a network connection to be established to or from a computer. For example, Client A is being used on a LAN with a local IP address 192.168.233.100 and wishes to request a web page from Server B which has IP address 192.168.233.2. As a web page is being requested, port 80 on the server will be used so Client A will send its request to the socket address 192.168.233.2:80. Server B will be listening for web page requests on port 80.

In the request sent to the server, Client A will include its own IP address and a port number that has been temporarily generated for this communication, such as 50272. This is included so that the server knows to send the data back to socket address 192.168.233.100:50272. The client will listen on this port number for a reply. The transport layer of the TCP/IP stack uses the port number to direct packets to the correct application.

Subnet masking

> **KEYWORD**
> **Subnet masking**: a method of dividing a network into multiple smaller networks.

IP addresses are split into a network identifier and a host identifier. For example, the IP address 120.176.134.32 could be split up with the first part identifying the network and the last part being the actual device (or host) that is being used. So 120.176.134 is the network ID and 32 is the host ID. The network may be a local network, or it could be a remote network and the device could be a computer, printer or router, etc. Network IDs can also be written with zeros in the parts of the IP address that would be used to identify the host, e.g. the network ID 120.176.134 could alternatively be written as 120.176.134.0.

Addresses are split up in this way to make networks easier to manage and to make it more efficient when routing data. For example, in a LAN split across two buildings, the administrator may find it useful to allocate IP addresses according to which building each computer is in. Where a network is separated in this way, each part is known as a subnet or subnetwork. Data sent to a particular computer will only travel around the parts of the network that it needs to, making the network more efficient.

> **KEYWORD**
> **Gateway**: a node on a network that acts as a connection point to another network with different protocols.

When a computer on a network sends data to another computer, it needs to identify whether it is on the same subnet as the other computer. If it is, it can send data to it directly. If it is not, it will send data to the relevant router or **gateway**, which will in turn send the data on to the correct subnet and computer.

A gateway is a node on a network that acts as a connection point to another network with different protocols. For example, in an organisation, a gateway may be used to connect two different company networks together. For a home user, a gateway may be used by their Internet Service Provider (ISP) to provide access to the Internet.

The gateway carries out all of the protocol conversion required to enable the two networks to work together.

To identify whether the destination computer is on the same subnet, the sending computer needs to look at the network portion of the destination IP address to see if it is the same as its own. In the example above, if sender and receiver are on the same subnet, they would both have 120.176.134 as the first part of their IP address.

To do this a subnet mask is used. To understand this you need to remember your binary conversions as the IP address needs to be converted to its binary equivalent so that a bitwise logical AND can be performed. For example, the address 120.176.134.32 in binary is

01111000.10110000.01000110.00100000

Each device on a subnet is programmed with the same subnet mask. Within the subnet mask, a value of 1 is assigned to all the bits that are part of the network ID and a 0 to all of the parts that identify the host. In our case, the first three octets are the network address so the mask would be 11111111.11111111.11111111.00000000.

If the computer with IP address 120.176.134.32 has data to send to the computer with IP address 120.176.134.75 then the following operation will be carried out on the sending computer to check if the two computers are on the same subnet:

- 01111000.10110000.10000110.00100000 Full IP address of sending computer
- 11111111.11111111.11111111.00000000 AND Subnet mask
- 01110000.10110000.10000110.00000000 = Network ID of sending computer
- 01111000.10110000.10000110.01001011 Full IP address of destination computer
- 11111111.11111111.11111111.00000000 AND Subnet mask
- 01110000.10110000.10000110.00000000 = Network ID of destination computer

As the sending computer and destination computer both have the same network ID, the data can be sent directly from the sending computer to the destination. Otherwise, the data would be sent to a router that could forward the transmission to the subnet that the destination computer is on.

IP address v4 and v6

It soon became apparent that as the Internet was growing at such a rate, the original 32-bit code was not going to provide enough permutations for the number of devices that would be present on the Internet. Consequently a new system, known as IPv6 was created, which uses 128 bits represented as eight groups of four hex numbers separated by colons as follows:

13E7:0000:0000:0000:51D5:9CC8:C0A8:6420

This massively increases the range of numbers available as there are more digits in the number, and hex is being used rather than decimal allowing for a greater range within each group of numbers.

The v6 IP addresses are slowly replacing the original v4 format, although both systems are still in use at the time of writing. All of the concepts relating to IP addresses are the same regardless of the format. For example, subnetting can still be used on the v6 format.

There is more in TCP/IP in Chapter 41.

Dynamic Host Configuration Protocol (DHCP)

IP addresses are defined as either static or dynamic. Static IP addresses are ones that are assigned and then never change. Dynamic IP addresses are allocated every time a device connects to a network and this is perhaps the most common approach. The allocation is done automatically by an application as you log on. For home users, this is typically assigned by your ISP.

In simple terms, the application looks for an available IP address from its pool of addresses and allocates it to your device. Where there are hundreds of users logging on and off a network all the time, this is a very efficient system as it means the administrator does not have to do it manually.

> **KEYWORD**
>
> **Dynamic Host Configuration Protocol (DHCP)**: a set of rules for allocating locally unique IP addresses to devices as they connect to a network.

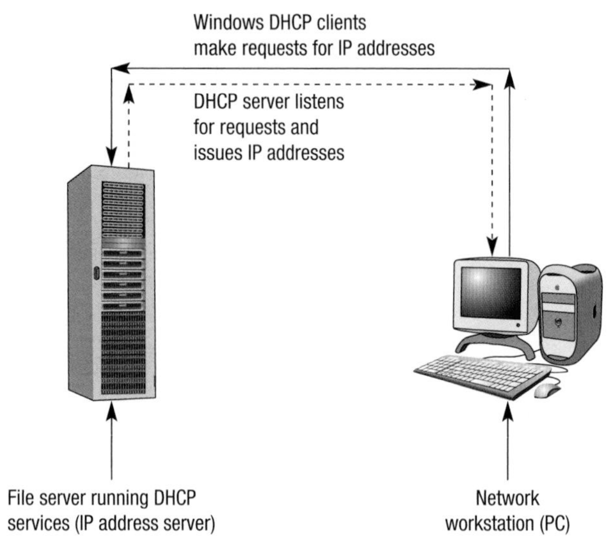

Figure 39.4 The DCHP server

A dedicated DHCP server is used on the network and handles the requests by managing a pool of available IP addresses, usually within a defined range of numbers depending on how the network is physically configured. In simple terms a user attempting to log on is making a request and the server will then offer that device a particular IP address, which may be the last used address for that device, or the next available address within the pool. When the user logs off, the reverse process takes place, freeing up the IP address for the next user.

Domain name server (DNS) system

Some estimates suggest that there are as many as a billion websites and the number is growing all the time, as is the number of users. This presents a complex problem in terms of allocating unique domain names to all of these sites. This job is carried out by large international organisations and they use the domain name server system.

Once allocated, the domain names are mapped to a unique IP address and this information is stored in databases on large servers called domain name servers (DNS). Humans use domain names as they are easier to remember than IP addresses. It is the DNS that maps the domain names to the IP addresses.

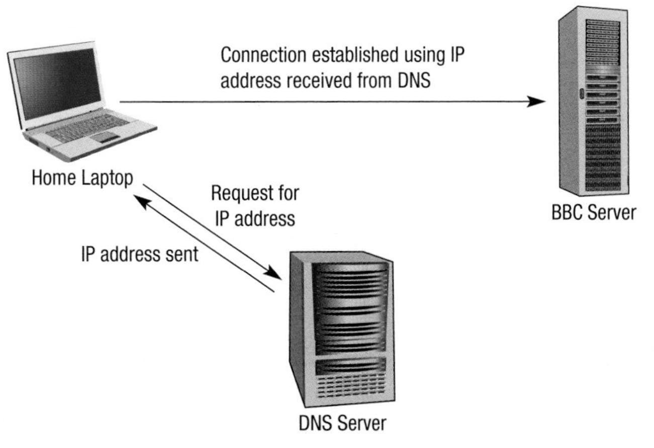

Figure 39.5 Process for connecting to the DNS server

Figure 39.5 shows how this works. If a user on their laptop wants to connect to the BBC website, a request is sent to the DNS to establish the IP address. The DNS looks in its database and sends the IP address to the laptop. A connection can then be established between the laptop and the BBC server.

As there are millions of addresses to be stored, all of this information cannot be stored on a single server. Consequently there are hundreds of DNSs in use around the world, all of which are connected to each other. Where that particular DNS does not store the IP address, it will send a request to other DNSs until the relevant information is found. The entire DNS system therefore has to be carefully organised and controlled across the whole world to ensure that no two domain names or IP addresses are duplicated and to make sure that these addresses are available whenever they are requested. In simple terms, you could compare it to a massive directory enquiries system like we have for telephone numbers.

Internet registries

The organisation that oversees the allocation of domain names and IP addresses is called ICANN – the Internet Corporation for Assigned Names and Numbers, who are based in the USA. They have a department called the Internet Assigned Numbers Authority (IANA), who at the time of writing, manage a further five large organisations around the world called **Regional Internet Registries (RIR)**. Each of these has a defined region of the world and therefore a defined set of IP addresses that they are responsible for allocating.

In Europe, the RIR is called RIPE NCC. In turn, each RIR has several members called National Internet Registries (one per country) who in turn have members called Local Internet Registries, all of whom have responsibility for allocating the IP addresses in specific geographical areas. It is organised into this hierarchical structure to make it easier to manage and to ensure that domain names and IP addresses are not duplicated.

KEYWORDS

Internet registries: organisations who allocate and administer domain names and IP addresses.

Regional Internet Registry (RIR): one of five large organisations that allocate and administer domain names and IP addresses in different parts of the world.

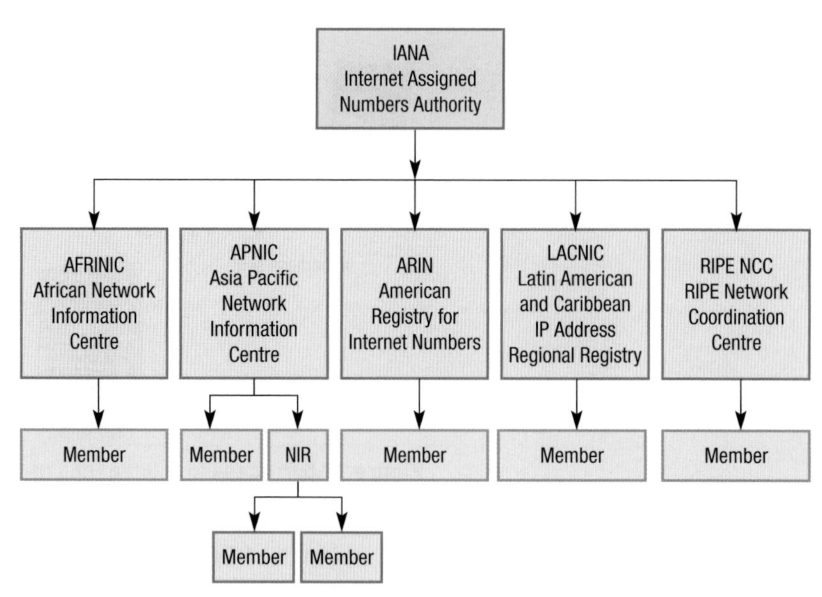

Figure 39.6 The global structure of Internet registries

Routing and gateways

In any communication there must be a sender and a receiver and a connection must be established between the two. There are a number of ways in which this connection can be made. When you connect to the Internet, a connection is established between your computer and the website that you are visiting. You probably realise that this is not a direct link. In the first instance, you connect to your Internet Service Provider (ISP) which in turn connects to the ISP hosting the website.

In fact, there may be many more connections in the circuit. Data being transmitted around a WAN will be sent via a number of nodes. A node is one of the connections within the network. In old-fashioned telephone terms we would call them exchanges. In Internet terms, there are thousands of nodes and, therefore, thousands of routes that a communication may take to reach its destination. Figure 39.7 shows the basic idea.

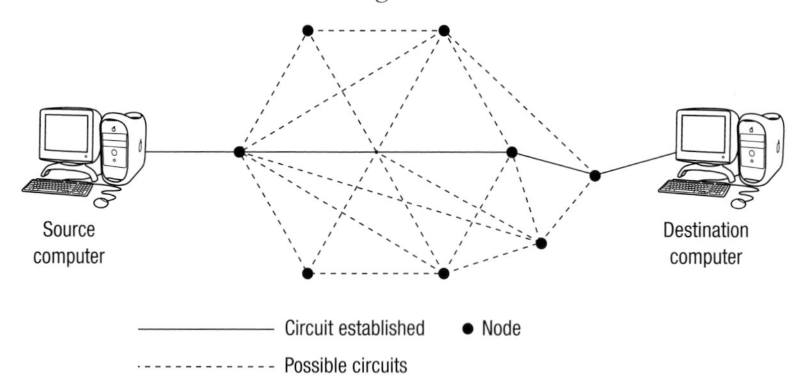

Figure 39.7 Circuit switching

KEYWORD

Packet: a block of data being transmitted.

The transmission will be routed through a number of nodes before a connection is established between sender and receiver. The router is used to send the data to the appropriate node on the network. It knows where to send it as each piece of data is sent as a **packet**, and it will read the header information in each packet of data being sent.

As the Internet is often referred to as a 'network of networks', data packets often need to be transmitted between networks as well as around them. Sometimes the networks will be dissimilar in that they use different protocols. When data are passed between two networks that use different protocols, a gateway must be used to convert between the two protocols.

KEYWORD

Routing: the process of directing packets of data between networks.

As the name implies, **routing** finds the optimum route between sender and receiver, which may be made up of many nodes. At each stage of the routing process, the data packets are sent to the next router in the path, often with reference to a routing table. The routing table stores information on the possible routes that each data packet may take between nodes on its path from sender to receiver. Routing algorithms are used identify the next best step.

Packet switching

KEYWORD

Packet switching: a method for transmitting packets of data via the quickest route on a network.

One of the methods used to send data across networks is called **packet switching**. Data sent over the Internet are broken down into smaller chunks called packets. Each packet of data will also contain additional information including a packet sequence number, a source and destination address and a checksum. Packets of data are normally made up of a header, body and footer. For example, in a 1 KB packet, it might contain:

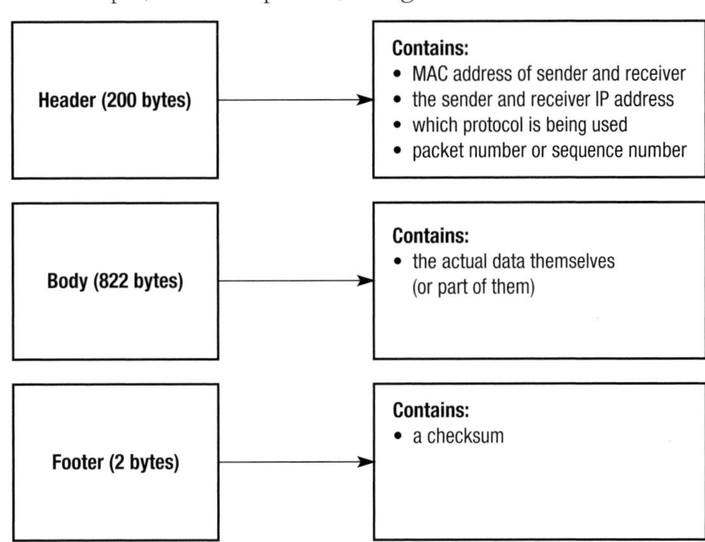

Figure 39.8 The contents of a typical packet of data

The packets are sent to their destination using the destination address. They are re-assembled at the other end using the packet sequence number.

KEYWORD

Checksum: a method of checking the integrity of data by calculating a sum based on the data being sent.

The **checksum** will identify any errors. It works by adding together the values of all the data held in a packet and transmitting those data along with the packet. For example, if the data contain number, it could add all of those numbers up and send the total value as the checksum. When the packet is received, the values could be added up and compared to the checksum. If it is the same then the chances are that the data have been received correctly. Where the checksum is different, the packet will be sent again.

Each packet can take a different route to its destination as it can be re-assembled at the other end regardless of the sequence in which packets are received. Therefore, the packets are routed via the least congested and therefore the quickest route. Data are transferred quicker using this method and are more secure as the packets are taking different routes. This optimises the use of each connection compared to circuit switching.

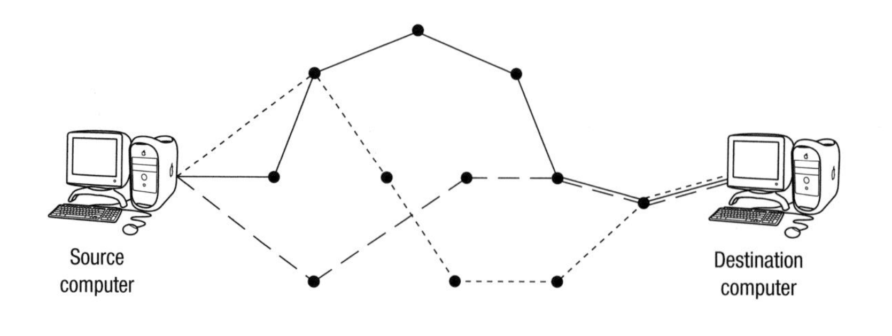

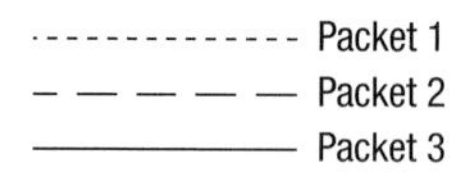

Figure 39.9 Packet switching showing three packets

Practice questions can be found at the end of the section on pages 360 and 361.

KEY POINTS

- The Internet is a global network of networks.
- The WWW is a resource available on the Internet made up of millions of websites and web pages.
- A uniform resource locator (URL) is the unique address of a resource on the Internet.
- A domain name identifies a domain on the Internet and is usually registered to a particular organisation or individual.
- The IP address is a numeric identifier that maps to the domain name.
- IP addresses are allocated to devices on a network either statistically or dynamically.
- A port identifies a specific application that can be accessed via a network. Common applications have dedicated port numbers which are called well-known ports.
- A socket is an endpoint of a communication flow across a computer network.
- Subnet masking is a method of identifying different subsets of a network.
- Internet domain names and IP addresses are administered by Internet Registries using the Database Name Server (DNS).
- Data is sent in packets and routed around using packet switching.

TASKS

1. Explain how packet switching works and how it enables fast transmission of data.
2. What are the components of a data packet?
3. Explain how a subnet mask can be used to route data to a particular part of a network.
4. Write out a URL and label each part.
5. Explain the relationship between a domain name and an IP address.
6. There is a one-to-one relationship between an IP address and a domain name. Why are both needed?
7. How is it possible to ensure that no two websites have the same name or IP address?
8. Describe the difference between a socket and a port.
9. What are 'well-known ports'? Give some examples.
10. What is the relationship between an IP address and a socket address?
11. Explain how port forwarding works.
12. Give an example of how you could calculate a checksum from a data packet containing text.

STUDY / RESEARCH TASKS

1. IP addresses are 32-bit numbers made up of four groups of numbers: ###.###.###.###. What are the first and last IP addresses available using this format? How many IP addresses are possible?
2. Which ISP currently has the highest number of users? Identify all the factors that have led to their success.
3. There are stories on the Internet that there are only 13 main DNS servers and that most of these are in America. How likely is this to be true? What would be the impact on the security of the Internet if there were just 13 DNS servers?
4. Some organisations have tried to create a worldwide map of Internet connections, e.g, the 'Internet Mapping Project'. Look at one of these maps. Explain why they all show clusters or hotspots of activity.

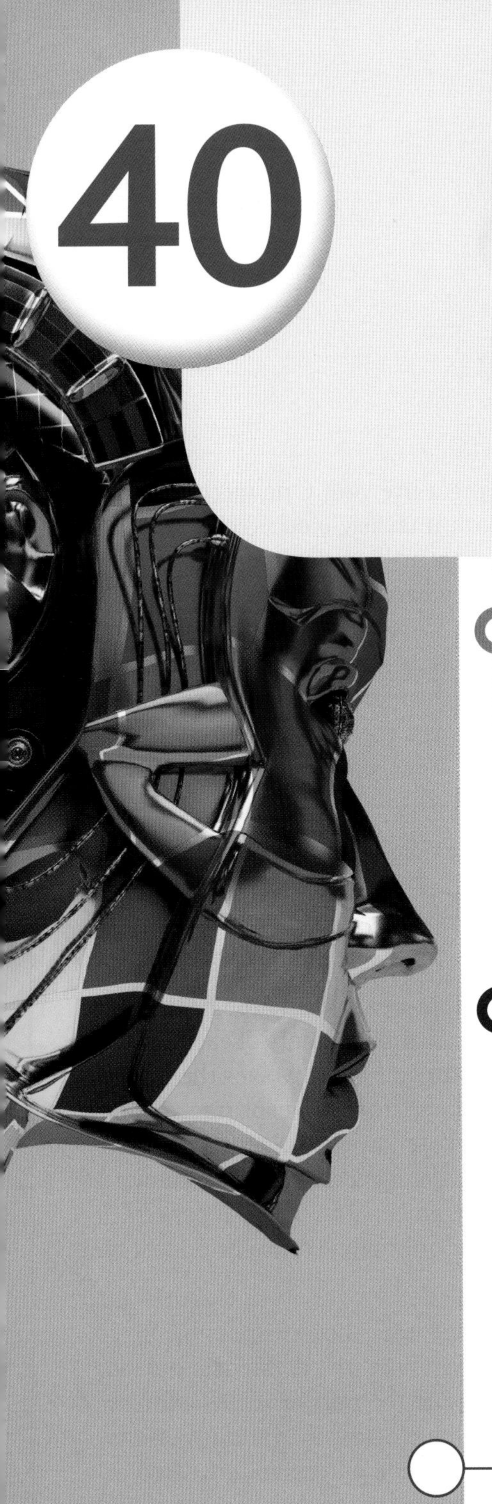

40 Internet security

A level only

SPECIFICATION COVERAGE

3.9.3.2 Internet security

LEARNING OBJECTIVES

In this chapter you will learn:

- how a firewall and proxy server can be used to protect a network
- how public/private key encryption is used to protect data
- how digital certificates and signatures are used to authenticate websites
- to understand the risk posed by Trojans, viruses and worms and how to protect against them.

INTRODUCTION

There are some inherent risks when using the Internet. These often relate to the potential threat of someone discovering personal or sensitive information about individuals and organisations and the information being misused. There is also an increasing risk from worms, Trojans and viruses which can cause network failure, corruption of files or denial of service leading to serious damage or significant problems for the individuals and organisations who are the victims. There are a number of measures that can be employed to either prevent or minimise the risks from these threats.

Firewall

KEYWORD

Firewall: hardware or software for protecting against unauthorised access to a network.

A **firewall** describes the technique used to protect an organisation's network from unauthorised access by users outside the network. A firewall can be constructed using hardware, software or a combination of both. The most secure firewalls tend to be those constructed from both hardware and software.

An organisation that has a local area network will have a number of computers linked together which will all have access to internal information. The LAN may also allow users to access the Internet and this is where the network becomes vulnerable. By utilising the LAN's connection to the Internet it is possible for hackers to use techniques and tools to access information stored on the LAN.

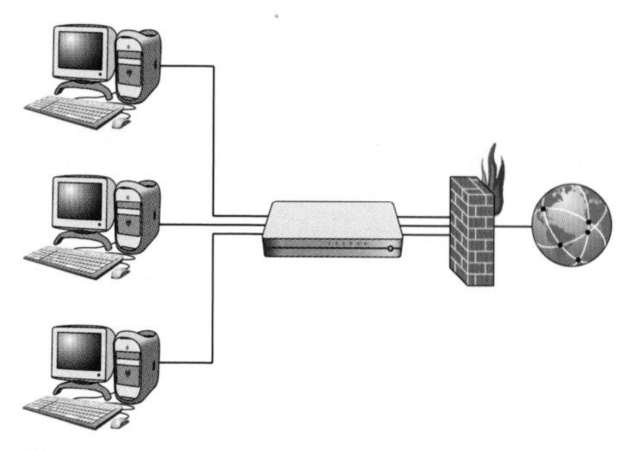

Figure 40.1 A firewall

> **KEYWORD**
> **Packet filtering**: a technique for examining the contents of packets on a network and rejecting them if they do not conform to certain rules.

There are a number of ways of creating a firewall. One method, known as **packet filtering**, uses two network interface cards (NICs) – one for the LAN and one for the Internet. When data packets are received through the Internet NIC, they can be examined before being passed around internally via the LAN NIC. Firewall software is used to examine the packets to ensure that they do not contain any unauthorised data. At a basic level, the header of each packet can be examined to check that it has come from a recognised source. If it has then it can be routed around the LAN. If it hasn't then that packet can be rejected.

Firewall software may also have a facility that keeps a log of all the data being transmitted so that it can be traced. The IP address of the computer sending each packet can be recorded. It may also generate automatic warnings if it identifies that the server is being attacked by hackers.

> **KEYWORD**
> **Stateful inspection**: a technique for examining the contents of packets on a network and rejecting them if they do not form part of a recognised communication.

At a more sophisticated level, rather than just examining the header information, it is possible to examine the actual contents of each data packet. This is called **stateful inspection** and also involves the firewall examining where each data packet has come from. It keeps track of all open communication channels and therefore knows the context of each packet it receives. For example, if the packet is received from a known communication source and forms part of an existing series of packets, it will be accepted. When it comes from an unknown source or port, it may be rejected.

As firewalls have developed over the years, it has become possible to examine each packet in more detail, looking not just at the data but also the protocols being used, the IP address and the port or socket address of the source. This means that packets can be blocked for various reasons, for example if they contain malware, if they come from an untrusted or unknown source, or if they involve an unknown process.

Proxy server

One security measure that can be used at this stage is a proxy server. The word proxy means 'on behalf of' so in this context it is a server that acts on behalf of another computer. By routing through a proxy server there is no direct connection between the computer on the LAN and the Internet. Instead, all requests get passed through the proxy server and can be evaluated to ensure that they come from a legitimate source or to filter users so that they only have access to specific websites.

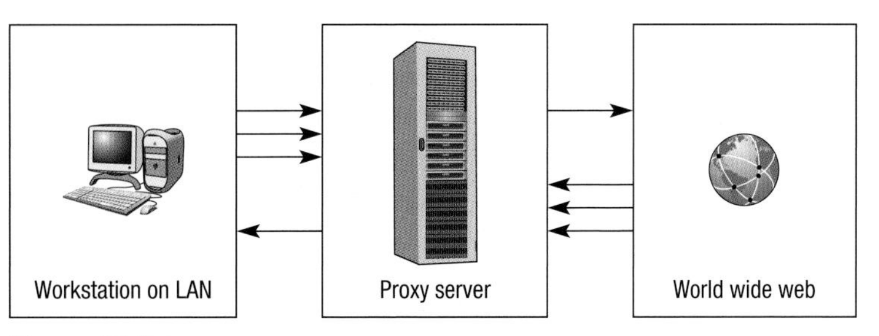

Figure 40.2 A proxy server

In Figure 40.2, the arrows represent requests which are then filtered through the proxy server, with only certain data being allowed through in each direction depending on how the proxy server has been set up.

Private/public key encryption

In Chapter 27 we looked at the way in which data can be encrypted so that if it were intercepted, it would make no sense to the person who intercepted it. Encryption techniques make use of a key, which is a string of numbers or characters that are used as a code to encrypt and then decrypt the message. Typically the key may be 128-bit or 256-bit enabling billions of permutations for the way in which data can be encrypted. Without the key, the message cannot be understood.

KEYWORD

Symmetric encryption: where the sender and receiver both use the same key to encrypt and decrypt data.

One method is **symmetric encryption** where one key is shared between the sender and the recipient as shown in Figure 40.3.

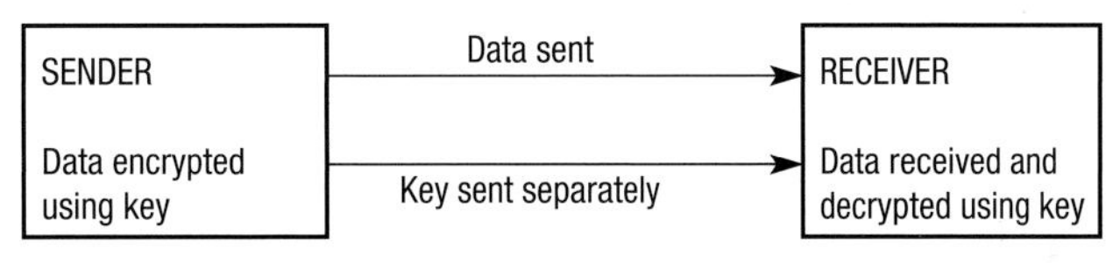

Figure 40.3 Symmetric encryption

In addition to the data, at some point the sender has to send the receiver the key. This is known as key exchange. It is possible but not advisable to send the key with the data. Other methods of key exchange include using digital signatures or certificates, or using password-protected systems. Once the key is sent and in the receiver's computer, it can then be used to encrypt and decrypt further messages.

KEYWORDS

Asymmetric encryption: where a public and private key are used to encrypt and decrypt data.

Private key: a code used to encrypt/decrypt data that is only known by one user but is mathematically linked to a corresponding public key.

Public key: a code used to encrypt/decrypt data that can be made public and is linked to a corresponding private key.

There is an inherent weakness with this system in that if the key is intercepted then it would be possible to work out what it is, therefore making all further communications vulnerable to unauthorised access. One way of getting round this is to use **asymmetric encryption**, which makes use of two related keys in combination, a private one and a public one. The algorithm used to create the two keys results in so many permutations that it is almost impossible to work out the combined key. Both sender and receiver have their own pair of public and private keys.

For example, with two computers A and B:

- A will have a **private key** known only to A.
- A will also have a **public key**, which is mathematically related to the private key. It is called a public key as anyone can access it.
- B will also have a private key and a related public key.
- For A to send a secure message to B, A will first encrypt the message using B's public key.

- As the private and public keys are related, the message can only be decrypted by B using B's private key.
- As no-one else knows B's private key, even if the message were intercepted, it could not be decrypted.

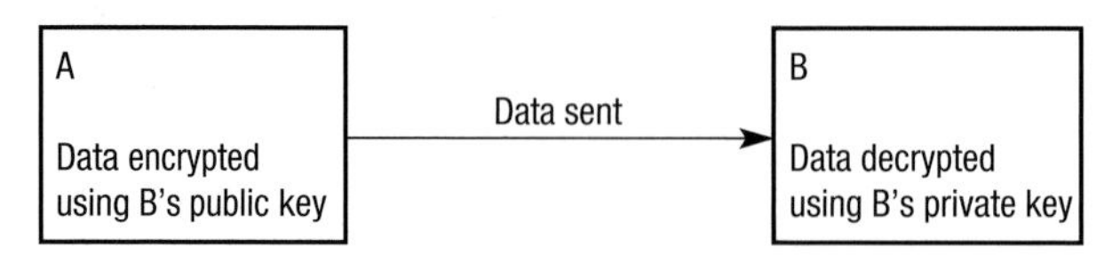

Figure 40.4 Public and private key encryption

An important factor that makes asymmetric encryption so secure is that although the two keys are mathematically related, it is virtually impossible to work out what the private key is from looking at the public key. This is mainly due to the large number of digits being used in the keys and the algorithms being used which will produce billions of possible permutations. At the time of writing, 1024-bit keys are common although as computing power increases, it becomes more likely that these codes are feasibly breakable and the number of bits used will increase.

Digital certificates and signatures

A **digital certificate** is a means of proving who you are when dealing with people and organisations on the Internet. It is usually used by businesses to authenticate that they are genuine and is important in the use of asymmetric encryption as a secure way of sharing public keys. Certificates are also used by some government agencies such as the Inland Revenue. The certificate typically contains the name of the organisation, their domain and server name and a serial number which is registered with a **Certification Authority** who issues the certificates.

KEYWORDS

Digital certificate: a method of ensuring that an encrypted message is from a trusted source as they have a certificate from a Certification Authority.

Certification Authority: a trusted organisation that provides digital certificates and signatures.

Digital certificates, sometimes referred to as SSL (Secure Socket Layer) certificates, were introduced to encourage people to do business on the Internet, as many consumers were, and still are concerned about fraud. If a hacker discovers your credit card number, then they could purchase items from the Internet using your card. Websites using digital certificates usually advertise the fact prominently on the site using the logo of the Certification Authority. Issuing organisations at the time of writing include Symantec (under the VeriSign and Thawte brands), Comodo Group, Go Daddy and Global Sign. You may see their names and logos on various websites.

A **digital signature** is another method of ensuring the authenticity of the sender. In the same way that a signature helps to prove someone's identity in real life, a digital signature does the same thing on the computer. However, rather than being an actual signature, a digital signature uses mathematical functions and the public/private key method.

KEYWORD

Digital signature: a method of ensuring that an encrypted message is from a trusted source as they have a unique, encrypted signature verified by a Certification Authority.

For example, if A wants to send a message to B with a digital signature:

- The message being sent has a publicly known hashing algorithm applied to it (see Chapter 10) to create what is known as a hash.
- The hash is encrypted using A's private key as described earlier in the chapter.

- The hash is appended to the message and becomes the digital signature.
- The message is sent to B who then uses A's public key to decrypt the hash.
- The hash is then put through the same publicly known algorithm and the result is compared to that in the original message.
- Where the two hashes are the same, the message is authenticated and where they are different then the message cannot be authenticated.

Trojans

KEYWORD

Trojan: malware that is hidden within another file on your computer.

A **Trojan** is a computer program designed to cause harm to a computer system or to allow a hacker unauthorised access. It is one of a group of malware programs, which is short for malicious software. The distinguishing feature of a Trojan is that it is hidden away inside another file and that it is not always obvious that a computer is infected. The Trojan does not replicate itself in the same way as other malware and therefore it can remain undetected for a long time.

This gives a hacker the opportunity to access a computer remotely without the knowledge of the user. Once access is achieved it is possible to carry out theft of data with a view to carrying out further crime. Alternatively, the Trojan may simply be used to cause damage to the computer or data stored on it.

Hackers are individuals or groups that gain or attempt to gain unauthorised access to individual computers or the networks of organisations. Hackers' motives vary enormously. At one end of the scale there is the amateur hacker who views hacking as a game and simply enjoys breaking into other people's systems. When they get in, they rarely do any damage. At the other end of the scale, there are professional hackers who can make a living by carrying out fraudulent acts.

There are also groups of 'ethical hackers' who enjoy the notoriety that hacking brings. These people tend to target large organisations such as Microsoft in order to expose weaknesses in their security measures. Their justification for this is to make big businesses take a more serious approach to Internet security. Other hackers have political or religious motivations and may target the websites of government agencies or religious groups in order to get their own views across. In some cases, hacking is used as a form of terrorism or sabotage against a particular nation.

CASE STUDY: FLAME

The Flame malware, first reported in 2012, is being classed as a new generation of superbug that is part Trojan, part worm and part virus. Unlike many other malware programs, Flame is quite large at 20 MB and its origin is currently unknown. Once it has installed itself it has the capability to monitor network traffic, access data and programs, take screenshots, record conversations and monitor keystrokes among other things.

The malware is proliferating in the Middle East and is so large and complex that some people believe it can only have been written by a state government for the purposes of collecting information and espionage.

Viruses

KEYWORD

Virus: a generic term for malware where the program attaches itself to another file in order to infect a computer.

A **virus** is a small malware program that is designed to cause damage to a computer system or the data stored on it. A computer gets infected when the malware installs itself on the computer from a number of sources including pop-ups, email attachments or file downloads.

The virus itself will be attached to another file but once installed on the host machine, it will activate. The defining feature of a virus is that it replicates itself and can therefore cause extensive damage to individual computers and networks as, like a human virus, it can spread anywhere.

Viruses are created for various reasons and have various impacts. At the lowest level a virus may simply display an unwanted message. At the other end of the scale, viruses can destroy whole networks and entire databases.

Worms

KEYWORD

Worm: malware or a type of virus that replicates itself and spreads around a computer system. It does not need to be attached to another file in order to infect a computer.

The nature of the Internet means that viruses now have the potential to spread very quickly around the world. Many of the world's most infamous viruses are classed as **worms**. Worms replicate themselves and are designed to spread, exploiting any weaknesses in a computer's defences. The defining feature of a worm is that it does not need to be attached to another file to infect the computer.

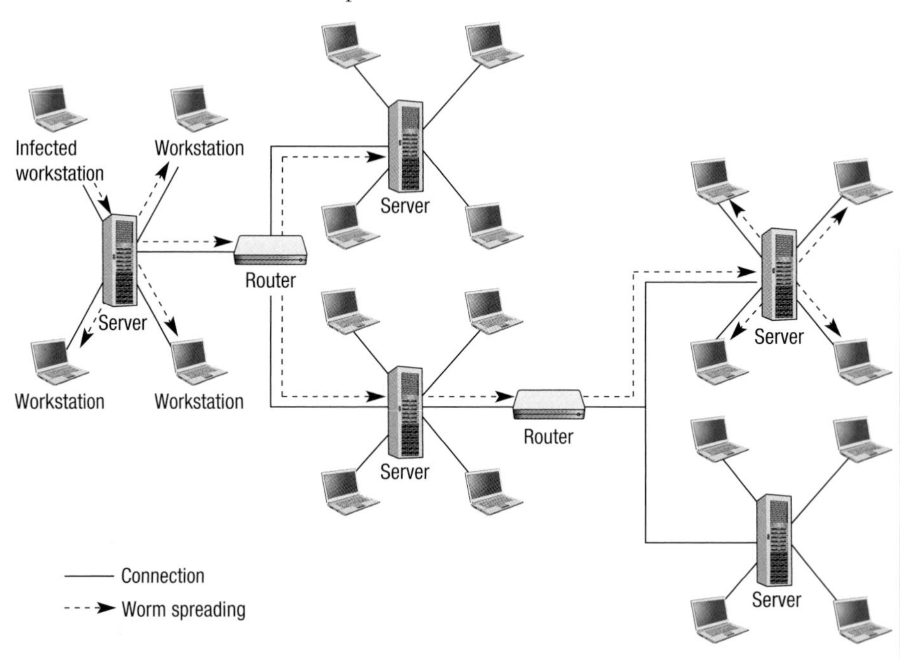

Figure 40.5 How worms spread

Figure 40.5 shows how a worm can quickly be passed around a network and any other network that connects to it. In this scenario, it is possible for every computer connected to get infected.

Well-known worms over recent years include 'Mydoom', 'Sobig.f', 'Iloveyou' and 'Melissa'. All of these proliferated via email and were able to spread quickly as the malware was automatically sent to the email address list of infected machines. Malware such as this infected millions of computers, caused billions of pounds worth of damage and forced some very large websites to close down temporarily.

Protecting against Trojans, viruses and worms

There are several ways of reducing the risks and the actions taken depend on what kind of user you are:

- as a user there are actions you can take to protect your own computer and data
- as a programmer there are steps you can take when writing your code that will make programs more secure
- as a system administrator, there are particular steps that can be taken to keep the whole system more secure.

As users there are many things we can do as individuals to protect our computers:

- use anti-virus software and other anti-malware software and keep it up to date
- keep operating system software up to date
- use a firewall
- do not open attachments or click on pop-ups from unknown senders
- operate a white list of trusted sites
- ensure sites use HTTPS, digital signatures and certificates
- use passwords on programs and files
- encrypt data files.

As programmers we can:

- select a programming language with in-built security features including tools that check for common security errors
- use recognised encryption techniques for all data stored within the program
- set administrative rights as part of the program and carefully control access and permission rights for different users
- don't load up lots of Internet services as part of your code unless they are needed
- thoroughly test your code as errors can be exploited, specifically testing for known security issues
- keep code up to date in light of new security threats
- never trust the user! Many threats are internal to an organisation and might not be malicious. Major problems can be caused through accidental misuse by a user.

As a system administrator we can:

- ensure that requests are coming from recognised sources
- use a network firewall and use the packet filtering and stateful inspection techniques as described earlier in this chapter
- use encryption techniques as described earlier and ensure digital certificates and signatures are used and are up to date
- keep anti-virus software up to date
- update the network operating system regularly.

Practice questions can be found at the end of the section on pages 360 and 361.

TASKS

1. Describe why symmetric encryption is not considered to be as secure as asymmetric encryption.
2. Explain how public and private keys are used in asymmetric encryption.
3. Explain how hashing is used to create a digital signature.
4. Describe how hardware and software can be used to create a firewall.
5. Describe three measures that a user can take to prevent unauthorised access to computer systems.
6. Describe three ways in which a computer system can be protected against viruses.
7. Discuss the following techniques, considering how effective they are against hackers:
   a) encryption
   b) digital certificate
   c) digital signatures.

KEY POINTS

- A firewall can be implemented using software and hardware and protects networks from unauthorised access.
- Data can be routed through a proxy server so that a network's server is not connected directly to the Internet.
- Symmetric encryption is where the sender and receiver both use the same key to encrypt and decrypt data.
- Private/public key encryption or asymmetric encryption uses two related keys, a public key to encrypt data and a private key to decrypt data.
- Digital certificates and signatures are verified by trustworthy organisations and ensures that data is coming from a trusted source.
- Viruses, Trojans and worms are all examples of malicious programs that can cause damage to your data or systems.

STUDY / RESEARCH TASKS

1. There are organisations that exist to help programmers write more secure code. One such organisation is the Common Weakness Enumeration (cwe.mitre.org). Find out about the most common security weaknesses listed on their website.
2. Three common errors in writing secure code are:
   a) buffer overflows
   b) cross-site scripting
   c) SQL injection.

   Find out what these mean and how as a programmer you could prevent these errors occurring.
3. New viruses come out every day. Identify a recent virus. What does the virus do? What was the intention of the person who wrote it?
4. 'No system is 100 per cent safe.' Discuss this statement considering all the methods available to an organisation to protect their computer systems and data.
5. Identify an organisation that provides digital certificates. Explain the level of security they provide for users.

41 Transmission Control Protocol / Internet Protocol (TCP/IP)

A level only

SPECIFICATION COVERAGE

3.9.4.1 TCIP/IP

3.9.4.2 Standard application layer protocols

LEARNING OBJECTIVES

In this chapter you will learn:

- how the TCP/IP stack is used to transmit data packets around networks
- about the four layers of the TCP/IP stack
- how sockets and ports are used in IP networks
- about the common transmission protocols including HTTP, FTP and SSH
- how email and webs servers are used.

INTRODUCTION

You have already come across the concept of a protocol, which is essentially a set of rules. In communications terms, it refers to the various rules that govern how data is sent around networks. In this chapter, we will be looking specifically at protocols that are relevant to networks and in particular the various layers of the Transmission Control Protocol and Internet Protocol (TCP/IP) stack which set the rules relating to the transmission of data in TCP/IP networks.

The TCP/IP stack defines the rules relating to transmission of data packets. IP controls the delivery of the packets and TCP keeps track of the packets and re-assembles them on receipt. **TCP/IP** is made up of a number of layers which are collectively referred to as a protocol stack. The TCP/IP stack is in line with the International Standard communication protocol stack called the Open System Interconnection (OSI) model. Within each layer there are a number of other **protocols**.

KEYWORDS

TCP/IP: a set of protocols (set of rules) for all TCP/IP network transmissions.

Protocols: sets of rules.

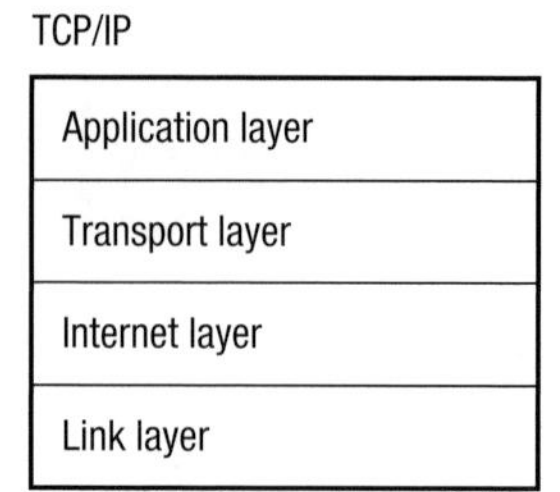

Figure 41.1 The layers of the TCP/IP stack

TCP/IP is made up of four main layers, as shown in Figure 41.1:

- **Layer 4 – Application layer**: The application layer handles the Domain Name System and a series of other protocols such as FTP, HTTP, HTTPS, POP3, SMTP and SSH, which are covered later in this chapter.
  For example: incoming and outgoing data are converted from one presentation format to another; presentation formats are standardised so that different types of data (sound, graphics, video, etc.) can be understood by the receiving device; data that have been compressed or encrypted can be interpreted.
- **Layer 3 – Transport layer**: This contains most of the configuration and coordination associated with the transmission that ensures that all the packets have arrived and that there are no errors in the packets. It also handles the way in which connections are made to create a path for data to travel between nodes. The sender and receiver are identified and authenticated and the communication is set up, coordinated and terminated. Network resources are identified to ensure that they are sufficient for the communication to take place.
  For example: connections can be opened and closed; **port** numbers are used to pass packets to the correct application in the application layer.
- **Layer 2 – Network or Internet layer**: Defines the IP addresses of devices that send and receive data and handles the creation and routing of packets being sent and received.
- **Layer 1 – Link layer**: This layer provides synchronisation of devices so that the receiving device can manage the flow of data being received. It identifies what network topology is being used and controls the physical signals that transmit the strings of bits around the network, that is, the actual transmission of the 0s and 1s. It also controls physical characteristics such as data transmission rates and the physical connections in a network. On wireless networks it handles the CSMA/CA protocol.

KEYWORD

Port: an addressable location on a network that links to a process or application.

The highest level is closer to the user in that the processes are usually handled using either the operating system or application software. The lower layers are handled using a combination of hardware and software including the physical or wireless connections between devices. It is referred to as a stack because of the way in which the request from the client machine passes down through the layers of the protocol and then back up through the layers of the server side, as shown in Figure 41.2. This means that the last action that takes place in the Link layer on the client computer becomes the first action in the Link layer in the server. This is an example of the last in first out (LIFO) structure that characterises a stack.

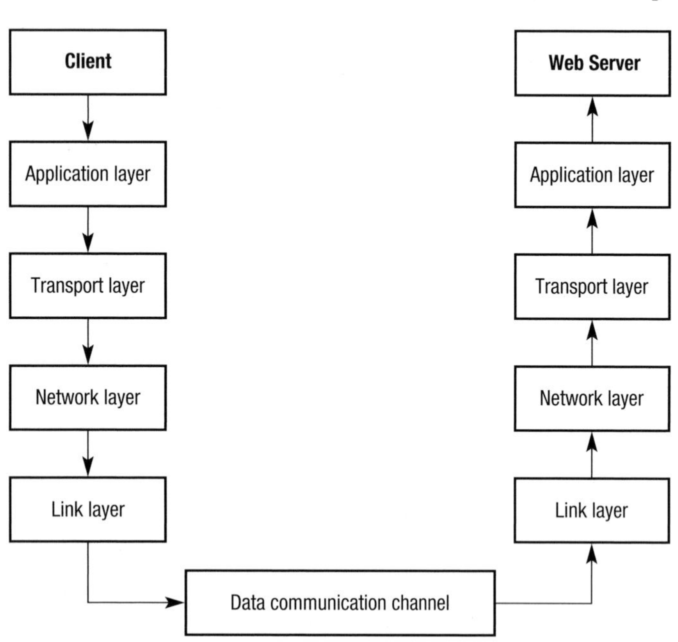

Figure 41.2 Simulation of the TCP/IP stack

Hypertext Transfer Protocol (HTTP) and Secure HTTP (HTTPS)

KEYWORDS

HTTP: a protocol (set of rules) for transmitting and displaying web pages.

Client–server model: a way of implementing a connection between computers where one computer (the client) makes use of resources of another computer (the server).

HTTPS: as above but with encrypted transmission.

Hypertext Transfer Protocol (HTTP) is the set of rules that govern how multimedia files are transmitted around the Internet. The content of the WWW is such that text, graphics, video and sound can all be transferred as part of a web page. **HTTP** ensures that the files are transferred and received in a common format. HTTP handles the transmission of this data. The formatting and display of web pages is handled separately, typically by HTML.

Hypertext refers to the fact that the web pages will have hyperlinks to other files. When you select a URL, either by typing it in or by clicking on a hyperlink, the HTTP protocol on your computer sends a request to the IP address of the computer that contains the web page. The HTTP protocol on this computer than handles the request and sends back the web page in the appropriate format.

This uses the **client–server model** which means that your web browser acts as a client, requesting the services of the computer that contains the web page, the server. Both client and server computers must use the same protocols so that the files can be sent and received in the same format.

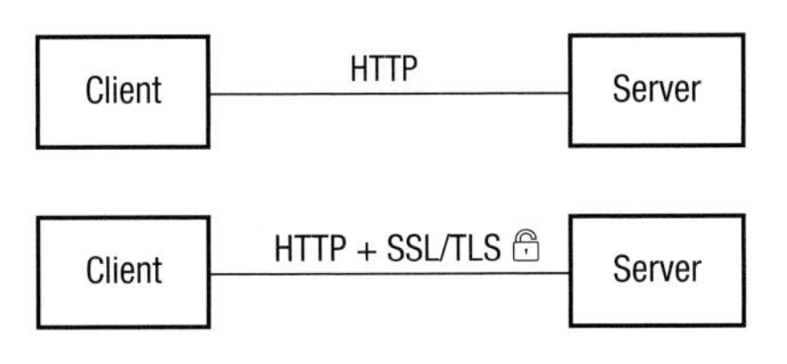

Figure 41.3 HTTP and HTTPS

HTTPS is an extension of the protocol with added security. This is commonly used on websites where personal information is used, such as banking websites. Additional security includes authenticating the web server and encrypting data that is being transmitted. It works by using either the Secure Socket Layer (SSL) protocol or the Transport Layer Security (TLS) protocol, both of which use data encryption.

File Transfer Protocol (FTP)

KEYWORD

FTP: a protocol (set of rules) for handling file uploads and downloads.

FTP is another set of rules relating to the transfer of files around the Internet. It is commonly used when a web page is uploaded from the computer of the person who created the site to the web hosting server. It is also used when software is downloaded from websites. When FTP is being used for this purpose, it will be shown as the prefix in the URL.

It is similar to HTTP in that it works using the standard layers of TCP/IP. However, HTTP tends to be used to transfer viewable content (web pages) whereas FTP is commonly used to transfer program and data files.

File transfer can be anonymous or non-anonymous (protected) depending on whether you need to identify yourself before the download. Where the site is protected a username and password is required. Anonymous sites do not require this.

Secure Shell (SSH) Protocol

KEYWORD

Secure Shell (SSH) Protocol: a protocol (set of rules) for remote access to computers.

The Internet is often used to enable a user to connect to a remote computer and execute programs and access resources on that computer. An example of this might be when a computer engineer fixes a problem remotely, or where you have access to school and college resources by logging on at home. This uses the client–server model whereby the computer that you use acts as the client and the computer that you control is the server. The server computer is more commonly referred to as the host.

In these situations, SSH is used to improve the security of the connection. It does this partially by creating a secure network of nodes through which the access is made available. Encryption is used on the data being transmitted using public key encryption (see Chapter 27). In addition, password and username login details would normally be required.

As SSH is secure, it is a useful protocol through which to access other services. For example, if you wanted to access the email server remotely, it would be more secure to do this using an SSH protocol than to access the email server without it.

SSH commands are usually input using a command line interface. This means that you have to know specific command words and the syntax (or format) in which you need to type the words in. This is similar to the old Disk Operating System or DOS interface that was the only way to operate computers until graphical user interfaces (such as Windows) were developed.

For example, to move around folders in Windows you would just click on the folder you wanted. In SSH folders are called directories and you need to remember various mnemonics. For example:

- cd change directory
- cd/windows/programfiles change to the windows/program files directory
- rm essay.doc delete the file essay.doc
- mv essay.doc essay1.doc rename essay.doc to essay1.doc
- cp essay.doc essay1.doc create a copy of the file essay.doc called essay1.doc
- vi essay.doc create a file called essay.doc

Simple Mail Transfer Protocol (SMTP) and Post Office Protocol (POP3)

KEYWORDS

SMTP: a protocol (set of rules) for sending emails.

POP3: a protocol (set of rules) for receiving emails.

SMTP and **POP3** are protocols used for sending and receiving emails. SMTP is a specific protocol for sending emails and works through a series of SMTP servers which store the email addresses of senders and recipients. By linking with DNS servers, the IP address of the recipient is identified and a connection can be established between sender and receiver. The data in the email can then be transmitted. SMTP uses ports 25 and 587.

Where the data cannot be sent for any reason, SMTP uses a queuing system to hold on to the email and then attempts to send it at a later time. It will continue to do this for a set number of times. If it still fails to send, it will send a message back to the sender indicating that delivery has failed. You also get a message if the SMTP or DNS server fails to identify the email address or IP address.

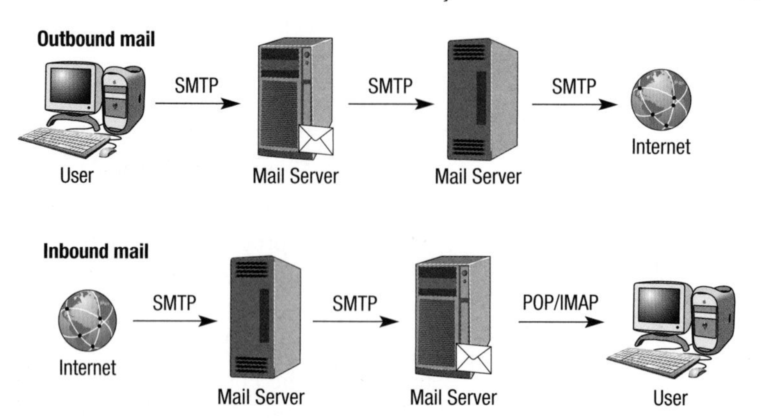

Figure 41.4 The SMTP and POP protocols

In order to receive mail, the client must first connect to the email server. POP3 is a protocol which checks for incoming mail using port 110. It works by creating a text file of any incoming messages associated with your email address. If a message is received, it will append it to your text file and the next time you log on, you will be able to access it in the form of an email message using an email client, which is the chosen email application being used on the network, such as MS Outlook.

Email and web servers

KEYWORDS

Email server: a dedicated computer on a network for handling email.

Web server: a dedicated computer on a network for handling web content.

Within a network, there may be one or more servers providing access to applications, storage space and other resources. Often servers are set up to perform specific functions on the network. Two examples of this are **email servers** and **web servers**.

- Email server: This is typically a high specification machine with large storage capacity that stores a database of all the network users and their email addresses. It also stores all outgoing and incoming mail. Specific software on the server is used to handle the storage and transmission of emails, allowing users to access their emails regardless of what other services are available. For example, accessing email would not be dependent on having a particular ISP.
  Typically an email application would be chosen within the organisation and all users would have access to it as an email client. When the user started to use the application, if POP3 was being used, port 110 would be used to retrieve incoming emails while port 25 would be used to send emails by SMTP.
- Web server: This is a server that hosts a website and handles traffic from users to the site. For a home user, the web server will typically be provided by their ISP. For business users it would be common to have one or more servers dedicated as web servers. This is of particular importance where the website is critical to the success of the organisation. For example, an online retailer would need enough web servers to ensure that users can get quick access to the website at all times. There have been many examples of servers crashing or slowing down due to unexpectedly large numbers of people accessing them.
  Data stored on a web server may be in various formats including text, scripts, and multimedia content. Web servers will make use of various protocols including HTTP to ensure that all of these data are correctly handled and formatted so that they appear correctly when viewed over the Internet regardless of the hardware and software being used by the user.

Web browsers

KEYWORD

Web browser: an application for viewing web pages.

A **web browser** is an application that allows users to view web pages and other resources and is critical in ensuring that websites appear exactly how they were designed. In simple terms a browser needs to retrieve resources via the URL, format them so that they display correctly on the screen and allow some form of navigation and other user features such as bookmarking and searching. This process may require several requests being made to the server in order to load the various resources that make up a web page including scripts, image, and style sheets. Not surprisingly, the main browsers such as Internet Explorer, Google Chrome and Mozilla Firefox all have similar features.

When a web page is loaded, a request is made to the domain name server (DNS), which translates the URL into an IP address. This IP address is then used to access the web page host. The host then serves the web page to the browser on the client computer as shown in Figure 41.5. All browsers must work within many of the protocols that we have looked at in this chapter, mainly the hypertext transfer protocol, which defines how data is transmitted.

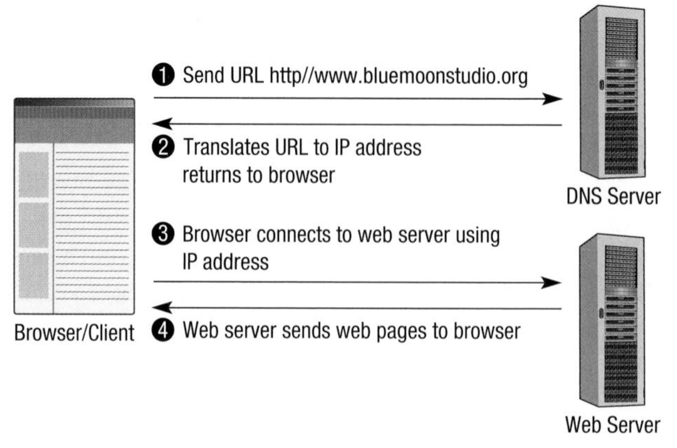

Figure 41.5 The process of requesting and receiving a web page

When a web page is loaded, the browser may cache it, which means that it is able to store it temporarily. The advantage of this is that if the user wishes to revisit the same page then it can be retrieved from the local cache rather than having to make the request to the DNS and host servers again. This renders the page much more quickly and reduces the dependence on bandwidth. However, if the page has changed on the host server, the version of the page held in the cache may be out of date and therefore will need refreshing.

Practice questions can be found at the end of the section on pages 360 and 361.

TASKS

1 What are the four layers of the TCP/IP stack?
2 Explain the need for protocols such as FTP and HTTP.
3 What are the key differences between HTTP and HTTPS?
4 How is it possible to complete access to a computer remotely using SSH?
5 Why might a network administrator choose to allocate servers to specific tasks?

STUDY / RESEARCH TASKS

1 Describe the process that Nominet uses to allocate domain names.
2 The WWW is one service available via the Internet. Identify two other services that are available.
3 Identify three browsers that are available and explain why you might choose to use one rather than the others.
4 Identify other ways in which information can be downloaded from the Internet without accessing the WWW.
5 What is Telnet protocol and why is it no longer in widespread use?
6 Telnet is sometimes associated with hacking. Why is this the case, and what security measures could prevent its use for unauthorised purposes?
7 What are the alternatives to POP3?

KEY POINTS

- Protocols are sets of rules. There are several protocols that related to the transmission of data around networks.
- The TCP/IP stack is a four-layered set of protocols for computer networks, including the Internet.
- A socket is an endpoint of a communication flow on a computer network that uniquely identifies an application and device.
- Hypertext Transfer Protocol (HTTP) is the set of rules that govern how multimedia files are transmitted around the Internet.
- FTP is a set of rules relating to the transfer of files around the Internet.
- SSH is a set of rules relating to the remote access of computers on a network.
- SMTP and POP3 are protocols relating to email.

42 The client–server model

A level only

SPECIFICATION COVERAGE

3.9.4.10 Client–server model

3.9.4.11 Thin- versus thick-client computing

LEARNING OBJECTIVES

In this chapter you will learn:

- that the client–server model is one where a high specification server provides resources to any number of lower specification clients
- how an application programming interface provides a common way for programs to work together across networks
- what CRUD and REST are and how they are used
- what JSON and XML are and how they are used
- what thick- and thin-client computing is and how they compare.

INTRODUCTION

The client–server model is a methodology for connecting computers together, usually over a network where one computer provides access to resources for other computers that are connected to it. Typically this might involve having a main server with large amounts of processing power and storage capacity with any number of other computers (clients) attached to it that then use the resources of the server.

The client may have few resources of its own and therefore has to request the services of the server. In a typical star topology like the one shown in Figure 42.1, each client has its own physical cable connection to the server.

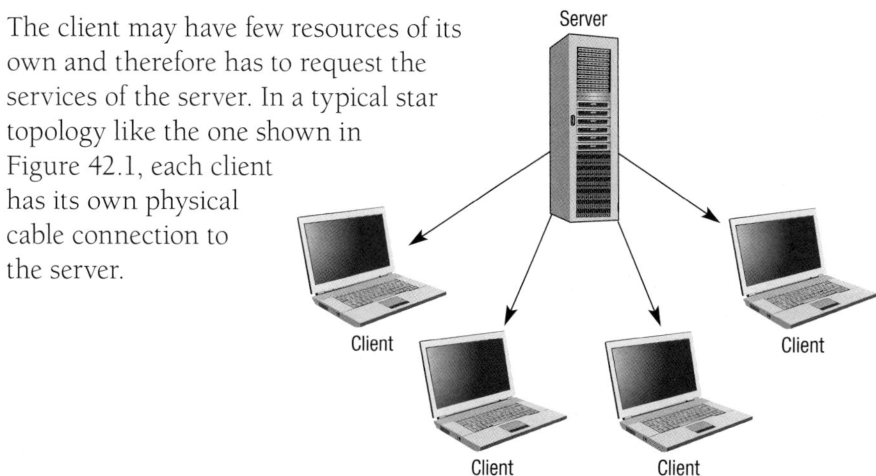

Figure 42.1 A typical star network

In a small network, there may be one server where all of the programs and data are stored. Each client then requests a service as and when it needs it. This might be to run a particular application, access a particular data file, or gain access to the Internet. In practice, there are more likely to be several servers in the network serving hundreds of clients. Servers and clients can take on different roles depending on what tasks are completed by the users. Examples of different servers include:

- File server: In a traditional network, the file server contains any type of computer file, which could be programs or data.
- Web server: A server is used to serve up web pages for an Intranet.
- Proxy server: Each client computer is provided with a gateway to the Internet through the server.
- Print server: All client print requests are sent to the same server where they are prioritised, buffered and then printed.
- Database server: The server will store the contents of the databases and access to the data will come from the individual clients.
- Application server: The server executes all of the procedures needed to run applications.

The client–server model works on the basic principle of sender and receiver. To initiate any communication and sharing of resources, the client must make a request to the server. In turn the server responds to that request and then provides the service that is being requested.

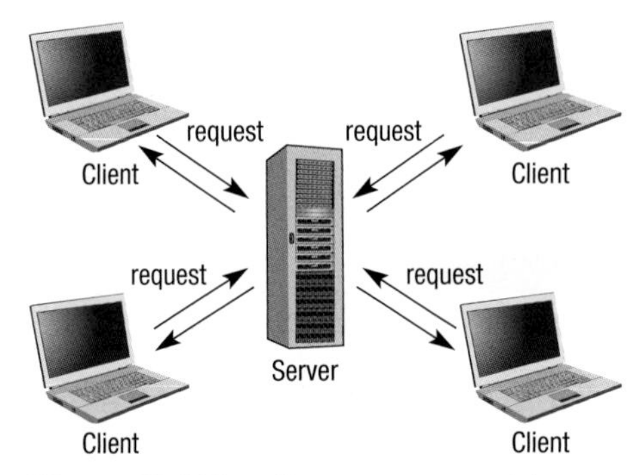

Figure 42.2 The client–server model

The client–server model is also used on the Internet. For example, email is built on this model with each user being the client, sending requests to their email provider, who responds to each request as it is received. FTP services also use the model with the client making a request to upload a file, which is then handled by the FTP server.

Application Program Interface (API)

As we saw in Chapter 41, in order to communicate with each other effectively, protocols are needed to define the rules by which the communication will take place. All of these requests and the corresponding transfer of services from server to client take place in the application layer of the TCP/IP stack, which we looked at in detail in Chapter 41.

In addition to the protocols, an **application program interface (API)** is used. An API defines the way in which programs can work together. They are usually made up of standardised subroutines that can be customised to provide an interface between one program and another. When using web services, an API can also define the protocols that will be used.

KEYWORD

Application program interface (API): a set of subroutines that enable one program to interface with another program

KEYWORD

Websocket protocol: a set of rules that creates a persistent connection between two computers on a network to enable real-time collaboration.

One of these is the **websocket protocol**, which creates a connection between a client and a server. The client first sends a handshaking request to the server in order to establish the connection. In response, the server creates a full duplex connection on a single socket. This allows simultaneous exchange of data in both directions enabling the client and server to communicate on an ongoing basis without the need to constantly refresh a full web page. Effectively, the websocket has created a dedicated link between the two computers. This is routed through port 80, the dedicated HTTP port, meaning that it will work in situations where non-web Internet connections have been blocked using a firewall.

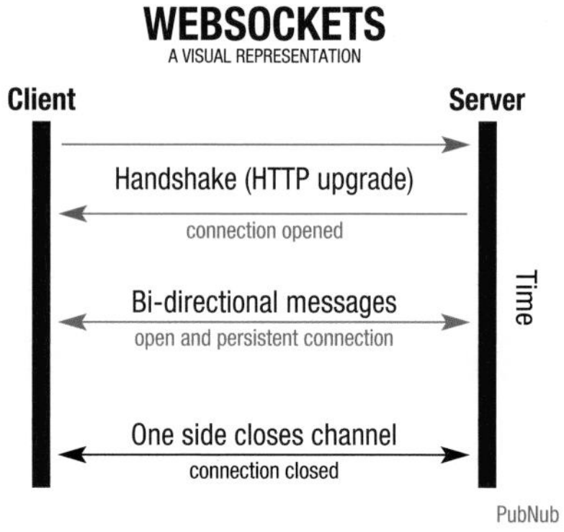

Figure 42.3 A websocket handling data passing between client and server

KEYWORDS

Packet: a block of data being transmitted.

Message: the name given to a packet of data being transmitted using the websocket protocol.

Data is sent in **packets** called **messages**, with minimal header information, allowing for very fast transfer of data in both directions. As these are all going down a persistent connection it allows for real-time collaboration between client and server. This is of particular value for certain web applications, where data is constantly changing and time is critical. For example, an online travel agent needs to respond instantly to ensure that one holiday is not sold to two different people at the same time; a share dealer requires an instantaneous share price in order to make a trade.

CRUD and REST

Many users access databases through a network or over the Internet. In these situations, there are conventions and styles that are used to ensure that data is stored, managed and represented correctly in the database.

KEYWORD

CRUD: an acronym that explains the main functions of a database: Create, Read, Update, Delete.

The four main processes required with databases can be defined by the acronym **CRUD**:

- C: Create
- R: Retrieve
- U: Update
- D: Delete

As well as representing the main database functions, it also refers to the way in which data is actually displayed and reported on via the user interface. Without these four functions it is not possible to have a complete database. All databases will conform to the CRUD principle regardless of

KEYWORDS

SQL (Structured query language): a programming language used to manage data within a relational database.

REST (Representational State Transfer): a methodology for implementing a networked database.

HTTP (Hypertext transfer protocol): the protocol (set of rules) to define the identification, request and transfer of multimedia content over the Internet.

how they are built. Relational databases, which are covered in Chapter 43, conform to the CRUD standard. In fact there is a one-to-one relationship between CRUD and **SQL** commands as shown in the table below:

CRUD	SQL
Create	INSERT
Retrieve	SELECT
Update	UPDATE
Delete	DELETE

REST stands for Representational State Transfer and is a design methodology for networked database applications. It uses the hypertext transfer protocol (HTTP) to carry out each of the four CRUD operations on a networked database.

HTTP uses request methods which define the way in which data will be handled. In the same way that CRUD can be mapped to SQL statements, it can also be mapped to the HTTP request methods.

CRUD	HTTP
Create	POST
Retrieve	GET
Update	PUT
Delete	DELETE

REST is an efficient way of implementing database applications over a network as it makes use of existing protocols within **HTTP**, which has already been adopted as the standard way of transferring data. This means that it will work on any type of local machine architecture, with any operating system and can be run through a firewall for added security. The basic process is shown in Figure 42.4.

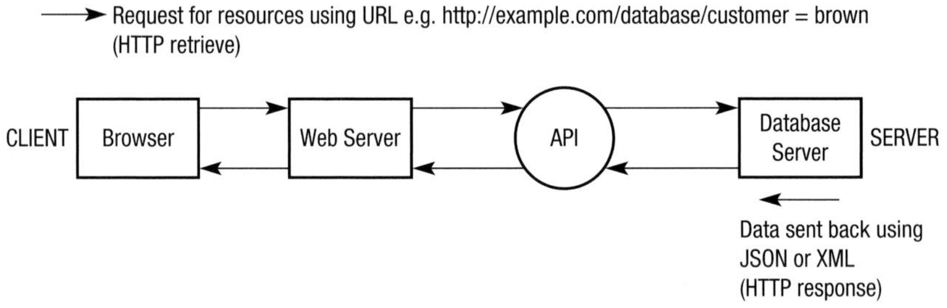

Figure 42.4 The REST model

- The client makes a request to the server from the browser of the local machine.
- The service requested is the database identified by its Uniform Resource Locator (URL), which uniquely identifies the resource on the server and contains the database query.
- The API is run from the server and accessed by the browser to coordinate processes between client and server applications.
- HTML files are used to ensure data is displayed in the correct format on the client side.
- Requests and data are transferred using HTTP.
- JSON (JavaScript Online Notation) or XML (Extensible Markup Language) are used to return the results of the query.

For example, to create a query to find all customers named 'Brown' might look like this:

http://www.example.com/customers/brown

This query, which looks like a normal URL, is sent to the server using the GET statement in HTTP. The result of the query will then be returned in JSON or XML format as shown in the next section.

JSON (JavaScript Online Notation) and XML (Extensible Markup Language)

KEYWORDS

JSON (JavaScript object notation): a standard format for transmitting data.

XML (Extensible markup language): a method of defining data formats for data that will be transmitted around a network.

JSON and **XML** are two alternative methods for formatting data objects that are being transferred across servers and web applications. Both have become standard methods.

JSON is a data format originally created as part of the JavaScript programming language, but now available as a standalone format that can be implemented using most programming languages. It is defined as human-readable and is made up of an object and values. For example, a database of names might look like this:

```
{"customers":[
    {"firstName":"Alan", "lastName":"Brown"},
    {"firstName":"Asif", "lastName":"Javid"},
    {"firstName":"Mary", "lastName":"Smith"}
]}
```

In this case the objects are `firstName` and `lastName` and the values of each are shown in speech marks. In standard database terms an object is a field and a value is record.

JSON is a compact code that is very easy to understand from a human point of view and therefore easy to implement. Similarly it is easy for computers to parse (or interpret) as each object and value is clearly defined and described on each line.

XML is a markup language that defines how data is encoded. Formatting data in XML is similar to writing code in a programming language and therefore requires more knowledge than producing a JSON file. The example below shows the same database of names written in XML:

```
<customers>
    <customer>
        <firstName>Alan</firstName>
        <lastName>Brown</lastName>
    </customer>
    <customer>
        <firstName>Asif</firstName>
        <lastName>Javid</lastName>
    </customer>
    <customer>
        <firstName>Mary</firstName>
        <lastName>Smith</lastName>
    </customer>
</customers>
```

JSON vs XML

> **KEYWORD**
>
> **Client-server database**: a way of implementing a database where the database is put into a server and various users can access it from their workstations. The processing, for example, running a query, will take place on the server.

JSON and XML have developed as the two main methods of sharing data on networked **client-server databases**. They have many things in common but there are also differences. Table 42.1 summarises the main similarities and differences:

Table 42.1 Comparison of JSON and XML

	JSON	XML
Human readable	Very easy to read as it is based on defining objects and values.	Slightly less easy to read as data is contained within markup tags.
Compact code	Less code is created than XML.	Requires more code then JSON.
Speed of parsing	Quicker than XML as data is clearly defined as object and value.	Slower than JSON as the data has to be extracted from the tags.
Ease of creation	Easier to create as the syntax of the coding is easier.	Similar to programming so therefore more knowledge is required.
Flexibility and extendibility	Works with a limited range of data types, which may not be sufficient for all applications.	Provides complete freedom over what data types are created and therefore allows greater flexibility.

Thin- vs thick-client computing

> **KEYWORDS**
>
> **Thin client**: in a network where one computer contains the majority of resources, processing power and storage capacity, which it distributes to other clients.
>
> **Thick client**: in a network where resources, processing power and storage capacity are distributed between the server and the client computers.
>
> **Terminal**: a computer that has little or no processing power or storage capacity used as a client in a thin client network.

A **thin client** is a computer that depends heavily on a more powerful server to fulfil most of its requirements and processing. The server would be a large powerful computer with lots of processing power and storage capacity that stores the main programs and datasets. The client then taps into these resources, which are not available on the local machine, using a much lower specification computer. In this scenario the majority of hardware and software resources are at the server end. In this scenario the server actually runs the software with the client machine simply acting as a '**terminal**' with very little processing power and no hard disk.

A **thick client** is a fully specified computer like the ones most people have at home. They do not need servers to carry out their processing most of the time. In thick- client computing the resources are allocated between client and server in a different way giving the client greater processing power, more local storage and access to software that is installed and run from the client machine. In this scenario more of the hardware and software resources are at the client end.

The decision on whether to configure a network using a thin- or thick-client model depends largely on what tasks users are completing and what resources they need. Also, with many applications being hosted on the Internet and accessed via browsers, organisations can move more towards thin clients with services available via the Internet rather than via the LAN.

There are advantages and disadvantages with each system as shown in Tables 42.2 and 42.3.

Table 42.2 Advantages and disadvantages of the thin-client model

Advantages	Disadvantages
Easy and cheaper to set up new clients as fewer resources are needed.	Clients are dependent on the server so if it goes down, all clients are affected.
The server can be configured to distribute all the hardware and software resources needed.	Can slow down with heavy use.
Hardware and software changes only need to be implemented on the server.	May require greater bandwidth to cope with client request.
Easier for the network manager to control clients.	High-specification servers are expensive.
Greater security as clients have fewer access rights.	

Table 42.3 Advantages and disadvantages of the thick-client model

Advantages	Disadvantages
Reduced pressure on the server leading to more uptime.	Reduced security if clients can download software or access the Internet remotely.
Clients can store programs and data locally giving them more control.	More difficult to manage and update as new hardware and software need installing on each client machine.
Fewer servers and lower bandwidth can be used.	Data is more likely to be lost or deleted on the client side.
Suitable for tablets and mobile phones that require more of the processing and storage to be done on the server side.	Can be difficult to ensure data integrity where many clients are working on local data.

Practice questions can be found at the end of the section on pages 360 and 361.

KEY POINTS

- The client–server model is a methodology for connecting computers together, usually over a network where one computer provides access to resources for other computers that are connected to it.
- An application program interface (API) defines the way in which programs can work together.
- CRUD stands for Create, Retrieve, Update, Delete.
- Representational State Transfer is a design methodology for networked database applications.
- JSON and XML are two alternative methods for formatting data objects that are being transferred across servers and web applications.
- A thin client is a computer that depends heavily on a more powerful server to fulfil most of its requirements and processing
- A thick client is a fully specified computer like the ones most people have at home.

TASKS

1 What are the principles of the client–server model? Give examples of where it might be used.
2 What is thin- and thick-client computing?
3 Give three advantages and three disadvantages of thin-client computing.
4 What is the purpose of an API?
5 Explain how the websocket protocol creates a persistent connection between client and server.
6 Explain the relationship between CRUD, REST, SQL and HTTP.
7 Explain where you might use JSON and XML and why you might choose to use one rather than the other.

STUDY / RESEARCH TASKS

1 Will all applications eventually be hosted on the Internet and accessed via a browser?
2 Research some commonly used APIs.
3 How is full duplex transmission achieved on a network?
4 What is SOAP and how does it compare to REST?
5 Research into other data transfer formats such as YAML or SXML.
6 JSON and XML have been adopted as standard. Why do some formats get adopted and other disappear?

Section Nine: Practice questions

1 When data is transmitted, additional bits of data are often added to each bit string.

a) Asynchronous data transmission uses start and stop bits. Use an example to explain the purpose of these.

b) The ASCII coding system uses seven bits to encode a character. The eighth bit can be used as a parity bit. Explain how a parity bit is used when transmitting ASCII codes using even parity.

c) Use an example to show how a check digit can be used. What is the purpose of the check digit?

2 A college uses a LAN (Local Area Network) to share software and printers between its students. The diagram shows the current topology.

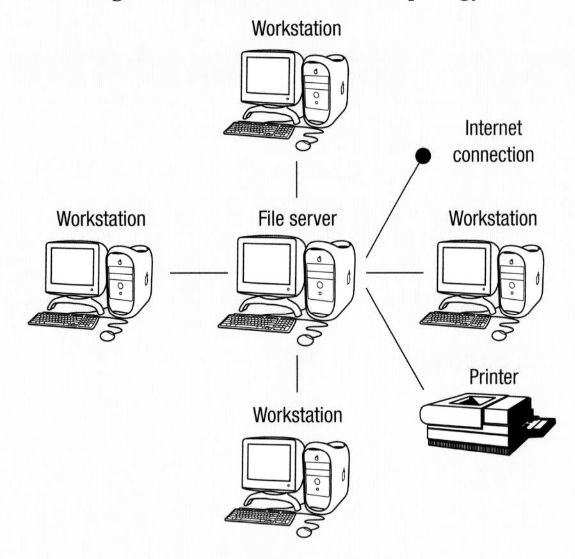

a) Name this topology.

b) Give one advantage of this topology.

c) Give one disadvantage of this topology.

3 Computers could be connected in a topology such as a star or bus. State one advantage of a bus topology over a star topology.

4 An example of a fully qualified domain name is www.aqa.org.uk. Using this example, explain each part of the address:

a) www **b)** aqa **c)** org **d)** uk

5 The domain name is referred to as a look-up for the IP address.

a) What does this mean?

b) Why do users prefer to use a domain name rather than an IP address to access a server?

6 The Internet is one example of a WAN (Wide Area Network).

a) Describe a WAN.

b) Identify two protocols that are used on the Internet. Why are protocols needed?

c) Explain how security of data transmission could be improved with the use of a digital signature.

d) Describe how public and private keys are used to create asymmetric encryption.

7 A computer connects to a server using port 60. The IP address and port number create a socket address.

a) What is a socket?

b) What is a port number?

c) What is an IP address?

d) How can computers connected to a network be identified?

8 Port A of the router in the diagram is assigned the IP address 192.168.1.1. Port B is assigned the IP address 213.208.10.146. Which of these IP addresses needs to be registered with an Internet Registry and why?

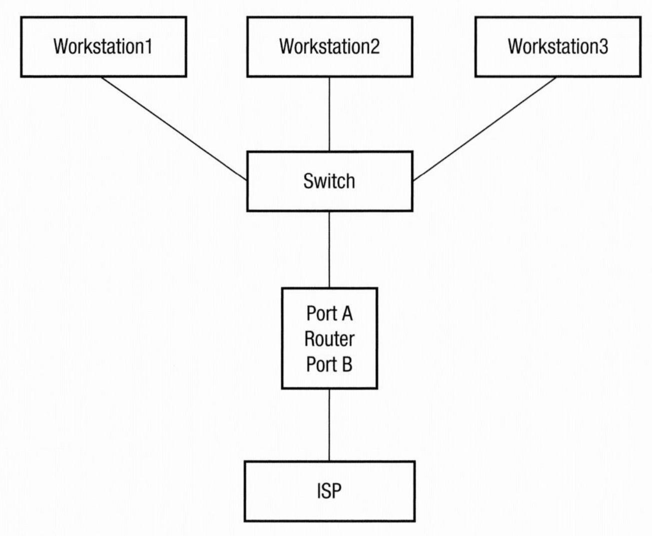

9 Some governments put pressure on ISPs to monitor and control the way in which users access the Internet. Give two reasons why governments may want to control users' access to the Internet.

10 Explain how the collision detection system called Carrier Sense Multiple Access with Collision Avoidance (CSMA/CA) CMSA/CA protocol works in a WiFi network.

11 What is the relationship between baud rate, bit rate and bandwidth?

12 The star topology in Figure 38.2 is configured as a client–server network.

a) Explain the term client–server.

b) Under what circumstances might a peer-to-peer network be used?

c) Explain the difference between a thick and a thin client.

13 The full IP address of a workstation in a network is shown in binary as:

01110101.00110010.10010100.10101000

The administrator uses a subnet mask to identify the first 24 bits of the IP address: 255.255.255.0. Show how the subnet mask would be applied and what the resulting address would be.

14 What are the main features of the Secure Shell (SSH) protocol?

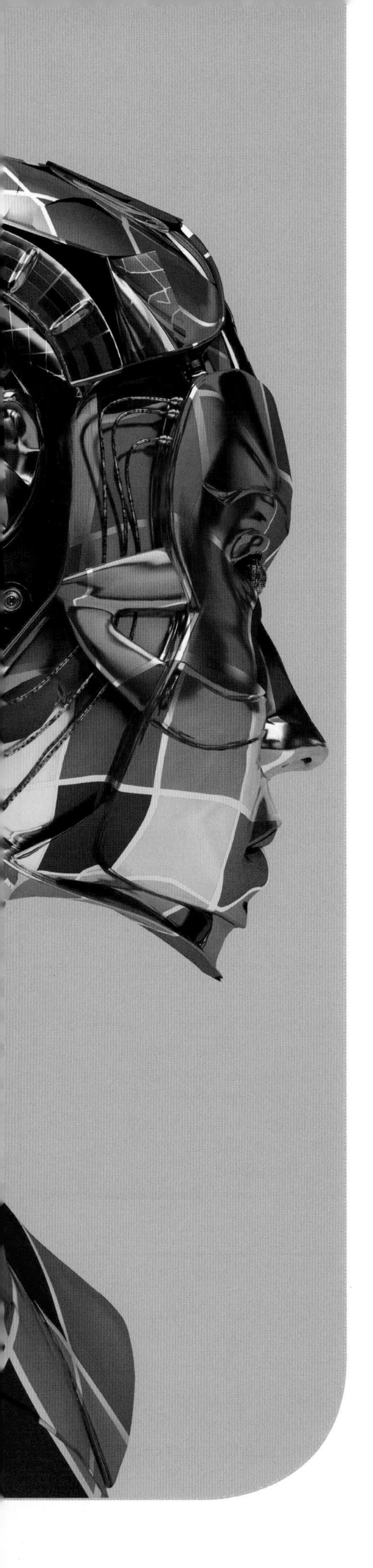

Section Ten: Fundamentals of databases

43 Relational databases

A level only

SPECIFICATION COVERAGE

3.10.1 Conceptual data models and entity relationship modelling

3.10.2 Relational databases

3.10.3 Database design and normalisation techniques

LEARNING OBJECTIVES

In this chapter you will learn:

- how relational databases store related data in linked tables
- how to define relationships between tables and link them together using primary and foreign keys
- how to represent relationships using Entity Relationship (ER) diagrams
- how to define the data within a relational database
- what the main components of a relational database are and how to set one up
- how to normalise a relational database.

INTRODUCTION

A **relational database** models data as mathematical relations, with each relation being composed of tuples of data, and the data within each tuple being related in some way. When a relational database model is implemented by software, a relational database program will represent each relation as a table, and each tuple will be a record within a table. The data in different tables can be linked together to express relationships that exist between the data in the tables.

KEYWORD

Relational database: a method of creating a database using tables of related data, with relationships between the tables.

Relationships

If you were to set up a database for a movie download site, you may create one **table** to store customer data, one to store data about the movies and one to store download details:

- The `CUSTOMER` table contains data all of which are related to the customer, for example, their name and address.
- The `MOVIE` table contains data related to the movie, for example, the title and genre.

KEYWORD

Table: a method for implementing on entity and attributes as a group of related data.

- The **DOWNLOAD** table contains data related to the actual download itself, for example, when it was downloaded, how much the customer paid, what file type was downloaded.

There are some real-world relationships between these three tables. For example:

- one-to-many: one customer may have many downloads
- many-to-many: one customer could download many movies and one movie could be downloaded by many customers.

It is also possible, albeit less common, to have a one-to-one relationship although there are none in this particular example.

Entities

KEYWORD

Entity: an object about which data will be stored.

In a relational database, a relation stores information about an **entity** and its attributes. An entity is an object about which data will be stored. In our movie example, a customer may be an entity that has attributes such as name and address. These relations are described and stored within tables. In this example, we have already identified three tables: **CUSTOMER**, **MOVIE** and **DOWNLOAD**.

One of the first tasks when creating a relational database is to decide on how many tables are needed to solve the problem. To do this, you must use a technique called normalisation, which ensures that databases are truly relational and are organised effectively. In view of this a further table called **MOVIEFORMAT** is added to the database. The reasons for this are explained in more detail later in this chapter in the section on normalisation.

Note that it is common practice to identify tables with capital letters and this has been adopted in this book.

Attributes

KEYWORD

Attribute: a characteristic or piece of information about an entity, which would be stored as a field in a relational database.

An **attribute** is a piece of information about an entity, which is implemented as a field in a relational database. With the movie download database example, we will store different items of data relating to each entity in a table. Possible attributes, only some of which we will use, include:

CUSTOMER: Customer Name, Address, Phone Number, Date of Birth

MOVIE: Movie Title, Age Classification, Genre

DOWNLOAD: Date of Download, Price, Method of Payment

MOVIEFORMAT: File type

Entity relationship diagrams

A relationship is the link created between two entities. Each entity is likely to be related to at least one of the other entities. There are three types, or degrees, of relationship, two of which exist in the movie database:

- One-to-many: One customer will have many downloads.
- Many-to-many: Many customers could have many downloads.

KEYWORD

Entity relationship diagram: a visual method of describing relationships between entities.

Entity relationship diagrams are used to show these relationships as shown in Figure 43.1.

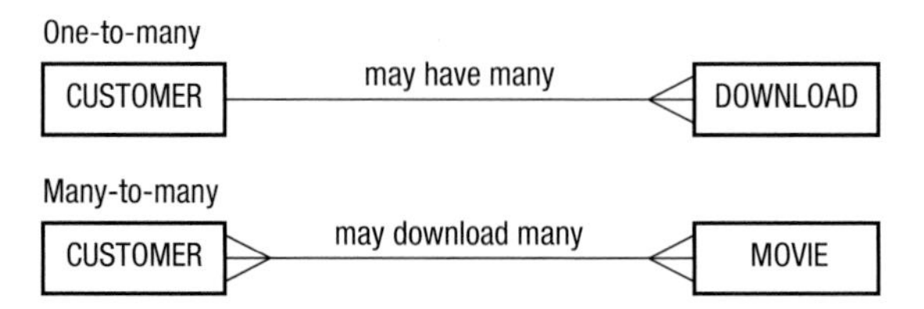

Figure 43.1 Entity relationship diagram

Notice that the name of the entity is shown in the box with the lines indicating the nature of the relationship. Labels are usually added above the lines to clarify the relationship.

The nature of relationships is sometimes hard to define. You should choose the one that best describes the relationship in logical terms. In our example you could say that the relationship between `CUSTOMER` and `DOWNLOAD` could be any one of the three:

- One customer has one download.
- One customer has many downloads.
- Many customers have many downloads.

However, the most accurate way to describe it is that one customer could have many downloads because this best describes the nature of the relationship in a real-life context.

When creating a relational database, you should replace any many-to-many relationships with one-to-many relationships. In the example, we replace the many customers to many movies relationship setting up the `DOWNLOAD` entity as a link as shown:

Figure 43.2 Resolving a many-to-many relationship

The third type of relationship, which does not exist in the movie database, is a one-to-one relationship. In a school, if a teacher only taught in one classroom and that classroom was only used by the one teacher, then this would be a one-to-one relationship. This relationship is shown in Figure 43.3.

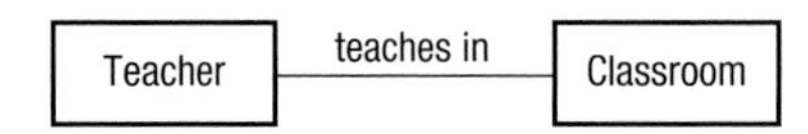

Figure 43.3 Entity relationship diagram for a one-to-one relationship

Primary key and entity identifier

KEYWORDS

Primary key: an attribute that can be used to uniquely identify every record within a table.

Entity identifier: an attribute which can uniquely identify each instance of an entity.

The **primary key** is the attribute in each table that uniquely identifies each record. It is related to the concept of an **entity identifier**, which is an attribute that can be used to uniquely identify each instance of an entity at the conceptual level. There must be a way of ensuring that every record in an entity table can be identified individually, otherwise the relationships between the tables cannot be made. For example, in the `CUSTOMER` table we will need to store hundreds of names and addresses. We could use the customer's name as the primary key but if there were two customers with the same name, then we would not be able to tell one from the other. There are three possible solutions:

- Use a unique attribute: Sometimes there is an attribute that is already unique. For example, if you were storing personal details you could use the National Insurance number as this is unique to every person in the country. If you were storing data about cars, you could use the Vehicle Identification Number (VIN) as this is unique to a particular car.
- Create a unique attribute: We could invent a unique code or identifier (ID) for each customer. Then if two people had the same name, the ID would be used so that you knew which was which. This could be used in our example as we could create a customer ID. Some relational database programs such as Microsoft Access have a facility called 'AutoNumber' which automatically allocates a unique number to each record.
- Use a composite key: Two or more attributes could be used in combination. For example, using name and address as a composite key may ensure that each record is unique as it is unlikely that you will have two customers with the same name at the same address. However, it is still possible, for example, a father and son who are both called John Smith who live at the same address.

Foreign key

KEYWORD

Foreign key: an attribute in a table that is a primary key in another table and is used to link tables together.

A **foreign key** is an attribute that appears in more than one table and is used to create the link between tables. The foreign key in a table must be a primary key from another table. For example, if one customer can have more than one download, how do we create the one-to-many relationship between the CUSTOMER table and the DOWNLOAD table?

The answer is to put the CustomerID in the DOWNLOAD table as a foreign key. The relationships in our movie download case study could be shown as in Figure 43.4.

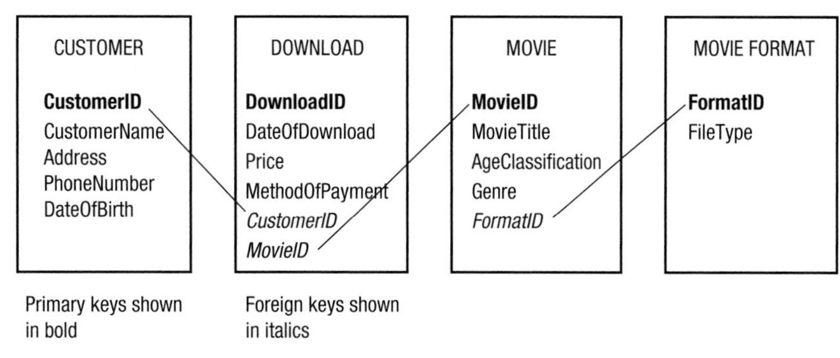

Figure 43.4 Links between tables

- Primary keys have been added for each entity in the form of unique IDs.
- CustomerID appears on the CUSTOMER table as the primary key and on the DOWNLOAD table as a foreign key.
- MovieID appears on the MOVIE table as the primary key and on the DOWNLOAD table as a foreign key.
- FormatID appears on the MOVIEFORMAT table as the primary key and on the MOVIE table as the foreign key.

Now that the relationships have been created, the four tables become one database. Users of the database will be unaware of the structure in the background. As far as they are concerned they are dealing with one database that contains all the information they need. It is common practice to write out the details of relational databases in standard database notation as shown:

CUSTOMER (<u>CustomerID</u>, CustomerName, Address, PhoneNumber, DateOfBirth)

MOVIE (<u>MovieID</u>, MovieTitle, AgeClassification, Genre, FormatID)

DOWNLOAD (<u>DownloadID</u>, DateOfDownload, Price, MethodOfPayment, CustomerID, MovieID)

MOVIEFORMAT (<u>FormatID</u>, FileType)

Note:

- the name of the table is shown in capitals
- all the attributes are placed between brackets
- primary keys are underlined.

Normalisation

KEYWORD

Normalisation: the process of ensuring that a relational database is structured efficiently.

Normalisation is the process of ensuring that a relational database conforms to certain rules that ensure that the data within it is stored in the most efficient way. In simple terms a database is normalised when there is no redundant data and when each item of data is stored in the correct table and at an atomic level.

- Redundant data occurs when the same field is unnecessarily duplicated in two or more tables. For example, many different customers may download the same movie. If all the movie details were stored every time it was downloaded, much of the data would be redundant as we only actually need to store the movie details once and then link those details to each customer who downloads it.
- Storing the same data multiple times can also lead to the problem of data inconsistency, for example we might store the same customer's details several times but the telephone number stored might differ. How would we know which was correct?
- Storing data at an atomic level means that they cannot be further decomposed. For example, a table may contain an attribute called `Address` that stores the full address of the customer. At an atomic level, this could be decomposed into several attributes, for example: `HouseNumber, Street, Town, County,` or `AddressLine1, AddressLine2` etc.

When a database is constructed according to these rules it is said to be in normal form.

There are various levels of normal form and the level that a programmer needs to go to depends on the complexity of the database. For A level, you should be able to develop a database to 'third normal form' (3NF). The process of normalisation is shown below for the movie download database.

First normal form (1NF)

First normal form is achieved by ensuring that a table does not contain repeating attributes or repeating groups and that all of the data in the table is atomic. For example, a first attempt at creating a database for the movie download system might produce Table 43.1.

Table 43.1

CustomerID	CustomerName	Address	DateOfDownload	MovieID	Movie Title	Genre	Format	FileType
1	John Smith	1 High Street	19/03/15 19/03/15	1 2	The Hangover 22 Jump Street	Comedy Comedy	LowRes LowRes	MPEG-2 MPEG-2
2	Mary Jones	14 Acacia Avenue	19/03/15 19/03/15 19/03/15	3 4 2	The Hunger Games Robocop 22 Jump Street	Sci-Fi Sci-Fi Comedy	HiRes HiRes LowRes	MPEG-4 MPEG-4 MPEG-2
3	John Smith	23 Maple Drive	19/03/15 19/03/15	5 4	How to Train Your Dragon Robocop	Children Sci-Fi	HiRes HiRes	MPEG-4 MPEG-4

- This is not in first normal form (1NF) because there are repeating groups, which are shown in the **`DateOfDownload`**, **`MovieID`**, **`MovieTitle`**, **`Genre`**, **`Format`** and **`FileType`** columns. A repeating group is when a group of values is stored in a particular row/column intersection in a database table instead of a single value.
- To satisfy first normal form, the repeating groups should be replaced by creating one record for each download as shown in Table 43.2.

Table 43.2

CustomerID	CustomerName	Address	DateOfDownload	MovieID	MovieTitle	Genre	Format	FileType
1	John Smith	1 High Street	19/03/15	1	The Hangover	Comedy	LowRes	MPEG-2
1	John Smith	1 High Street	19/03/15	2	22 Jump Street	Comedy	LowRes	MPEG-2
2	Mary Jones	14 Acacia Avenue	19/03/15	3	The Hunger Games	Sci-Fi	HiRes	MPEG-4
2	Mary Jones	14 Acacia Avenue	19/03/15	4	Robocop	Sci-Fi	HiRes	MPEG-4
2	Mary Jones	14 Acacia Avenue	19/03/15	2	22 Jump Street	Comedy	LowRes	MPEG-2
3	John Smith	23 Maple Drive	19/03/15	5	How to Train Your Dragon	Children	HiRes	MPEG-4
3	John Smith	23 Maple Drive	19/03/15	4	Robocop	Sci-Fi	HiRes	MPEG-4

- If a customer downloads more than one movie then there will be multiple records for the same customer. In the case of Mary Jones, she has three records in the table as she has downloaded three movies.
- Each download can now be uniquely identified with a composite key made up of **`CustomerID`** and **`MovieID`**, so this could be made into the primary key. It is possible that one customer could download the same movie again, in which case this primary key would not be adequate, but for this example we will assume that a customer will only download the same movie once.

Second normal form (2NF)

Second normal form is achieved by ensuring the database is in first normal form and then removing attributes that depend upon part but not all of the primary key by creating additional tables.

The non-key attributes are all the other attributes apart from the primary key. For example, **Address, DateOfDownload, Genre** and **FileType** are non-key attributes. To be in second normal form, any non-key attributes that depend upon part but not all of the primary key should be removed to another table. For example, the **Address** of the customer is dependent on the **CustomerID**, but not on the **MovieID**. Similarly, the **Genre** is dependent on the **MovieID** but not on the **CustomerID**. So, both **Address** and **Genre** are examples of attributes that depend on part but not all of the primary key. This means that separate tables are needed to store the customer data and the movie data.

To start with, we will separate the information about customer downloads and movies into two tables:

Table 43.3 CUSTOMER DOWNLOAD

CustomerID	CustomerName	Address	DateOfDownload	MovieID
1	John Smith	1 High Street	19/03/15	1
1	John Smith	1 High Street	19/03/15	2
2	Mary Jones	14 Acacia Avenue	19/03/15	3
2	Mary Jones	14 Acacia Avenue	19/03/15	4
2	Mary Jones	14 Acacia Avenue	19/03/15	2
3	John Smith	23 Maple Drive	19/03/15	5
3	John Smith	23 Maple Drive	19/03/15	4

Table 43.4 MOVIE

MovieID	MovieTitle	Genre	Format	FileType
1	The Hangover	Comedy	LowRes	MPEG-2
2	22 Jump Street	Comedy	LowRes	MPEG-2
3	The Hunger Games	Sci-Fi	HiRes	MPEG-4
4	Robocop	Sci-Fi	HiRes	MPEG-4
5	How to Train Your Dragon	Children	HiRes	MPEG-4

Notice that when we split the initial table up into two tables, we have kept an attribute which is common to both tables (**MovieID**) so that we can link the information in the two tables together.

Each movie in the **MOVIE** table (Table 43.4) is now identified by the primary key **MovieID**. Every non-key attribute in the **MOVIE** table depends on the whole of this primary key, so this table is now in second normal form. The **MovieID** field also exists in the **CUSTOMER DOWNLOAD** table, as a foreign key.

The **CUSTOMER DOWNLOAD** table (Table 43.3) is more problematic as the primary key for this table would be a composite key made up of the **CustomerID** and the **MovieID**. Together, these two fields form a primary key as they would be unique to each record because we have assumed that a particular customer will only download the same movie once. It is still the case that this table is not in second normal form as it contains attributes that depend upon part, but not all, of the primary key. For example, **CustomerName** depends upon the **CustomerID** but not the **MovieID**. The solution is to split this table up further into a **CUSTOMER** table and a **DOWNLOAD** table:

Table 43.5 CUSTOMER

CustomerID	CustomerName	Address
1	John Smith	1 High Street
2	Mary Jones	14 Acacia Avenue
3	John Smith	23 Maple Drive

Table 43.6 DOWNLOAD

CustomerID	MovieID	DateOfDownload
1	1	19/03/15
1	2	19/03/15
2	3	19/03/15
2	4	19/03/15
2	2	19/03/15
3	5	19/03/15
3	4	19/03/15

The **CUSTOMER** table (Table 43.5) now has the attribute **CustomerID** as the primary key. The two other attributes in this table depend on the whole of the primary key so this table is now in second normal form. In fact, as the primary key for this table consists of only one attribute, it would not be possible for an attribute to depend upon part but not all of this.

As we have assumed that each customer will download a movie only once, the **DOWNLOAD** table (Table 43.6) can have a composite key made up of **CustomerID** and **MovieID**. The only other attribute in the table, **DateOfDownload**, depends on both of these, so this table is also in second normal form.

Third normal form (3NF)

Third normal form is achieved by ensuring the database is in second normal form and then removing non-key attributes that depend upon other non-key attributes by creating additional tables.

If we look at each of the three tables in turn:

- Table 43.4 **MOVIE**: It can be noted that the **FileType** depends upon the **Format**. All LowRes films are recorded in MPEG-2 format and all HiRes films are recorded in MPEG-4 format. Therefore, the non-key attribute **FileType** depends upon the non-key attribute **Format** so we can split the format information off from the movie information to create two new tables:

Table 43.7 MOVIE

MovieID	MovieTitle	Genre	Format
1	The Hangover	Comedy	LowRes
2	22 Jump Street	Comedy	LowRes
3	The Hunger Games	Sci-Fi	HiRes
4	Robocop	Sci-Fi	HiRes
5	How to Train Your Dragon	Children	HiRes

Table 43.8 MOVIEFORMAT

Format	FileType
LowRes	MPEG-2
HiRes	MPEG-4

Both of these tables are now in third normal form. Table 43.7 **MOVIE** has **MovieID** as the primary key and all the non-key attributes in this table depend upon **MovieID** and no other non-key attributes. Table 43.8 **MOVIEFORMAT** has **Format** as the primary key and only contains one other attribute, **FileType**. As there is only one non-key attribute in this table, it must be in third normal form as it is not possible for a non-key attribute to depend on another non-key attribute.

- Table 43.5 **CUSTOMER**: The **Address** and **Name** both depend on the **CustomerID** and not on each other, so this table is already in third normal form.
- Table 43.6 **DOWNLOAD**: There is only one non-key attribute, so this cannot possibly depend upon another non-key attribute, so this table must already be in third normal form.

Fully normalised design

The final, fully normalised design of the database is as follows:

CUSTOMER

CustomerID	CustomerName	Address
1	John Smith	1 High Street
2	Mary Jones	14 Acacia Avenue
3	John Smith	23 Maple Drive

MOVIE

MovieID	MovieTitle	Genre	Format
1	The Hangover	Comedy	LowRes
2	22 Jump Street	Comedy	LowRes
3	The Hunger Games	Sci-Fi	HiRes
4	Robocop	Sci-Fi	HiRes
5	How to Train Your Dragon	Children	HiRes

MOVIEFORMAT

Format	FileType
LowRes	MPEG-2
HiRes	MPEG-4

DOWNLOAD

CustomerID	MovieID	DateOfDownload
1	1	19/03/15
1	2	19/03/15
2	3	19/03/15
2	4	19/03/15
2	2	19/03/15
3	5	19/03/15
3	4	19/03/15

- The `CustomerID` is the primary key of the `CUSTOMER` table and a foreign key in the `DOWNLOAD` table.
- The `MovieID` is the primary key of the `MOVIE` table and a foreign key in the `DOWNLOAD` table.
- The `Format` is the primary key of the `MOVIEFORMAT` table and a foreign key in the `MOVIE` table.
- The `DOWNLOAD` table has a composite key made up of `CustomerID` and `MovieID`.
- At this point, a database designer might choose to add an additional field, `DownloadID`, to the `DOWNLOAD` table which would be unique for each download. This would mean that the composite key of `CustomerID+MovieID` could be replaced by a primary key that was just one field. This might be considered to be an improvement, but it is not required for normalisation.

In summary, the characteristics that a relation database design must have to be fully normalised are:

- All of the data must be atomic / there must be no repeating groups / no repeating attributes.
- There should be no partial dependencies, where a non-key attribute depends upon part but not all of the primary key.
- There should be no non-key dependencies, where a non-key attribute depends upon another non-key attribute.

Practice questions can be found at the end of the section on pages 390 and 391.

TASKS

1 What is a relational database?

2 Explain the following terms:
  a) entity **b)** attribute **c)** relationship.

3 Describe how primary keys and foreign keys are used to create relationships between tables.

4 A company employs engineers to fix faults on photocopiers. The company has 20 engineers and over 200 clients. The engineers will travel to client sites in order to fix their machines and can fix between one and five machines a day. The company employs a 'work controller' who takes the call from the client and then allocates an engineer to the job. She uses a relational database to keep track of all the details relating to the clients, engineers and jobs. Part of the database is shown below in standard notation:

```
CLIENT (Name, Address)
ENGINEER (Name, Address)
JOB (Date, Nature of Problem)
```

  a) Suggest a suitable identifier for each table.
  b) Draw an entity-relationship diagram to show the relationship between the three entities.
  c) Suggest foreign keys that could be used to create the relationships.
  d) Identify four other attributes that the company may store.
  e) Now complete the standard notation.
  f) How could attributes be combined to form a composite key the `ENGINEER` table?

5 A garage uses a database to store details about its customers, their cars and the repairs that are carried out. The system is currently stored as a single table, an extract of which is shown below.
   a) Identify three problems with the way the database is currently stored.
   b) A relational database is to be designed using three tables: `CUSTOMER`, `CAR` and `REPAIR`. Normalise the database to 3NF showing your answer in standard notation:
   `ENTITYNAME (Primary key, attribute 1, attribute 2 …)`
   This will involve identifying suitable primary and foreign keys.

Customer	Address	Reg no	Make of car	Date repaired	Repair carried out
John Brown	1 High Street	M222 HGG	Ford Escort	11/03/15	Replace exhaust
Mary Jones	10 Low Road	K222 HKK	VW Golf	11/03/15	Electrical fault
John Brown	1 High Street	P333 AAA	Citroen Saxo	11/03/15	New tyres
Jane Fox	2 New Lane	J123 AAA	VW Polo	11/03/15	Starter motor

KEY POINTS

- Relational databases are made up of related data stored within a series of linked tables.
- Entity relationship diagrams can be used to show the relationships that exist between tables.
- The main relationships are: one-to-one, one-to-many and many-to-many.
- Primary keys uniquely identify each record within a table.
- A foreign key in one table is the primary key in another table and is used to link the tables together.
- Normalisation is the process of ensuring that data is stored efficiently to eliminate data redundancy and ensure data consistency.

STUDY / RESEARCH TASKS

1 There are further stages of normalisation beyond third normal form (3NF) including fourth normal form (4NF) and Boyce-Codd normal form (BCNF). Research these and find out why they are necessary.

2 What is a flat-file database and why is a relational database more suited to storing complex data?

3 Create a relational database for the movie download database used in this chapter, using a programming language or database application available within your centre.

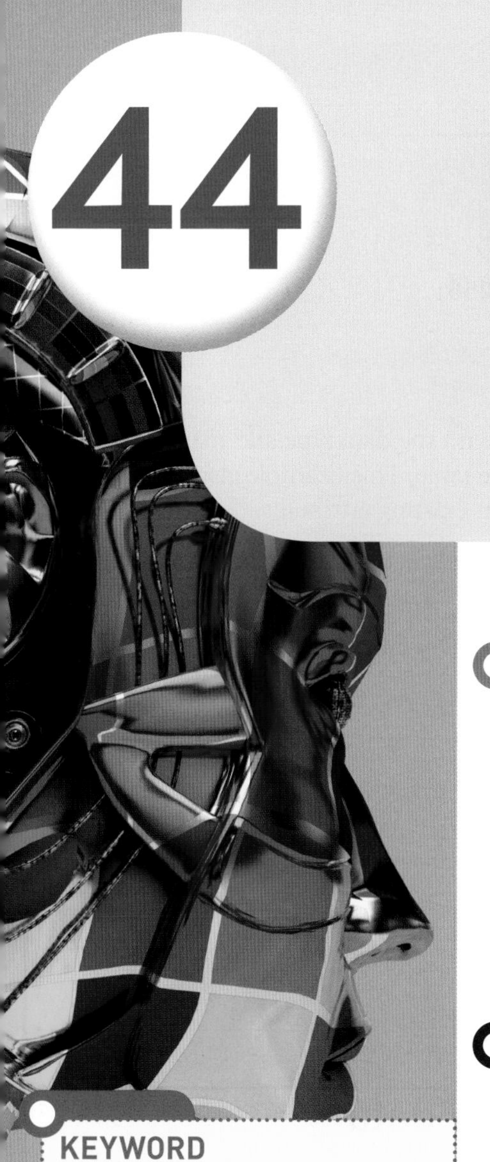

44 Structured query language (SQL)

A level only

SPECIFICATION COVERAGE

3.10.4 Structured query language (SQL)

3.10.5 Client–server databases

LEARNING OBJECTIVES

In this chapter you will learn:

- how to use SQL to define a table, enter, update and delete data
- how to use SQL to query data, including searching and sorting
- what a client–server database is and what problems are caused when there is shared access to a database
- what a database management system is.

INTRODUCTION

Structured Query Language (SQL) is a specialised programming language that is used for managing relational databases. Its functions allow users to define tables, insert, update and delete data and to carry out queries on data to produce and output subsets of the main data. In common with other programming languages, SQL works by typing in lines of code. Examples are shown below for the main functions using the movie download database from the previous chapter as an example.

KEYWORD

Structured query language (SQL): a specialised programming language for manipulating databases.

Defining a table

To create a **table** the user needs to define the name of the table and each of the attributes including the data type and length. If you have used MS Access or any other proprietary database package, you will notice that this is a very similar process, but is achieved through typing code in the correct **syntax** rather than using a graphical user interface.

KEYWORDS

Table: a method for implementing an entity and attributes as a group of related data.

Syntax: the rules of how words are used within a given language.

```
CREATE TABLE Customer
(
CustomerID varchar (5),
CustomerName varchar (255),
CustomerAddress varchar (255),
PRIMARY KEY (CustomerID)
);
```

Notice the syntax of the commands with the data type and maximum length shown for each attribute in the table. This example also shows how the primary key can be assigned from existing attributes.

There are a number of supported data types, with examples shown in Table 44.1.

Table 44.1 Examples of supported data types in SQL

Character (n)	character string with fixed length (n)
Varchar (n)	character string variable length with maximum field length (n)
Boolean	true or false
Int	short for integer and is a whole number
Decimal (p,s)	decimal number with number of digits before and after the decimal point
Real	any number up to 7 decimal places
Date	in the format day, month, year
Time	in the format hour, minutes, seconds

Entering and updating data

To enter data into a table, you need to specify the name of the table and the column where you want to enter it:

```
INSERT INTO Customer (CustomerID, Name, Address)
VALUES ("1", "John Smith", "1 High Street");
```

This will create a row of data in the Customer table entering the details for John Smith. An alternative syntax can be used where every field is being entered as follows:

```
INSERT INTO Customer
VALUES ("1", "John Smith", "1 High Street");
```

As every field is being entered it is not necessary to input all the field names.

To update data, you need to specify the table and column and identify the item of data that needs updating. For example, to update John Smith's address:

```
UPDATE Customer
SET Address = "29 Wellington Street"
WHERE CustomerID = "1";
```

Notice that the **WHERE** command is being used to make a selection using the **CustomerID** rather than the name to ensure the correct record is updated.

Deleting data

Deleting data is a similar process to updating data as you have to define the table and then use a selection statement to identify the data that you want to delete. To delete John Smith from the database:

```
DELETE FROM Customer
WHERE CustomerID = "1";
```

Note that John Smith's unique ID is used to identify which record to delete to avoid the possibility of there being more than one customer named John Smith and the wrong record being deleted.

You can use wildcards within a selection statement. For example, to delete all records:

```
DELETE * FROM Customer;
```

This will delete all of the records while keeping the structure of the table intact. You can also delete all records as follows:

```
DELETE FROM Customer;
```

Querying data

KEYWORD

Query: a search or sort carried out on data that retrieves the answer to a question.

In simple terms a **query** is a search and/or sort. An extract of code is shown below relating to a query that extracts the name and address of all customers called John Smith in the database:

```
SELECT CustomerName, CustomerAddress
FROM Customer
WHERE CustomerName = "John Smith"
ORDER BY CustomerName DESC;
```

- `SELECT`: Identifies the columns that you wish to extract. In this example, all the columns are from one table. Where the columns are from more than one table and if the field name is used in more than one table, it is necessary to include the name of the table followed by a full stop and then the name of the column. For example, `Customer.CustomerName` indicates that we wish to extract the `CustomerName` column from the `Customer` table.
- `FROM`: Indicates the table or tables that are needed to extract the data. In this example, there is only one table. If there were more, you would have to list them all, separating them with commas.
- `WHERE`: Indicates the condition that must be met. In this case, the condition is `CustomerName = "John Smith"`. There may not be any conditions, for example, you may simply want to print a list of customer names. If this is the case, the `WHERE` statement can be left out altogether.

More complex statements can be used within the `WHERE` structure including AND and OR statements. For example, the condition could be `CustomerName = "John Smith" OR CustomerName = "Mary Jones"`. To extract all the data for such records, the following SQL would be written:

```
SELECT *
FROM Customer
WHERE CustomerName = "John Smith" OR CustomerName =
"Mary Jones";
```

Note that the * is used as a wildcard which means that all attributes are extracted when the query is run.

Example

Suppose we wanted to perform a query to produce the name and address of all customers who have downloaded movies on a particular day, perhaps to identify the total value of downloads on that day. The extract should be sorted in ascending order by customer name. This involves querying three tables.

CUSTOMER

CustomerID	CustomerName	Address	PhoneNumber	DateOfBirth
1	John Smith	1 High Street	01555354354	30/03/67
2	Mary Jones	12 Acacia Avenue	01555564333	23/04/78
3	John Smith	23 Maple Drive	01555653535	23/08/72

MOVIE

MovieID	MovieDownloaded	GenreClassification	Age	Price
1	Robocop	SciFi	PG	12.99
2	How to Train Your Dragon	Kids	PG	10.99
3	22 Jump Street	Comedy	15	9.99
4	The Hunger Games	Drama	PG	9.99
5	The Hangover	Comedy	18	10.99

DOWNLOAD

DownloadID	DateOfDownload	CustomerID	MovieID
1	19/03/15	1	1
2	19/03/15	1	2
3	19/03/15	1	3
4	19/03/15	2	3
5	20/03/15	2	4
6	20/03/15	3	1
7	20/03/15	3	1

```
SELECT CustomerName, Address, DateOfDownload,
Movie.MovieID, MovieDownloaded, Price
FROM Customer, Download, Movie
WHERE Download.DateOfDownload = "19/03/15" AND
Customer.CustomerID = Download.CustomerID AND
Movie.MovieID = Download.MovieID
ORDER BY CustomerName DESC;
```

Note that the table names are being shown in the **SELECT** statement as more than one table is being used to perform the query. This would extract the following data:

CustomerName	Address	DateOfDownload	MovieID	MovieDownloaded	Price
Mary Jones	14 Acacia Avenue	19/03/15	3	22 Jump Street	9.99
John Smith	1 High Street	19/03/15	1	Robocop	12.99
John Smith	1 High Street	19/03/15	2	How to Train Your Dragon	10.99
John Smith	1 High Street	19/03/15	3	22 Jump Street	9.99

This data may be extracted in the form of a table, or may be compiled directly into a report. Notice the use of the `DESC` command after the `ORDER BY` to ensure the data is sorted in descending order.

Client–server databases

Where databases are being used in a network environment it is likely that there will be a dedicated database server, particularly where the organisation has a large and complex database. The database server holds and manages the database itself so that all amendments, searches and so on are carried out at the server.

This would normally be done through a **database management system** or DBMS, which is a program that controls the data that is kept on the database. This will help to maintain the integrity of the data as it ensures that there is only ever one version of the data.

One of the biggest benefits of working with a database is that it can be accessed by many different users or programs and the data can in turn be used in many different ways. However this can generate problems – what happens if two programs try to access and update the same item of data at the same time? What happens if a program needs to add a new attribute to an entity, but other programs that access the same dataset don't 'know' about the extra attribute?

The DBMS controls the data that are kept on the database. It also manages how the data are stored, whereabouts in the system they are kept, and it can control access rights as well. The diagram shows how the four departments in a company all access the data files via the DBMS – none of them have direct access to the data.

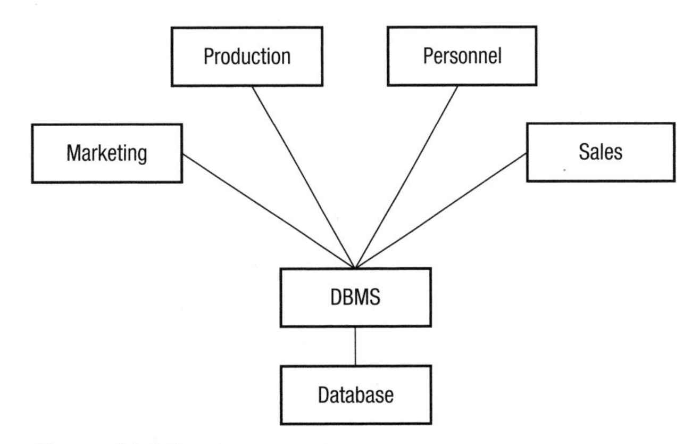

Figure 44.1 Database management system

KEYWORDS

Client–server database: a way of implementing a database where the database is put into a server and various users can access it from their workstations. The processing, for example, running a query, will take place on the server.

Database management system: software that enables the management of all aspects of a database including adding, updating and querying data.

Issues with concurrent access on shared databases

One common problem with a database that is accessible to a number of users is what to do if a number of users are trying to access the same data at the same time. As long as the users are only reading the data this is not a problem, but if two or more users want to write data there will be problems.

Imagine two users both want to put some data in the same location. The first person presses the save button and their updated data get saved to the file. A few seconds later the second presses their save button. Their data now overwrite the first person's data but how do you know which is the most up-to-date version?

KEYWORDS

Record locks: a technique to temporarily prevent access to certain records held on a database.

Serialisation: a technique to ensure that only one transaction at a time is executed from multiple users on a database.

Timestamp ordering: a technique to ensure multiple users can execute commands on a shared database based on the timestamp of when the data was last written to or read from.

Commitment ordering: a technique to ensure concurrent transactions on a shared database are executed based on the timestamp of when the request is made and also the precedence the request takes over other simultaneous request.

- **Record locks**: One solution to these problems is to put a lock on the data. As soon as a user with write access takes an item of data a lock is put on that data item so that no other user can save to that location, and the second person will not be able to save their version of the data without acknowledging that the data may have been altered by another user. You can see a simplified version of this process in use across a network. Two users can both load the same word-processed file, but only the first person to load will be given write access – the other person will see a read-only version of the file.
- **Serialisation**: This is the process of only allowing transactions on a particular database to take place one at a time, that is, in serial format. This process, which is managed by the DBMS, ensures that each transaction is carried out in the correct sequence to avoid compromising the integrity of the data. For example, a typical transaction maybe to read data to or from a record. If another transaction is taking place on that record at the same time, then serialisation ensures that the two transactions take place one after each other in the correct order. The DBMS will be used to identify which transactions need serialising and also provide a schedule for dealing with them.
- **Timestamp ordering**: Every transaction that takes place on the database will have a read and write timestamp that indicates the last time the record was written to or read from.
  To ensure serialisation, the timestamp can be used as a method of sequencing the transactions. To avoid concurrency issues where two transactions are taking place on the same record at the same time, the DBMS can use the timestamp to identify the last action that took place and it will use a protocol to decide whether to execute the current transaction or not based on the timestamp.
- **Commitment ordering**: This system looks at each command it has been asked to execute on the database in terms of when it was made but also in terms of whether it should take precedence over other commands. This depends on the nature of the command and what the impact of it is on the database. For example, where a command from one user could cause deadlock, it would be blocked until the dependent action was completed. This can be implemented by building a graph of transaction dependencies.

Practice questions can be found at the end of the section on pages 390 and 391.

KEY POINTS

- SQL is a query language usually used with relational databases, which works in a similar way to a programming language in that you need to type in lines of code.
- SQL can be used to define tables and add, update and remove data.
- SQL can be used to search and sort data held within one of more tables.
- A database management system (DBMS) is used to control and administrate databases.
- Databases can be made available to users using the client-server model where the database is stored on a database server.
- Problems can arise where several users are accessing the data at the same time.
- There are various techniques for dealing with these problems including record locking, serialisation, timestamp ordering and commitment ordering.

TASKS

1 Write SQL code to create the following queries from the movie download database example used in this chapter.
  a) Select all kid's movies.
  b) Select all customers who have downloaded comedy films.
  c) Select all customers born before 1972.
  d) Select all customers who downloaded movies on 20/03/13 sorted in descending order on CustomerName.

2 Give an example of a how a wildcard can be used when using SQL.

3 What issues might arise when multiple users have access to the same database at the same time? What can be done to avoid these problems?

STUDY / RESEARCH TASKS

1 A database can either be held on a database server or it can be distributed to the users on a local basis. Explain how these two systems work.

2 What is an object-oriented database and how does it differ from a relational database?

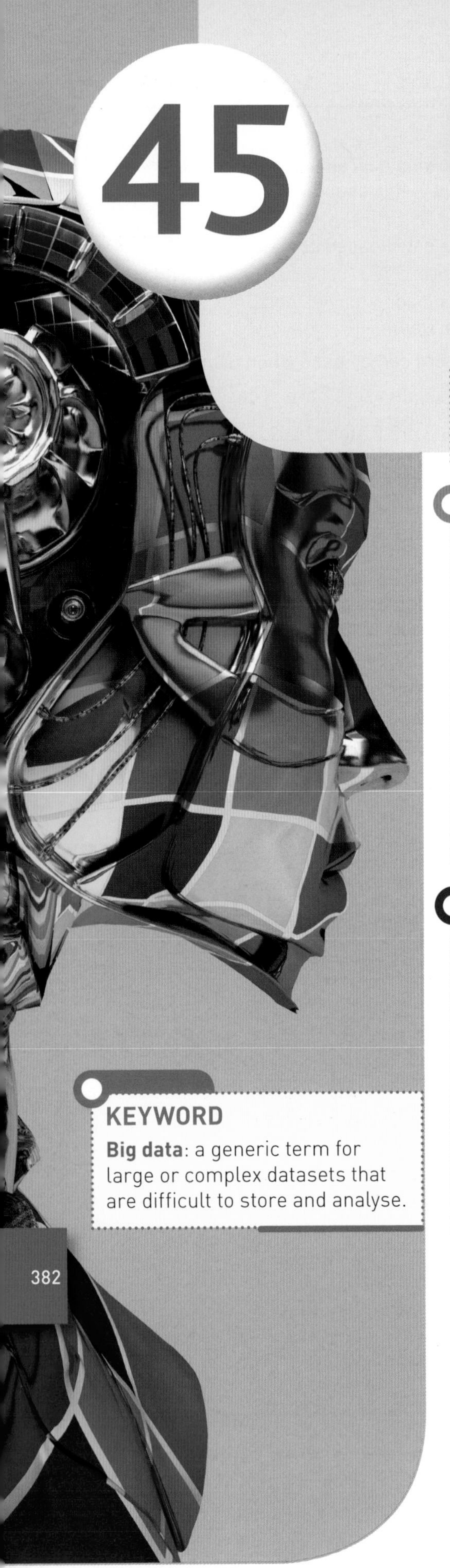

45 Big data

A level only

SPECIFICATION COVERAGE

3.11 Big data

LEARNING OBJECTIVES

In this chapter you will learn:

- how to define big data and what specific issues programmers face when dealing with it
- in what circumstances big data is created and used
- how data can be structured or unstructured
- how machine learning techniques can be applied to big data
- how to model big data
- how distributed processing techniques can be used with big data
- how functional programming techniques are useful on big data.

INTRODUCTION

Big data is a generic term given to datasets that are so large or complicated that it is difficult to store, manipulate and analyse them. The difficulty partly comes from the fact that the dataset is so massive that you would need multiple servers to physically store and provide access to it within a timescale that is useful. Another complication is that standard database software would not be able to cope with the quantity of data generated so it would be difficult to structure the data and produce any meaningful analysis of them. A further complexity is the speed at which the data are changing as many big data projects use data that are being updated in real-time.

KEYWORD

Big data: a generic term for large or complex datasets that are difficult to store and analyse.

Data are usually collected for a specific purpose. Data are given structure to turn them into information and information then has to be used for it to become knowledge. If the quantity of data cannot be structured and analysed, then they cannot produce any useful results. This is the challenge of working with big data – if it can't be turned into useful information, there is little point collecting it in the first place.

Figure 45.1 The data, information, knowledge process

Big data is a difficult term to define as there is no accepted quantifiable definition. The three main features of big data are:

- volume: the sheer amount of data is on a very large scale
- variety: the type of data being collected is wide-ranging, varied and may be difficult to classify
- velocity: the data changes quickly and may include constantly changing data sources.

Examples of big data

Big data is used for different purposes. In some cases, it is used to record factual data such as banking transactions. However, it is increasingly being used to analyse trends and try to make predictions based on relationships and correlations within the data. For example, scientists use data to predict the impact of climate change; business analysts use big data to predict sales; healthcare specialists use it to predict the spread of diseases; meteorologists use it to predict major weather events.

Big data is being created all the time in many different areas of life. Examples include:

- Scientific research: Scientists generate large volumes of data that could be measured in terms of petabytes or exabytes. These may be from readings from weather sensors, data collected from telescopic observations, biological results of experiments or global statistics on health issues. In all of these cases, the data are being collected and analysed for scientific purposes, usually to improve the quality of people's lives. For example, the human genome project uses masses of data to try to find the causes of genetic illnesses, with a view to eradicating them.
- Retail: All large businesses make use of data. Online retailers in particular can have millions of customers generating billions of sales. These data are used to improve the performance of the business. Sales data can be collected and analysed to help spot trends in consumer behaviour enabling businesses to become more profitable.
- Banking: The banking sector has to handle billions of transactions on an annual basis. They need to keep these data secure and have an audit trail of every single transaction to prevent against loss and fraud.
- Government: Most government departments and agencies have massive datasets. For example, the NHS records every single patient, appointment and operation. These data are critical to the successful treatment of patients and in many cases are a matter of life and death.
- Mobile networks: There were an estimated 4.6 billion mobile phone contracts around the world at the time of writing. All of the customer and call data are recorded to enable bills to be generated.
- Security: Legislation allows for mobile phone calls, texts, email messages and other online communications to be recorded. This represents billions of items of data every day and can be used by the security services to spot terrorist threats.
- Real-time applications: Many applications, particularly online and mobile, make use of real-time data. For example, weather apps take data readings from sensors, city traders use software that enables trading based on second-by-second share price fluctuations.

- The Internet: A lot of big data gets created through everyday use of the Internet. For example, data from social media websites could be analysed to understand social attitudes and trends. Data from search queries can be used to understand how people use the web.

As you can see from the range of examples, one of the challenges of working with big data is that the dataset is constantly changing as new data are arriving. For example, weather data may be collected from remote sensors every few seconds. To be useful, the data may need to be stored, processed and transmitted to data users within a few seconds or minutes.

KEYWORD

Latency: the time delay that occurs when transmitting data between devices.

The concept of **latency** is critical here as it is in other areas of computing. In this context, latency could be described as the time delay between a user making a request for data and those data being received, or the amount of time it takes to turn the raw data into meaningful information. With big data there may be a large degree of latency due to the amount of time taken to access and manipulate the sheer number of records. For some applications a large amount of latency may be acceptable, for others, the data may need to be processed within seconds.

CASE STUDY: H1N1 AND GOOGLE

When there is a flu pandemic, health services around the world have to respond quickly to introduce vaccination programmes, ensuring that they have enough vaccine and that it gets to the right people. There is a major problem in targeting resources to the neediest places and predicting where the outbreak will cause the greatest loss of life. Various health organisations such as the NHS in the UK and the Center for Disease Control (CDC) in the USA use big data, which is fed into them from health care professionals around the country and around the world.

Historical data can be used to understand how viruses might spread and more up-to-date data are used to understand what is happening in the immediate term. The data are not real-time as there is a delay in receiving and capturing them.

To help with this issue, Google set up their own big data project called Google Flu Trends (GFT), which analysed real-time data of searches being made relating to flu. The theory is that there is a correlation between people searching for information about flu, with the actual incidence of flu symptoms. By matching the geographical location of the searches, it would be possible to predict areas where flu is prevalent or spreading on a daily basis.

Structured and unstructured data

Big data provides major challenges not just in terms of the volume of data and frequency of change, but also the nature of the data being collected. Most databases work on the model that the data will fall into columns and rows, otherwise referred to as fields and records. This makes data easy to organise and store as they can be entered into the appropriate fields. When data are analysed, it is relatively easy to carry out searches and sorts to query the data as we saw in the previous chapter.

However, some data do not fit into this model. Data can be defined as either structured or unstructured.

KEYWORD

Structured data: data that fit into a standard database structure of columns and rows (fields and records).

- **Structured data**: These are data that can be defined using traditional database techniques using fields and records. This means that it is possible to give each data item a field name and type. For example, customer data would fall into this category as it is possible to identify a field for name and address. Banking data would be structured as it would be possible to define sort code and bank account number as fields.

KEYWORD

Unstructured data: data that do not fit into a standard database structure of columns and rows (fields and records).

- **Unstructured data**: These are data that cannot be defined in columns and rows. These might include multimedia data, web pages and the contents of emails, documents, presentations. It is important to note that although each of these types of data has its own structure, they do not fit easily into the standard database structure therefore making it difficult to analyse the contents of the data.

Machine learning techniques

Where quantitative data are stored in standard relational database format it makes it relatively simple to query data to produce results. For example, if an online retailer wanted to know how many of a particular product they sold this week, they can do a simple query to find this information. Even on a large dataset, this could be produced relatively quickly and accurately.

However, qualitative data are much harder to analyse and it is this type of data that is most likely to be unstructured. For example, if the online retailer asks for feedback from their website in the form of customer comments, they could receive millions of items of data, all written in free text. It would take a long time for someone to read all of these, so techniques collectively known as machine learning can be used to automate the process.

Machine learning covers everything from pattern recognition to artificial intelligence systems. In this context, at a simple level, the machine could learn to look for patterns of words within a text in determining the nature of the feedback. For example, it could look for positive or negative words to classify the feedback as positive or negative. In order to do this, it would need to be programmed with the words or phrases to look for.

A more advanced form of machine learning is where the computer is able to develop its own knowledge based on the data it is manipulating. This is particularly valuable with big data as there may be patterns and correlations that exist within the data that are not immediately obvious. One technique, called predictive analytics, is widely used in the financial and insurance sectors to predict risk.

CASE STUDY: CAIS

CAIS (pronounced keys) is the Credit Account Information Sharing database that stores over 400 million records of credit transactions, everything from bank loans, to credit cards and overdrafts. This information is used by banks and other lenders to determine whether a customer has a good or bad credit history and to identify the level of risk associated with that customer. Different lenders can use the data to determine a credit score, which in turn helps them to set credit limits and interest rates.

It is an example of big data, where the data need to be very accurate and provided quickly, normally within a few seconds or minutes during the loan application process.

Issues with big data

There are a number of issues with big data.

- Datasets are so large that they are too difficult to store and analyse.
- Unstructured data can be very difficult to analyse in an automated way.
- Specialised software is needed to manage and then extract meaningful information from the data.
- Massive storage and processing power is needed, meaning that many big data applications may only be carried out on supercomputers or large dedicated networks.
- Data are constantly changing so it is difficult to keep track of every change.
- Finding a correlation in a dataset does not necessarily mean you have found the answer to a problem. In other words, it is possible to infer the wrong conclusion from the data.
- There is an issue with concurrency where several users are working on the data at the same time.

Modelling big data

Most big datasets are stored in what are called data warehouses, which as the name suggests are very large. In common with a normal warehouse, there needs to be some system for understanding what is stored within the warehouse. One method is to use fact-based **modelling** which attempts to identify fundamental facts within the data that in turn identify all of the entities within the data. These are represented as diagrams or expressed in natural language.

For example, a large online retailer may have millions of items of data. They will record uniquely identifiable, time-stamped facts such as 'We sold 1 million rechargable batteries' or 'We sold 2 million clock radios'. Facts such as these can be defined in a more abstract form such as 'We sold `QUANTITY PRODUCTS`'. In this format, it is possible to model other data that the retailer records.

KEYWORDS

Modelling: recreating a real-life situation on a computer.

Graph schema (database): a method of defining a database in terms of nodes, edges and properties.

Node: in database modelling, it is an entity.

Properties: in database modelling, it is items of information stored within each entity.

Edge: in a database graph schema, it refers to the link and relationship between two nodes.

To represent this in graphical form, **graph schema** can be created based on the graph data type that we looked at in Chapter 9. Graphs are made up of **nodes**, **properties** and **edges**. A graph schema for a dataset for an online retailer might look like that shown in Figure 45.2.

- A node is an entity such as a customer, product or picker.
- A property is relevant data relating to the node, such as the customer or product name.
- An edge shows the link and describes the relationship between the two nodes, for example it shows which customer bought which product, or which picker has been assigned to collect and post the items.

These are considered to be more efficient models for dealing with large amounts of constantly changing data. In the example above, the online retailer might want to link two orders made by the same customer at the same time to the same picker in the distribution centre.

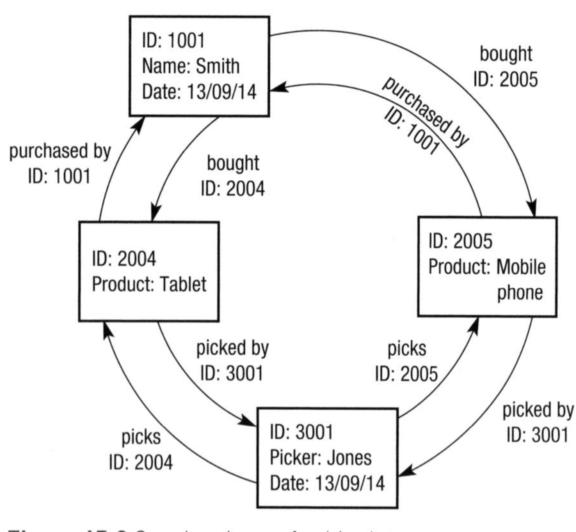

Figure 45.2 Graph schema for big data

Distributed processing

The main issue with big data is that one computer cannot store all of the data or analyse it quickly enough. One solution is to split the work over several computers, typically by adding more servers or workstations to a network and then distributing the processing between the processors of each.

KEYWORD

Distributed processing / computing: the principle of spreading large and complex tasks over a number of computers or servers.

This is known as **distributed processing** or **distributed computing**. For big data, often a dedicated network is set up to work on the same main task. Typically, one computer will be allocated as the master computer within the network and will control the others through the operating system and specialist software. Each computer on the network is allocated its own subtask and messages are then passed between the computers in order to meet the overall goal. The network can be implemented on a client–server or peer-to-peer basis.

This can be implemented as a distributed network with each of the main computers being a server and then further workstations being attached to the server. Similarly, the network can be extended to include Internet services, such as the use of online data storage. This is an example of cloud computing where the Internet is used to provide a service that you would normally get from the LAN.

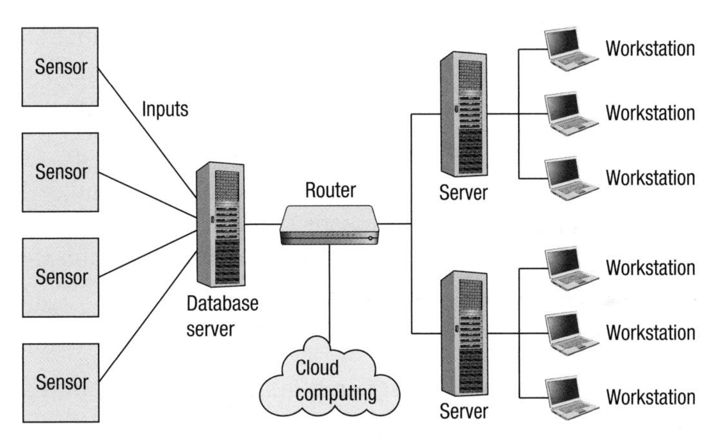

Figure 45.3 Distributed processing

Big data is almost bound to require distributed computing of some sort as the data entry source is usually remote to the computer where the data are stored and analysed. For example, customer data will come from the Internet or meteorological data will come from remote sensors.

KEYWORDS

Distributed program: a program specifically written to be used in a distributed processing environment.

Functional programming: a programming paradigm that uses functions to create programs.

Where specialist software is used to tackle a task using distributed computing, it is referred to as a **distributed program**. There are implications for programmers writing distributed programs in terms of dealing with multiple users and ensuring the integrity of large datasets. One resolution to this is to use **functional programming** techniques.

Functional programming for distributed programs

In Chapter 5 we looked at programming paradigms, which describe the way in which any particular programming language works. For example, Visual Basic is described as an imperative or procedural language as the programmer has to define the list of steps required to implement the program. Key features of imperative languages are the need to declare variables and to identify the correct sequence of events to achieve the desired outcome.

KEYWORDS

Variable: a data item whose value will change as the program is run.

Mutable: changeable.

State: in programming it refers to the state that the variables are in, i.e. the value that is currently stored.

Side effects: in programming it refers to the fact that the value contained within a variable will change as the program is run, which has implications for other parts of the program.

Imperative language: a language based on giving the computer commands or procedures to follow.

Object-oriented language: a programming paradigm that encapsulates instructions and data together into objects.

Function: a subroutine that returns a single value.

Concurrence: the concept of two users trying to access the same data item at the same time.

Functional programs differ in that they use expressions that are similar to normal mathematical expressions in order to evaluate data. There are no **variables** as such, as every item of data is treated as a function. A feature of a variable in an imperative language is that it is **mutable**, which means it will change as the program runs. This means that it will have a **state**, which is the current value of the variable at any given point when the program is being run.

This is said to have **side effects**, which means that the programmer has changed the state of a variable and this may impact on how the program runs. This is a normal consequence of using **imperative** or **object-oriented** languages. Where the side effect is known it is then possible to write future lines of code in the knowledge of what has gone before.

This presents a problem where many computers might be working on the same data at the same time, as each computer will need to know the state of the variable at any point. With functions, the values and the expressions are all that is needed make up the lines of programming code. This means that the value produced by any one line of code is entirely the result of the **function** and is not dependent on the state of any of the variables.

This makes this type of coding particularly suited to analysing big data using distributed processing, as there will be multiple users all accessing the data at the same time from different computers. This is known as **concurrence** and it is problematic because it can cause data locking as we saw in the previous chapter. Also, where there are several users, the side effects of any previous coding may not be apparent.

Because functional programming does not use variables as such and as the value of the variable is not changing, there will be few or no side effects. This means that the user always gets the original value of the variable to put into their own functions and the output of the function will be local to their machine.

There is more on functional programming in the next two chapters.

Practice questions can be found at the end of the section on pages 390 and 391.

TASKS

1 Define 'big data'.
2 What problems are associated with big data?
3 How can machine learning techniques be used to analyse qualitative data?
4 Give three scenarios where big data could be used.
5 Give three examples of how big data could be used to predict future events.
6 Create a database graph scheme that shows how data about you might be stored and used by your school/college.
7 How does distributed processing help with storing and analysing big data?
8 What is a programming paradigm?
9 What features of functional programming make it particularly useful for writing programs to be used in a distributed processing environment?

KEY POINTS

- Big data is data that are either too large or too complex to be handled using traditional database techniques.
- Big data is becoming more common, particularly with the volumes of data being generated via the Internet.
- Some data cannot be structured into the columns and rows of a traditional database.
- Unstructured data are more difficult to query and techniques such as machine learning are used to interrogate unstructured data.
- Data models can be produced to try and understand how big data is structured.
- Big data is often spread across a number of servers in order to cope with its size. Where this is the case there is an added complication of working on data split across two or more servers.

STUDY / RESEARCH TASKS

1 How successful has the Google Flu Trends (GFT) program been in predicting outbreaks of flu?

2 How is big data being collected and used by scientists working on the Large Hadron Collider at CERN?

3 Will single processor computers simply become a thing of the past?

4 'When analysing data, the more data you have, the more accurate your results will be.' Is this statement necessarily true?

5 What are the main features, advantages and disadvantages of cloud computing?

6 Can computers think for themselves?

Section Ten: Practice questions

1 A hospital stores details of its wards, patients and their medical conditions in a database.

- For each patient their unique patient number, surname, forename, address, date of birth and gender are stored.
- Wards have a unique name and a number of beds. The name of the nurse in charge is also recorded.
- Each medical condition is assigned a unique medical condition number and the name and the recommended standard treatment are recorded.
- Patients may suffer from one or more medical conditions and a particular medical condition may be attributed to more than one patient.
- The medical conditions of each patient are recorded.
- A patient can be assigned to only one ward at any one time.
- Each ward may have patients with different medical conditions but each patient can be assigned to only one ward at any one time.
- Four entities for the hospital database are **`Ward`**, **`Patient`**, **`MedicalCondition`**, **`PatientMedicalCondition`**.

a) Copy the partially completed entity relationship diagram and show three more relationships which exist between the given entities.

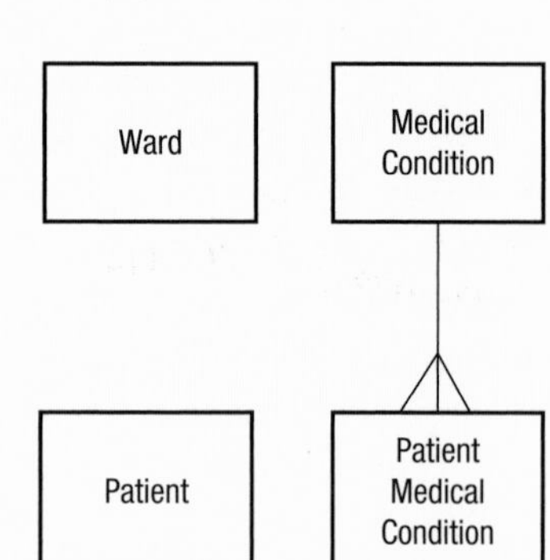

b) Using the standard format (**`TABLENAME [PrimaryKey, Non-keyAttribute1, Non-keyAttribute2,`** etc.**]**) to describe tables, state all attributes for the following entities underlining the primary key in each case.

i) **`Ward`**
ii) **`Patient`**
iii) **`MedicalCondition`**
iv) **`PatientMedicalCondition`**

2 In a multi-user database, concurrent access is allowed to each data item.

a) What is meant by concurrent access?

b) Why must concurrent access be controlled in a multi-user database?

3 Acme Ltd supply products to their customers. The data requirements for the database system are defined as follows.

- A unique **`ProductID`** and product description are recorded for each product.
- The quantity in stock of a particular product is recorded.
- A unique **`CustomerID`** is assigned and the name, address and telephone number of each customer is recorded.
- An order placed by a customer will be for one or more products.
- Acme Ltd assigns a unique code to each customer order, **`AcmeOrderNo`**.
- A customer placing an order must supply a code, **`CustomerOrderNo`**, which the customer uses to identify the particular order.
- A customer may place one or more orders.
- Each new order from a particular customer will have a different customer order code.
- A particular order will contain one or more lines with each line numbered; the first is 1, the second is 2 and so on.
- Each line will reference a particular product and the quantity ordered.
- A specific product reference will appear only once in any particular order placed with Acme Ltd.
- After normalisation, the database contains four tables based on the entities **`Customer, Product, Order, Orderline`**.

a) Show three more relationships which exist between the given entities by completing the diagram below.

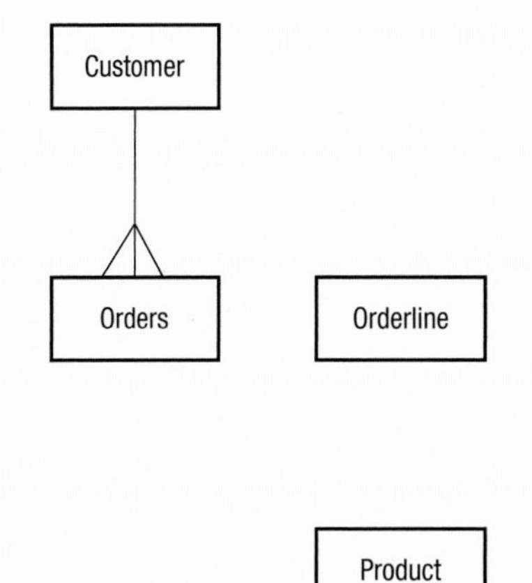

Using the standard format `TableName (PrimaryKey`, Attribute1, `Attribute2`, etc.) describe tables, stating all attributes, for the following entities underlining the primary key in each case.

i) `Product`

ii) `Customer`

iii) `Order`

iv) `OrderLine`

b) Using SQL commands **`SELECT, FROM, WHERE`** and **`ORDER BY`**, write an SQL statement to query the database tables for all customer names where the orders have been despatched in ascending order of `AcmeOrderNo`.

4 The widespread use of the Internet has led to a massive increase in the volumes of data being collected. Some of this falls into the category of 'big data'.

a) What is meant by big data?

b) Give two examples of big data.

c) Why might functional programming be suitable when working with big data?

d) How does distributed processing help with big data?

e) Why might some people have concerns about the way in which data is collected about them over the Internet?

Section Eleven: Fundamentals of functional programming

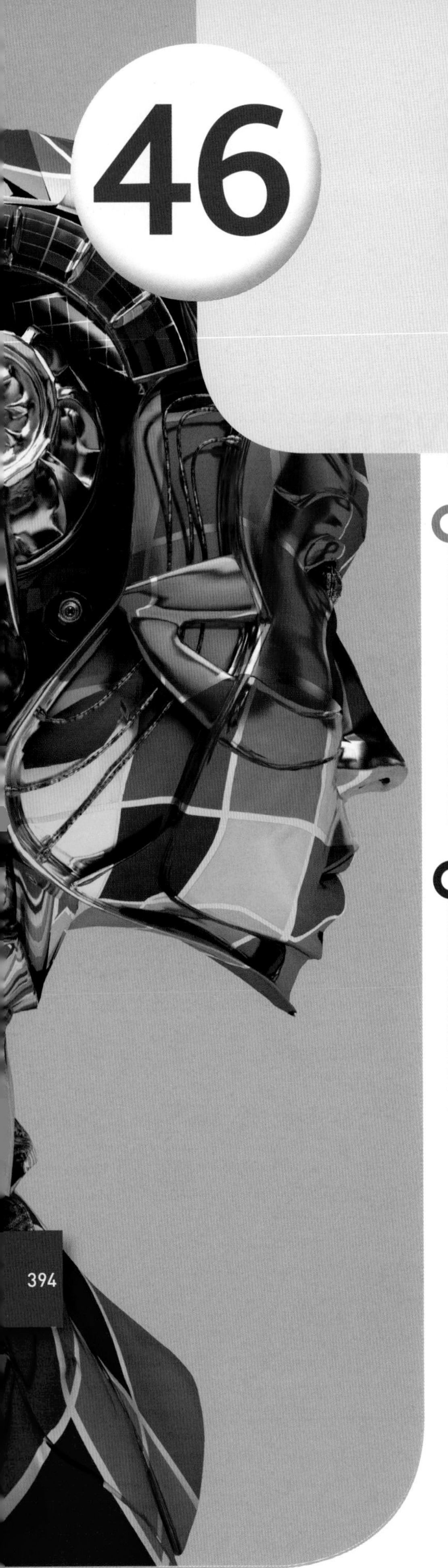

46 Basics of functional programming

A level only

SPECIFICATION COVERAGE

3.12.1 Functional programming paradigm

LEARNING OBJECTIVES

In this chapter you will learn:

- what the functional programming paradigm is
- how functions can be used to construct programming code
- what a function type is
- what a first-class object is and how they are used in code
- how to apply functions fully and partially
- how to combine existing functions to make new ones.

INTRODUCTION

In Chapter 23 you were introduced to the concept of a mathematical function, which is an expression or rule that takes an input value from a set and returns an output value from another set. For example, the function f could map the input set {1, 2, 3, 4, 5} to an output set {1, 4, 9, 16, 25}, in which case the function could be described as $f(x) = x^2$ as each value in the input set is squared to produce the corresponding value in the output set.

Figure 46.1 A basic function

In Chapter 4 we looked at using functions within programming languages where they are used as a type of subroutine that requires an argument to be entered, which is the value that the function needs to work on. It will then return a result. In the next two chapters we will look at functions in more detail. In particular we will examine the functional programming paradigm, which uses functions as a way of constructing algorithms.

The functional programming paradigm

A programming paradigm is a method of programming. In your studies so far you have come across two main paradigms:

- Procedural languages: Also known as imperative languages, these require the programmer to input lines of instructions in sequence, which the program then carries out.
- Object-oriented languages: The programmer creates self-contained objects that contain code and data.

The **functional programming paradigm** is an example of a **declarative programming language** where all the algorithms call functions. This means that the lines of code look and behave like mathematical functions, requiring an input and producing an output. A value is produced for each function call. The idea is that there is one main function, which in turn calls further functions in order to achieve a specific task. Each function may call another function, or call itself recursively in order to generate a result.

The concept is that where a function is being used, it will always return the same value if it is given the same input so there can be no unforeseen side effects. One of the problems with **procedural programming languages** is that the value of a variable can change throughout the program and that changes made to the value in one subroutine may have an impact within another subroutine. Programmers often spend a lot of time tracking the value of variables when debugging, trying to find out where a program has gone wrong. In theory, a well-designed functional program will not have this problem.

The motivations behind using the functional programming paradigm include:

- Program requirements may be better defined as a series of abstractions based on functions rather than a more complex series of steps.
- Broader abstractions can lead to fewer errors during implementation.
- Functions can be applied at any level of data abstraction making them highly re-usable within a program.
- Functional code is easier to test and debug as each function cannot have any side effects, so only needs testing once.
- There are no concurrency issues as no data is modified by two threads at the same time.
- In multi-processor environments the sequence that functions are evaluated in is not critical.

As functional programs use mathematical expressions they lend themselves to writing applications that require lots of calculations. For example they are used to improve the reliability of mobile telephone networks, to analyse large volumes of financial data, to create control systems in the field of robotics, as well as in the aerospace industry.

There are some languages that specifically use the functional programming paradigm such as Lisp, Haskell, Standard ML, Scheme and Erlang. Other languages provide support for functional programming including Python, Perl, C#, D, F#, Java 8 and Delphi XE.

KEYWORDS

Functional programming paradigm: a language where each line of code is made up of calls to a function, which in turn may be made up of other functions, or result in a value.

Declarative programming languages: languages that declare of specify what properties a result should have, e.g. results will be based on functions.

Procedural programming languages: languages where the programmer specifies the steps that must be carried out in order to achieve a result.

Function types

A **function type** refers to the way in which the expression is created. All functions are of the type A → B where it is defined with an argument type (A) and a result type (B). In our example, A is the set that contains {1, 2, 3, 4, 5} and B is the set that contains {1, 4, 9, 16, 25}.

A is also called the **domain** and contains objects within a particular data type, in this case integers. B is called the **codomain** and is the set from which the output values are chosen. As we saw in Chapter 23 there are many standard sets of values such as integers, reals etc. that a set can be drawn from. Note that not every value that exists in the codomain will necessarily be output. The values that are used are referred to as the range.

In functional programming, a value that is passed to a function is known as an argument. For example in the expressions $a = f(x)$ and $b = f(2, 4)$, x, 2 and 4 are arguments.

KEYWORDS

Function type: refers to the way in which the expression is created, for example, integer of the domain and codomain, where f: A → B is the type of function.

Domain: a set of data of the same type which are the inputs of a function.

Codomain: the set of values from which the outputs of a function must be drawn.

First-class objects and higher order functions

Within a functional programming environment, a function is a **first-class object**. This means that it has certain properties and can be used in particular ways within the program. A broad definition of a first-class object is any object that can be passed as an argument to or can be returned by a function. In functional programming, this means that a function can be passed as an argument to another function or can be returned from a function as the result. Other objects, such as integer values which can be passed as arguments to a function, are also first-class objects.

A function which can accept another function as an argument is known as a **higher order function**, the three most common of which are `map`, `fold` and `filter`. There is more on how these three functions work in the next chapter.

As an example of a first-class object, in Haskell you might write the function:

```
map (*2) [1,2,3,4,5]
```

`map` is a function, which takes in another function and applies it to every element in a list. `*2` (multiply by two) is the function that `map` is taking in. `[1,2,3,4,5]` is the list on which the function is applied.

The result of this higher order function would be `[2,4,6,8,10]`

In this example, `map` is a higher order function and *2, 1, 2, 3, 4 and 5 are all first-class objects.

KEYWORDS

First-class object: any object that can be used as an argument or result of a function call.

Higher order function: a function that takes a function as its inputs or creates a function as its output.

Function application

The process of providing the function with its inputs is known as **function application**. With our earlier example we had the function $f(x) = x^2$ and two sets – the domain and codomain. Set A (the domain) contained the inputs, which were all integers and set B (the codomain) contained the outputs, also integers.

KEYWORD

Function application: the process of calculating the result of a function by passing it some data to produce a result.

We input a single value from A, which can be described as the function taking its argument, or the argument being passed to the function. The function is then applied to the argument, which in this case means it squares it, to produce an output in B.

Consider a function to calculate the volume of a box:

```
volume = int x int x int → int
```

In this example the values 3, 4 and 5 are input to the function called **volume**. The argument is actually three values, which would be defined within the code as height, width and breadth.

```
let cuboidvolume height width breadth = height * width * breadth

cuboidvolume 3 4 5

60
```

Partial function application

In the **cuboidvolume** example, the function takes three values as a single argument as it needs them to perform the calculation. However, it is possible to pass any number of arguments into a function. Where this is the case, partial function application can be used to fix the number of arguments that will be passed.

KEYWORD

Partial application (of a function): the process of applying a function by creating an intermediate function by fixing some of the arguments to the function.

The idea of this is that when you have one function that takes lots of arguments, by partially applying the function you effectively create a new function that performs just part of the calculation. The **partial application of a function** can produce results that are useful in their own right in addition to the full application of the function.

For example, consider the two notations for a function that adds two integers together by taking two arguments:

```
add: int x int → int        add: int → int → int
```

- The first is a full application of the function which takes two integers as arguments and adds them together to create a result that is also an integer. Both values are passed as arguments at the same time.
- The second is a partial application that shows a new function being created, which always adds the first argument value onto a number. This new function is then applied to the second argument to produce the overall result.

If you applied this to a function as follows:

- Full function application would add *x* and *y* at the same time to create a result:
  `add (x,y): add (2,3) = 5`
- Partial function application would produce the function `add2`. This function would then be used with the second argument to produce the final result:
  `add (x,y): add (2,3) = add2 (3) = 5`

KEYWORD

Function composition: combining two or more functions together to create more complex functions.

Function composition

Function composition is the process of creating a new function by combining two existing functions together. This is one of the key principles of functional programming as the concept is to have complex functions that in turn are made up of simpler functions. As each component function produces its result, this is passed as an argument result to the calling function. This process continues for each of the component functions until a result is produced for the complex function as a whole.

For example, imagine you have two functions with f being applied to domain A and g being applied to the domain B:

f: A → B

g: B → C

Composition of these two functions would mean that the result of function f, becomes the input for function g.

In function f, A is the domain and B is the codomain, so A results in B.

In function g, B is the domain and C is the codomain, so B results in C.

Therefore a function composition of the two would result in A being the domain and C being the codomain, from which the range (output) is produced.

It can be visualised as follows:

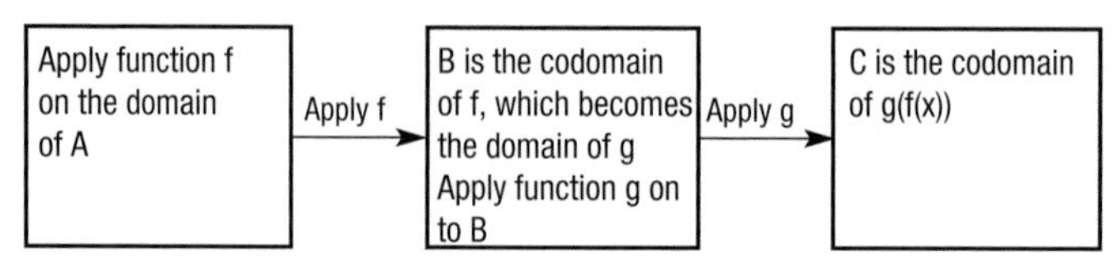

Figure 46.2 Composition of functions

Example

Let's assume that the function f is $f(x) = x^2$ and function g is $g(x) = x + 3$. To work out the composition of g and f:

- pass an argument with the value of, for example, 4
- function f squares it and produces the result 16
- 16 is the input for function g
- g adds 3 to the result
- the result of the combined function is 19.

This could be shown in the following notation:

g ° f where ° means the composition of functions g and f

$f(x) = x^2$

$g(x) = x + 3$

g ° f becomes $g(f(x)) = x^2 + 3$

Practice questions can be found at the end of the section on page 405.

KEY POINTS

- A mathematical function is an expression or rule that takes an input value from a set and returns an output value from another set.
- A programming paradigm defines a methodology for programming.
- Functions can be of different types defined by the way they are constructed.
- A function is an example of a first-class object.
- A broad definition of a first-class object is any object that can be passed as an argument or can be returned from a function.
- Functions can be applied partially or fully.
- Functions can be combined together to create new functions.

TASKS

1 Describe the basic properties of a function.

2 Define the functional programming paradigm.

3 What are the main reasons for using functional languages compared to procedural languages?

4 What is a first-class object? Give two examples.

5 Show the full and partial application of the equation $f(a, b, c) = a + b - c$ with a set of integers.

6 Use function composition to combine the equation $f(x) = x^3$ with $g(x) = x + 3$.

STUDY / RESEARCH TASKS

1 Research the way in which functional programming helps telecommunications create mobile networks that are highly reliable.

2 Research programming languages that are specifically designed around the functional programming paradigm.

3 Some programming languages such as C# and Python are procedural languages, but provide support for the functional programming paradigm. Pick one of these two languages and find out how this support is provided.

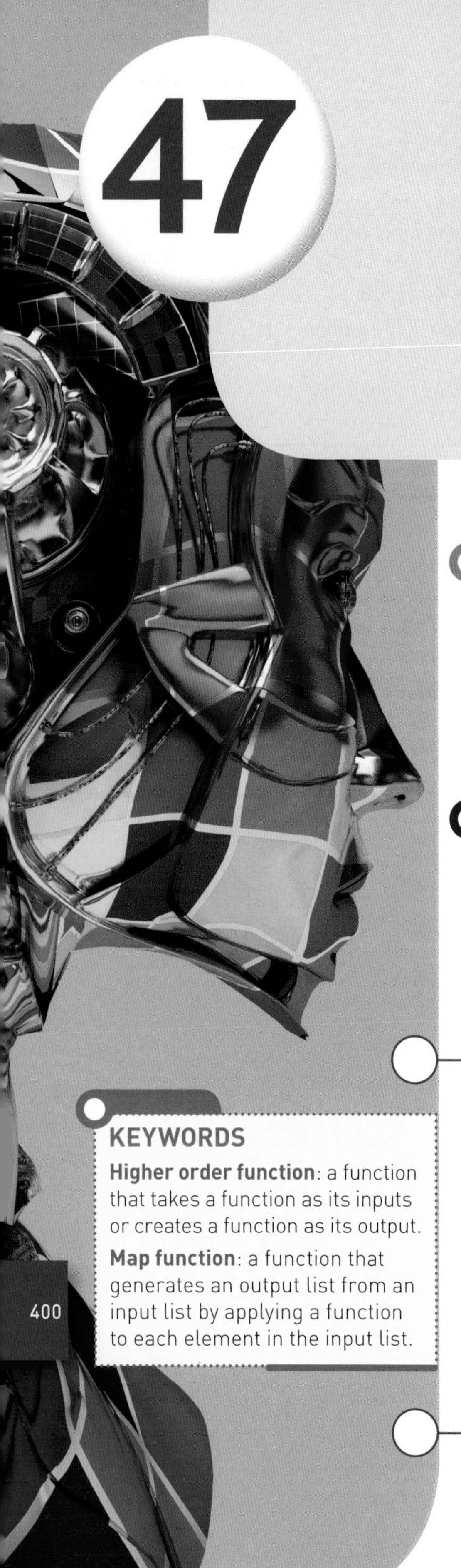

47 Writing functional programs

A level only

SPECIFICATION COVERAGE

3.12.2 Writing functional programs

3.12.3 Lists in functional programming

LEARNING OBJECTIVES

In this chapter you will learn how to:

- use higher order functions
- use the map, fold and filter functions
- manipulate lists using a functional programming language.

INTRODUCTION

In the previous chapter we looked at the basic principles and concepts behind functional programming. In this chapter we look at how to write simple functional programs focusing on handling lists and also looking at the higher order functions: map, filter and reduce.

KEYWORDS

Higher order function: a function that takes a function as its inputs or creates a function as its output.

Map function: a function that generates an output list from an input list by applying a function to each element in the input list.

Higher-order functions

In the functional programming paradigm, a **higher order function** is one that:

- takes a function or functions as its argument
- produces a function as the result of a function
- or does both of the above.

As we saw in the last chapter, the ability to combine functions is a key component in functional programming. Higher order functions are therefore those that can form part of larger more complex applications in order to produce a result.

The map function

The **map function** applies a given function to every element within a list and returns a corresponding list of results. In this chapter we will look at code that can carry out the function on every element of the list in one pass.

If we use a simple example f(*x*) = x^2, our data might look like this when represented as a list:

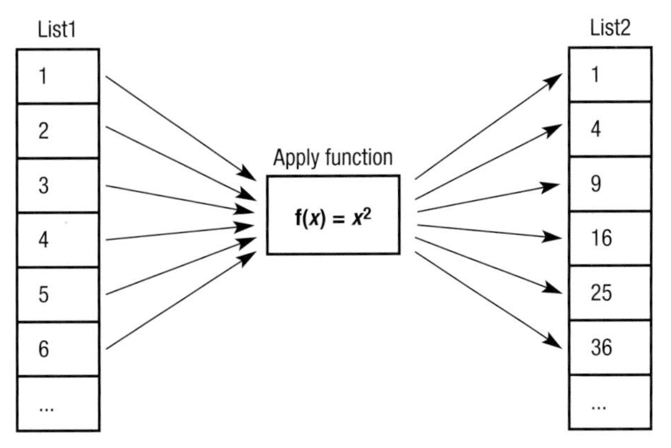

Figure 47.1 A function applied to a list that generates a new list

It is called a map function because it maps one element of the input list (List1) to the corresponding element in the output list (List2). The function it performs could be anything.

The code to carry our f(*x*) = x^2 function to a list of integers is:

```
square x = x * x
map square [1,2,3,4,5,6]
```

The filter function

KEYWORD

Filter function: a method of creating a subset based on specified criteria.

The **filter function** processes a list and then creates a new list that contains elements that match certain criteria. You may have come across filtering in database or spreadsheet programs where you filter a dataset to create a subset of the data of records matching a certain criteria. The operation is very similar to a search.

In order to create the filtered list, some kind of selection criteria needs to be applied to the list. This is sometimes called a predicate function and returns a Boolean value of either TRUE or FALSE. For example, a list could be filtered to include all values over 50, or all odd numbers. The way in which the statement is actually written depends on the programming language being used. Typical examples would be `If`, `Select`, `Remove_if`, `list.filter` and `where` statements.

For example, we might use the function **odd** to filter the odd numbers in a list (List1) into a new list (List2).

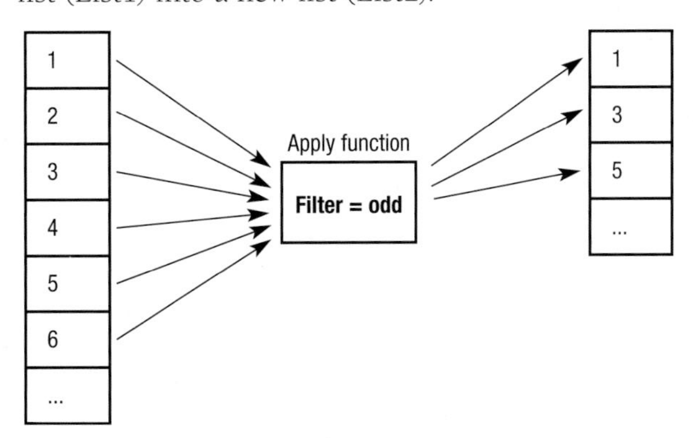

Figure 47.2 A function to filter odd values from a list

The code to carry the filter function is:

```
filter odd [1,2,3,4,5,6]
```

The reduce or fold function

KEYWORD

Reduce/fold function: a method of reducing a list to a single element by combining the elements using a function.

The **reduce or fold function** takes a list of values and reduces it to a single value. This is a recursive process as the function keeps processing until the list is empty.

The logic behind it is that if you can carry out a function on the first item of a list, then you can carry out the same function on every item of the list. For example, if you took a simple example where you wanted to add the items of a list together, you could reduce the list until you ended up with just one item in it. Let's assume the numbers 1–5 are in the list:

Table 47.1 The process of reducing/folding a list

Instructions	Original list	Apply function (add)	Result
Start with a list	[1, 2, 3, 4, 5]		
Take the first item out of the original list and apply the function	[2, 3, 4, 5]	1	1
Recurse	[3, 4, 5]	2 + 1	3
Recurse	[4, 5]	3 + 3	6
Recurse	[5]	4 + 6	10
Recurse	[empty]	5 + 10	15
List is now empty so the function will not recurse			

As you can see from Table 47.1 we have reduced or folded a list with five elements down to a single result. You can see that the instructions needed are the same each time. Therefore, we would only need to write one algorithm to carry out the process. This is a good example of efficient code as regardless of how many items were in the list, only one algorithm would be needed.

The code to carry out the reduce or fold function is:

```
foldl (+) 0 [1,2,3,4,5]
```

List processing

KEYWORDS

List: a collection of data items of the same type.

Identifier: the name of a list.

As we have seen, a **list** is a set of data items of the same type stored using a single **identifier**. It is made up of any number of elements and can contain any type of numeric or text strings. The only rule is that you cannot mix data types within a list. This means that you are then able to carry out operations on lists of data, which is much more efficient than trying to carry our operations on individual data items. In fact, we have already seen with the example in this chapter how useful and efficient the list structure can be.

Lists have a few components as shown in Table 47.2. Consider the following list of names called `FirstNames`: [Abdul, Brian, Chloe, Dave, Marlon, Nigel, Rafi, Sunil].

Table 47.2 Components of lists

List components	Explanation
Identifier	The name given to the list. In this case: `FirstNames`.
Data type	Identifies the data type being stored, e.g. text strings, integers, reals. In this case, it is text strings.
Elements	These are the individual values stored in the list. These are identified by their position in the list. For example, 'Dave' is element 3 assuming that the first element starts in position 0.
Head	The first element in a list. In this case it is Abdul.
Tail	All the other elements in the list apart from the head.
Length	The number of elements in the list. In this case there are eight.

KEYWORD

Tail: every element in a list apart from the head.

Head: the first element in a list.

Empty list: a list with no elements in it.

An important concept is that the **tail** of the list is not just the last item, but all of the items apart from the **head**. This is a useful concept as it allows lists to be defined in terms of a head and a tail, which can speed up processing.

It is important to remember that a list can be empty. This may be when it is set up and no data has been entered or it may be after a list has been processed and there is no data left to process. The **empty list** is often represented as brackets with nothing between them, e.g [ ].

There are various standard processes that can be carried out on lists as shown in Table 47.3.

Table 47.3 Standard processes that can be carried out on lists

Process	Description	Haskell code
Return the head of a list	Identifies the first element in the list.	`head [1,2,3,4,5]` `1`
Return the tail of a list	Identifies all of the other elements apart from the head.	`tail [1,2,3,4,5]` `[2,3,4,5]`

Test for an empty list	Checks whether there are any elements in the list.	`let MyList = [4,8,15,16,23,42]`
		`MyList`
		`[4,8,15,16,23,42]`
		`null MyList`
		`False`
		`let MyList = []`
		`MyList`
		`[]`
		`null MyList`
		`True`

Return the length of a list	Identifies how many elements there are in a list.	`length [1,2,3,4,5]` `5`

Construct an empty list	Creates a list that has no elements in it.	`let emptylist=[]`
		`emptylist`
		`[]`

Prepend an item to a list	Adds an item to the beginning of a list.	`Let SetA = [1,2,3,4]`
		`SetA`
		`[1,2,3,4]`
		`let SetB = [0] ++ SetA`
		`SetB`
		`[0,1,2,3,4]`

Append an item to a list	Adds an item to the end of a list.	`let SetC = SetA ++ [0]`
		`SetC`
		`[1,2,3,4,0]`

Practice questions can be found at the end of the section on page 405.

KEY POINTS

- A higher order function takes a function or functions as its argument and/or produces a function as the result of a function.
- The map function applies a given function to every element within a list and returns a corresponding list of results.
- The filter function processes a list and then creates a new list that contains elements that match certain criteria.
- The reduce or fold function takes a list of values and reduces it to a single value by applying a function.
- Functional programming uses lists of values. The functions can be applied to each element in the list.

TASKS

1 Using a suitable programming language, write your own code that will:

   a) Place the following values in a list: 2, 5, 3, 6, 4, 8, 9

   b) Identify the value at the head of the list

   c) Identify how many elements are in the list

   d) Apply a function to double every value in the list

   e) Filter values greater than 5

   f) Add all the values in the list together by reducing/folding it until there is just one element

   g) Create a message to indicate if the list is empty.

STUDY / RESEARCH TASKS

1 Using a functional programming language of your choice, create your own applications to solve other problems. For example you could create:

   a) a scientific calculator application that included common functions such as square, square root, sine, cosine etc.

   b) a username generator for a network manager that takes a user's real name and automatically generates and stores a username for them.

2 Research real-world applications of functional languages.

Section Eleven: Practice questions

1 A functional program is being used to process the following list.

item (0)	2
item (1)	4
item (2)	6
item (3)	8
item (4)	10

a) Write a function that would generate the set of numbers shown in the table.

b) Suggest a suitable identifier for the data.

c) Suggest a suitable data type for the elements of the list.

d) What value or values are at the head of the list?

e) What value or values are at the tail of the list?

2 Write Haskell code to reduce/fold a list.

3 Trace the reduce/fold algorithm in the table below.

Instructions	Original list	Apply function (add)	New list
Start with a list	{2, 4, 6, 8, 10}		

Use the table to explain what is meant by the term recursion.

4 Write Haskell code that will produce a new list (from an existing list) of values over 10 using the higher order function 'filter'.

5 What is a higher order function?

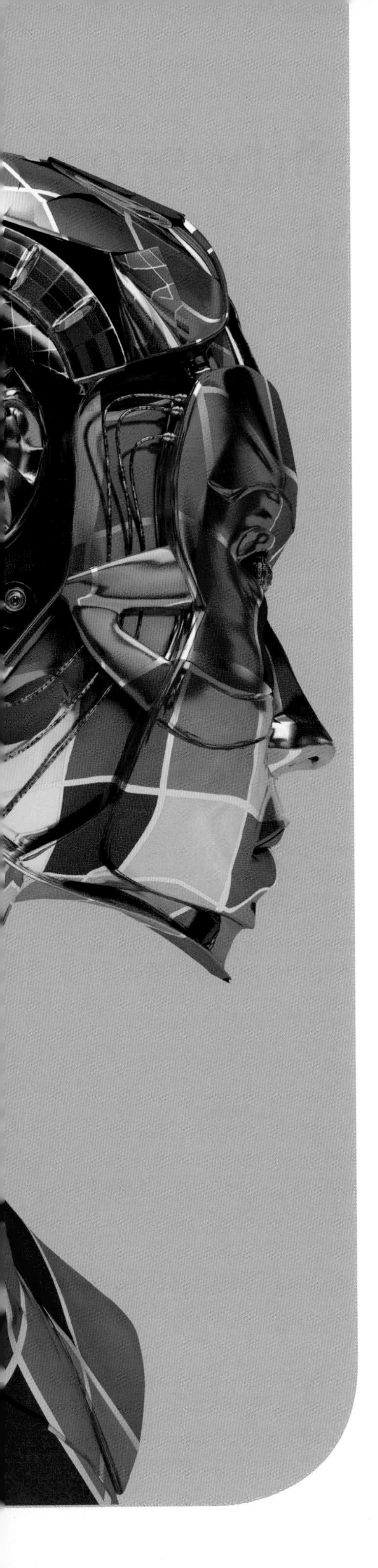

Section Twelve: Software development

48 Aspects of software development

SPECIFICATION COVERAGE

3.3.1 Aspects of software development

3.13.1 Aspects of software development

LEARNING OBJECTIVES

In this chapter you will:

- learn about the main stages of software development
- look in detail at the components of each stage: analysis, design, implementation, testing and evaluation.

INTRODUCTION

Software development is the process of creating and maintaining programs or applications. Software development is a process comprising a number of stages that a developer will work through to solve computer-based problems, typically: analysis, design, implementation, testing and evaluation. A system is more than just the programs that will be needed – it might include the file structures, hardware, operating system and people that will be working with it.

There is no requirement to work through every single stage as it will depend on the nature of the development. For example, to create a brand new application from scratch, a developer will work through every stage. However, to produce an update for an existing application may only need the developer to work through two or three of the stages. For AS level you need to be familiar with the processes that take place at each stage. For A level, you will work through each stage for your A-level project, which is worth 20% of your A-level grade. There is more on how to tackle the project in the next chapter.

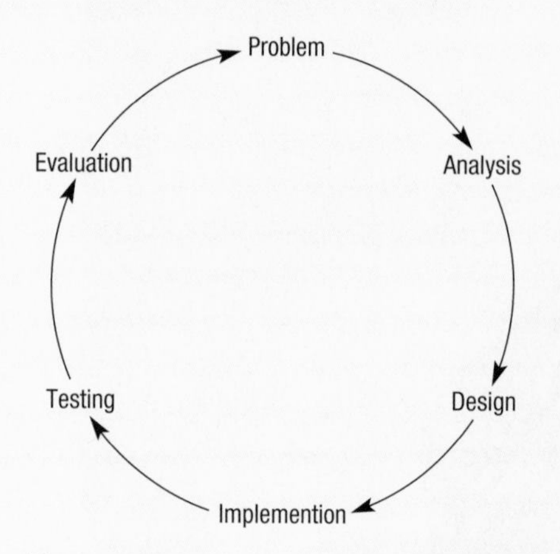

Figure 48.1 Stages of system development

Analysis

KEYWORD

Analysis: the first stage of system development where the problem is identified, researched and alternative solutions proposed.

The first stage is to identify and fully analyse the nature of the problem. We have already looked at some of the methods for defining problems in Chapter 17. **Analysis** also includes:

- understanding what has prompted the need for a new system
- gathering information
- carrying out a feasibility study.

Defining the problem

Having identified that there is a problem with an existing system or perhaps seen a new area that needs developing, it is very important to keep an open mind and concentrate on getting to grips with what the problem involves rather than looking for possible solutions. This involves identifying the scope of a problem and being realistic about how much of the problem a new system can solve. Any constraining factors may be identified here.

If you are creating a solution for someone else it is important that you agree the specification and scope of the work before you start. In general, the more you can involve the end users, the less likely you are to do something they do not want.

Prompts for a new system

There are many reasons for creating a new computer system. Many of these are based on the ever-increasing demands that are made of existing systems and the ever-changing advances in computer technology.

- Some existing systems simply cannot cope with the increased volume of data they are being asked to handle. For example, the way in which a bank maintains its data has changed beyond all recognition in the last 20 years, and banks have spent many millions of pounds updating their systems.
- New technology has meant that existing systems soon become outdated. For example, at one time it took a lot of time and effort to book a holiday. Now it can all be done in real time with the flights you want booked in a matter of minutes.
- The current system may be inflexible or inefficient. For example, the number of people flying is increasing every year and this has placed pressure on the airport immigration authorities. Computer-readable passports allow immigration staff far greater control and instant access to details of the person in front of them.
- New technology has created new opportunities. You have only to look at the way the use of the Internet has exploded to realise how many new opportunities such as e-commerce and e-banking there are. Other developments such as computer control and GPS have created other totally new fields. For example, cars are becoming increasingly automated. Current technology enables cars to park themselves.
- Commercial reasons. Many new systems are created in order to generate demand from customers. There isn't necessarily a need for a new system but companies introduce them because customers will buy them.
- New platforms and operating systems. Businesses create new systems to take full advantage of new platforms and operating systems. For example thousands of new apps have been created for iOS and Android devices.

- Increased processing power. As processing power increases, new software is written to take advantage of it. Some applications that are processor intensive may become feasible with the advent of faster processors.
- Increased network power. Many applications are developed to take advantage of ever increasing connectivity. For example, many of the leading applications written over the last few years have been based on social media.

Methods of gathering information

If the problem that you are going to tackle is based on an existing system then it may be worthwhile investigating how it currently works, though there are times when a completely fresh approach to a problem might lead to a better solution. Asking for opinions about a possible new system is a good idea too. There are several ways in which you can gather data about an existing system:

- Interview people who are involved with the current system. This will probably include the systems administrator, the people actually using the system and their customers or clients. Although this can be a time-consuming process, talking to the people who are actually using the existing system will give valuable first-hand knowledge and you can follow up on any comments that they might make. A drawback is that each person you talk to may give you a very personal view of the system.
- Unless they are carefully structured the data collected from an interview can be hard to make use of. Asking someone to fill in a carefully designed questionnaire or carrying out a survey will allow you to carry out a more accurate analysis of the responses, but these tend to restrict the data you gather to definite, closed answers. Questionnaires allow you to gather a lot of data relatively quickly.
- Although it can be very time consuming, observation of current practices will help to identify problem areas. It is objective rather than subjective and you may spot something that everyone else has missed.
- Examination of the current system including the files, paperwork and processes used will help to identify the data requirements of the new system. It will also help with the creation of the human–computer interface (HCI) of the new system. This process will also help you to see the overall scope of the problem.

Feasibility study

A **feasibility study** is a preliminary report to the person that asked for the new system in the first place. It will identify possible solutions and suggest the best way forward. The report will indicate how practical a solution is in terms of time and other resources, such as the availability of suitable software and hardware and the abilities of the end user to cope with the proposed method of solution.

Possible solutions might include doing nothing, having bespoke software written for you, writing the software yourself or buying an existing package 'off the shelf' and tailoring it to your needs.

KEYWORD

Feasibility study: an analysis of whether it is possible or desirable to create a system.

Design

Before work can start on the actual creation of the solution, the availability of appropriate hardware and software should be assessed. The choice of hardware will be driven by the users' needs and by the way in which data will be manipulated and stored. Work will begin on the file structures and algorithms that are going to be used.

Most big projects are far too big to be considered as one complete problem. The best solution is often to break them down into smaller modules. Each module will be self-contained, and the programmer will test it on its own. This approach will allow more than one person to work on the solution. Often the view a user has of the system will need to be defined so that all the modules will have the same 'feel'.

The process of looking at a big problem and breaking it down into smaller problems and then breaking each of the smaller problems down, and so on until each problem is manageable is known as the **'top-down' design** approach. The benefits of this approach are similar to the **modular design** system mentioned above, though there is the potential problem of getting too engrossed in small details such as the fine-tuning of the human–computer interface. It makes more sense to solve the overall problem first before you get too involved in screen layouts.

KEYWORDS

Design: the second stage of system development where the algorithms, data and interface are designed.

Top-down design: related to the modular approach, this starts with the main system at the top and breaks it down into smaller and smaller units a bit like a family tree.

Modular design: a method of system design that breaks a whole system down into smaller units, or modules.

Data flow diagram (DFD)

There are a number of ways of representing a problem and its possible solution. **A data flow diagram** is concerned purely with how data are moved round a system and as such it only needs four symbols.

KEYWORD

Data flow diagram (DFD): a visual method of showing how data passes around a system.

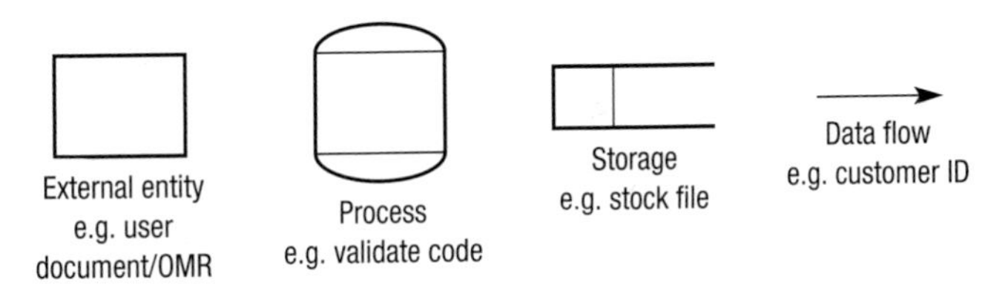

Figure 48.2 Data flow diagram symbols

The next diagram shows how a DFD might be used. It shows what happens after the electricity meter at a house has been read.

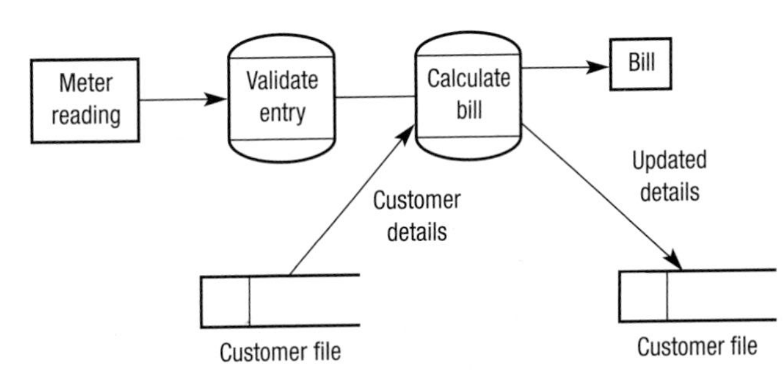

Figure 48.3 A DFD showing data flows for reading an electricity meter

Describing algorithms

At this stage, a description of the algorithms that will be used should be included. These will not be fully formed lines of code as it is only the design stage. For example, if search or sort algorithms are needed, they can be identified at this stage.

It may be appropriate to work the algorithms into pseudo-code that reflects the programming language being used. For example, if a functional programming language is being used, the main functions should be identified at the design stage. If a relational database is being used, the main SQL statements should be identified.

Data dictionary

KEYWORD

Data dictionary: a list of all the data being used in the system including name, length, data type and validation.

Where relevant, a **data dictionary** should be produced to show what data will be used and how it will be stored. Careful planning at this stage will reduce the number of problems that will be encountered later in the project. This is of particular relevance to database projects.

Details of the data to be stored, including the data type, length, title and any validation checks, will be stored in a data dictionary. This can be seen as a database about the database – it holds background details but not the data itself. Figure 48.4 shows part of the definition of a table in Microsoft Access. This forms part of the data dictionary that defines the whole solution.

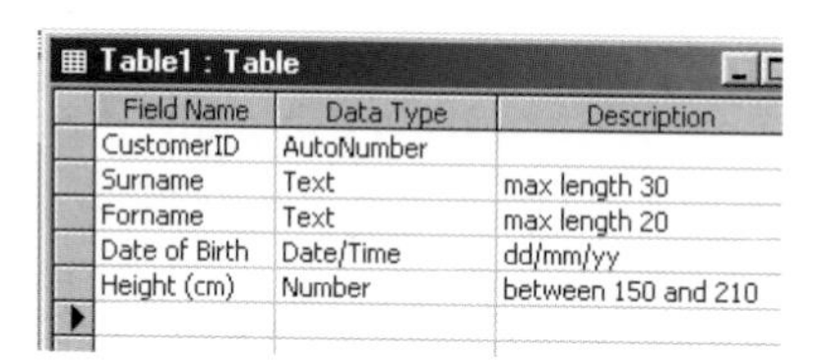
Table1 : Table

Field Name	Data Type	Description
CustomerID	AutoNumber	
Surname	Text	max length 30
Forname	Text	max length 20
Date of Birth	Date/Time	dd/mm/yy
Height (cm)	Number	between 150 and 210

Figure 48.4 Data dictionary definition of a table from MS Access

Variables table and data structures

KEYWORD

Variables table: a list of all the variables that a program will use, including names and data types.

It is important to decide what variables a program will need and what purpose they will serve. Some programming languages will only let the programmer use variables that have been declared. Declaring variables at the start of the program allows the programmer much tighter control of their program. The programmer will need to decide about certain characteristics of each variable – type, length, name and whether the variable is local or global. This example shows how three global variables might be set up in Visual Basic.

```
Public Age As Integer

Public Name As String

Public WearsGlasses As Boolean
```

In Chapter 4 we met the concept of using blocks or modules to make a project more manageable. Allocating names to these procedures and defining local variables used within them is also best carried out at this stage.

Volumetrics

The volume of data that a system will be asked to handle both now and in the future will have a bearing on how the programmer decides to store and handle the data. They will need to consider the throughput of data – how many transactions the system will need to cope with in a given time span – and also how much data the system will need to store at any one time. This will affect the storage media that is used, and it will also be a consideration when back-up strategies are being decided. The programmer will also need to consider how many users will be allowed to access the files and programs at one time.

It is important to realise that the databases that most students see in their time at school or college are generally small. Even a pupil database in a school with 1200 pupils is very small compared to the high street banks that have literally millions of customers each with hundreds of fields of data stored about them. The computers used by the DVLA store details of well over 20 million cars which means the planning needed for this project is very different to planning how the secretary of a swimming club will store their data.

The human–computer interface (HCI)

The human–computer interface is the term given to any form of communication between a computer and its user. For the majority of us this might seem limited to the computer screen with its familiar graphical user interface (GUI), but it can also include the layout of buttons on a mobile phone or house alarm system and the way information is presented on a tablet or smart phone or the flight controls of a new aircraft.

There are a number of aspects that need to be considered when designing an effective HCI. These include:

- Ease of use: There is no point creating sophisticated software if the functions are hidden behind a series of screens and button presses. A good HCI will feel almost intuitive.
- Target audience: What suits a child might not necessarily suit an adult so it is important that the programmer is aware of who the end user is going to be.
- Technology: There is little point in creating a screen layout that works comfortably on a 21 inch flat screen if the target audience mostly use tablets and smart phones.
- Ergonomics: The interface should be 'comfortable' to use. This is important if the user is likely to be sitting with the interface for a long period of time, for example, an airline pilot or a tele-sales operative.

Implementation

KEYWORD

Implementation: the third stage of system development where the actual code and data structures are created.

This stage is when the application is actually written creating the fully working system using the appropriate tools identified in the feasibility study. The process of implementing a system is based on the design, and the programmer(s) will need to be fully aware of the requirements set out in the design. It is important to note that many systems are modular and that the different modules do not all have to be written at the same time. For example, some parts of the system may still be at the design stage while other parts are being implemented. In some cases, one module is dependent on another one. In other cases, the design of one part of the system may change as a result of the implementation of another part.

It is important to realise that systems development has to be a responsive process and that the developer may need to respond to issues as they arise throughout the development process.

Prototypes

Creating a solution to a problem can be very costly both in terms of finance and time. There is little point in presenting the end user with a completed project if they are going to ask you to alter various details. In this case it would be a good idea to produce a **prototype**.

KEYWORD

Prototype: a stripped down version of a whole system built at the design stage to test whether the concept works.

The functionality of a prototype may vary depending on the nature of the project. For example, the human–computer interface may be very well developed in the prototype and quite closely reflect the finished system. However, the functionality behind it may be incomplete.

At this stage the end user is asked to comment on the product so far, and they will check to make sure all the major functions work as expected.

Testing

It is important that the system is tested to make sure it performs as expected. There are a number of test strategies that can be used and it is important to understand that any testing plan must include tests that are carried out as the code is written and not just at the end.

Test data are data that generate a known result, and test data will need to be devised that tests every aspect of the solution from the expected responses to the extremes that humans can subject a computer program to. As individual units or modules are completed they are tested to ensure they carry out the functions they contain. As the project proceeds, modules can be fitted together and at this stage integration testing takes place. This process makes sure that the modules work together. A lot of these processes make use of test data.

KEYWORDS

Testing: the fourth stage of system development that includes a range of tests using a variety of data.

Normal test data: test data that is within the expected range for the system and should therefore produce the correct result.

Boundary test data: test data on or close to the boundary of the acceptable range.

Erroneous test data: test data that is clearly incorrect and should produce an error.

Test data are data that will generate a known response. Typically three types of test data are used:

- **Normal**: Data that the system is expected to handle as they are within an acceptable range. For example, an age field could be tested with values between 0 and 110.
- **Boundary**: Data that are on the extremes of the acceptable range. This means testing the minimum and maximum values and those that are just inside and just outside the range. For example, with age, you might use 0, 1, 109 and 110.
- **Erroneous**: Data that are clearly incorrect and therefore you would expect the program to catch the error. For example, test data for age could be 1000 or an item of text instead of a value.

KEYWORDS

Development testing: testing that takes place during the development of the program.

Black box testing: using test data to test for an expected outcome.

White box testing: checks all pathways through the code, looking inside it and potentially adding extra commands to check what is happening.

Unit testing: testing carried out on just one module or component of the whole system.

Development testing

Black box testing involves entering test data into a routine or procedure and checking the resulting output against the expected outcome. Basically, it tests that an input produces the correct output without actually examining how it does it.

White box testing involves testing every aspect of a routine or procedure. Whilst black box testing is concerned with testing the data handling, white box testing considers all the other processes that are involved – for example, how the program reacts if it fails to find a suitable printer. White box testing checks all pathways through the code, looking inside it and potentially adding extra commands to check what is happening.

Unit testing makes sure each unit carries out the function it has been designed for. It incorporates both black box and white box testing.

Once all units have been tested, they are put together to form bigger sections. Integration testing is the process of making sure that the different modules that have been tested as individual units will work together.

KEYWORD

System testing: a range of tests carried out on the system once it has been completed.

System testing

System testing involves testing the system as a complete unit rather than as individual modules and making sure that it satisfies the specification agreed with the user.

Alpha testing is carried out on the finished system. This involves creating test data in-house. This test data will try to cover all the possible eventualities, so they will allow the system to be tested under normal conditions. The benefits of this process are that any problems that are found can be rectified before true live data are used by the end user. Another benefit of using a known set of test data is that if necessary the system can be stopped and restarted.

Some program developers will release an early or beta version of their new program to their potential users. At this stage the software is bound to have 'bugs' in it and the users are expected to send details of the problems they have encountered back to the programmers for them to resolve. This process is known as beta testing.

The benefits of this system are that passing your software to a number of people that have not been involved with the development will mean the testers will all use the system in slightly different ways and so highlight faults that might not have been found by normal means.

Although the developers will test the system they have developed as thoroughly as possible, it is the end users that need to be satisfied that the solution does what they wanted it to. Acceptance testing is carried out by the intended user. They enter their own live data and make sure the system matches the specification that was agreed with the program writers.

Some problems may only come to light some time after the system has been implemented by the user. These might include issues involving the volume of data the system is asked to cope with. Problems such as these will be resolved as part of the systems maintenance.

Evaluation

The final stage of the system development is **evaluation**. The solution that has been created was designed to match the specification that was agreed with the user. An evaluation compares the actual outcome with this specification. It should also contain suggestions for future improvements. It is these improvements and refinements that start off the whole cycle all over again. Evaluation may take place over an extended period of time. In fact, many systems are constantly being evaluated. There may be several criteria used to evaluate a system. For example:

- Functionality: Does it do what it is supposed to?
- Ease of use: This is not the same as 'easy to use' but it means that the level of complexity is appropriate to the user.
- Ease of implementation: How easy was it to transfer from the old system to the new system?
- Reliability: A measure of how much the system is 'up' or 'down'.
- Performance: Does the system meet its performance criteria, which might relate e.g. to the speed of operation or the amount of data it can handle?
- Cost effectiveness: Refers to how much it costs to implement the solution and whether the cost is justified.
- Ease of maintenance and adaptability: How easy is it to fix faults or add new modules?
- Longevity: How future-proof is the system?

KEYWORD

Evaluation: the final stage of system development where the system is judged according to certain criteria.

TASKS

1 You have been asked to gather data about an existing system from the employees that use it. You can either interview them or ask them to complete a questionnaire, or possibly even both. Compare the benefits and drawbacks of these two methods.

2 What are the benefits of using a top-down approach to solve a problem?

3 What is prototyping?

4 Explain why it is important for a programmer to define the variables they intend to use before they start writing the code itself.

5 What is meant by the term 'volumetrics'?

6 Explain the difference between white box and black box testing.

7 Software companies sometimes release beta versions of their software. What is a beta version, and what are the benefits to the company of using this technique?

KEY POINTS

- Software development is the process of creating and maintaining programs or applications. The first stage is analysis to identify and fully analyse the nature of the problem.
- The second stage is design where suitable algorithms, data structures and user interfaces are identified.
- The third stage is implementation when the program is written.
- The fourth stage is testing, where various methods are used to ensure the system works and does what it was designed to do.
- The final stage is evaluation where the system is assessed.

STUDY / RESEARCH TASKS

1 A company has developed an electronic system for registering pupils in a school. Draw up a table comparing the pros and cons of the various methods of delivering support for this program.

2 Describe five features of a good HCI (human–computer interface).

3 Describe the benefits and drawbacks of using off-the-shelf software, having bespoke software created or writing it 'in house'.

4 Draw a data flow diagram to show what happens when you take money out of your bank account via an ATM. Remember it only shows the flow of data.

49 Non-exam assessment (NEA)

A level only

SPECIFICATION COVERAGE

3.14 Non-exam assessment – the computing practical project

LEARNING OBJECTIVES

In this chapter you will learn:

- how to select an appropriate problem for your project
- how to tackle each section of the project
- how to interpret the mark scheme
- what programming skills and techniques you can use
- how to document your project.

INTRODUCTION

The A-level non-exam assessment constitutes 20% of the total A-level grade and requires you to work on a project either to identify a real problem that can be solved with a computer-based solution or to investigate a specific aspect of computer science. You are required to work through all of the stages of system development to produce a programmed solution. This is a major piece of work that should take at least 50 hours, with the majority of the marks available for the technical solution.

Selecting a project

The information given in this chapter relates specifically to the documentation required for the AQA A-level practical project now referred to as the non-exam assessment on the specification.

This project involves working through the stages of software development (see previous chapter) in the production of a solution to a chosen problem/investigation. Students should be aware of the agile approach to software development, the characteristics of which are:

- Planning is adaptive, which means that it may need to change during the stages of development.
- Developments may evolve as the technical solution is being written, for example, after a prototype has been created.

- Technical solutions should be completed well ahead of schedule to allow time for changes to be made.
- There should be a culture of continuous improvement, which may mean adding to or changing aspects of your code.

As a result of this you do not necessarily have to complete each stage of system development in order. However, your final write-up should be presented in the correct order.

You are expected either:

- to identify a realistic problem with a real end user and create a system that allows interaction with the user and involve the storage, manipulation and output of data, or
- to investigate a specific aspect of computing such as artificial intelligence or 3D graphical modelling with reference to a project supervisor.

You need to think about the areas of computing that interest you and that you are good at so that you can show off your programming skills. The solution must be completed using coding and 42 out of the 75 marks available are allocated to the technical solution.

The solution is assessed in five sections in accordance with the headings used for system development in the previous chapter.

Table 49.1 Breakdown of marks for A-level project

Analysis	9
Documented design	12
Technical solution	42
Testing	8
Evaluation	4
Total	75

Choosing a problem or investigation

Considerable thought should be given to the choice of project. Many students choose problems that are either too easy or too hard and as a consequence, they do not give themselves the opportunity to score highly in this area. AQA students must have a real end user or supervisor and this is sometimes the hardest part of the project. The project requires evidence of dialogue so you need to find a user who you can speak to or communicate with throughout the project rather than just at the beginning and end.

- Start with family and friends to see if there is anything they do on which you could base a project. For example, you could choose something based on their work.
- Many students base projects on their own work experience placements, particularly if they still have contacts at the organisation.
- Another source of projects is your own hobbies and interests. For example, you may be able to create systems related to gaming, social media, clubs or societies that you are involved in.

You should also base your choice on the tools and skills that you know are available in your centre, and on your level of expertise with different software. For example, it may be easier to produce your coursework using Visual Basic or Python if that is what you have been using throughout your course. You will probably find that there is plenty of help available on these packages from your school/college and from online forums.

The following is a sample of project ideas. This is not an exhaustive list.

- A simulation of a business or scientific issue. For example, a business issue such as modelling share prices, or a scientific issue such as modelling flu epidemics.
- An investigation of a well-known problem such as the game of life, the Towers of Hanoi or the travelling salesman problem.
- A solution to a data processing problem for an organisation, such as: membership systems (e.g. clubs, gyms), booking systems for organisations such as holiday companies or medical appointments; stock control systems; student timetabling and school reporting systems.
- The solution of an optimisation problem, such as production of a rota, shortest-path problems or route finding.
- A computer game.
- An application of artificial intelligence or investigation into machine learning algorithms.
- A control system operated using a device such as an Arduino board, Raspberry Pi or robotic arm.
- A website with dynamic content, driven by a database back-end. Note that the creation of a static website will not be sufficient for A level.
- Rendering a three-dimensional world on screen.
- An app for a mobile phone or tablet of a suitable complexity, perhaps chosen from the list above.
- Exploring large datasets, looking for and visualising correlations.

It is worth noting that the best projects are often the most realistic. It is much better if there is a real problem rather than a pretend one. Don't be scared to do something original even if it is very specific in what it will do.

Analysis

This is possibly the hardest part of the project as it involves identifying and interviewing a real user or working with a project supervisor. If your user has a genuine problem that needs solving then this section is much easier as they will provide much of the information you need. Your analysis should include:

- General background information on the organisation or person you are creating the system for. This should be sufficient for a third party to read and understand.
- A description of the problem with a clear statement that describes the problem area and specific problem that is being solved/investigated.
- An analysis of the critical path of the project in terms of identifying the main stages and the sequence which these should be done and the dependency between the stages.
- An outline of how the problem was researched, which might include an interview or questionnaire involving the user/supervisor.
- Source documents from the current system where relevant, or evidence of research into the chosen aspect of computing.
- Observation of the existing system where relevant.

- A list of the user's requirements and any limitations.
- A list of general and specific objectives that are realistic, achievable and measurable.
- Any modelling that helps inform the design stage, which may include graph models, entity-relationships models, data flow diagrams.

This section is marked according to four main criteria. These are:

- How well the problem has been scoped and whether it has been explained in a way that is easy to understand.
- Whether there is a fully documented set of measurable and appropriate specific objectives.
- Whether the requirements were identified though proper research and dialogue with the user.
- Whether the problem has been sufficiently well modelled to be of use in subsequent stages.

Design

AQA do not require full designs of every button, form, report, etc. They only need to see evidence of design for the 'key aspects' in a form that a third party could understand. This will typically be a mixture of written explanation and diagrams which could include:

- The overall system design, perhaps in the form of a top-down design diagram, system flowchart or entity relationship model.
- A description of the main modules that will make up the system.
- A description of the data items including data types and structures.
- A description of the file structures being used.
- Explanation of the main algorithms that will be used. It may be appropriate to use pseudo-code or specific code, for example SQL queries.
- A sample of rough designs of inputs and outputs including forms and reports. Examples of the design of the human–computer interface.
- An explanation of any library software that will be used, e.g. scientific or data visualisation libraries.
- An explanation of any database or web design frameworks being used.
- It is acceptable to show screenshots of any aspect of the design from the programmed solution even though these will not actually be produced until the technical solution is complete.

This section is marked by looking at how well your design describes how the key aspects of the solution are structured:

- fully/nearly fully explained
- adequately explained
- partially explained
- inadequately explained.

Technical solution

This is the main part of the project where most marks are awarded. It is split into two parts:

- The completeness of the solution: A total of 15 marks are available here and awarded depending on how well your solution meets the

requirements that you identified in your analysis. The three bands are: Meets almost all requirements / Meets many requirements / Meet some requirements. The total marks awarded will be a matter of judgement. Where you cannot meet all of the requirements, you are advised to identify the main requirements of the system and make sure that you meet these.
- The techniques used: 27 marks are available in this section and are awarded based on how proficient you are at using certain programming techniques. Note that to be proficient means that you have successfully implemented a particular part of the code. That is, your code must actually work.

You must include enough evidence to prove that you have fully implemented the design. This could include:

- Self-documenting code, which means code that uses meaningful identifiers, logical structures and annotation (comments) that allows a third party to understand it.
- An overview guide, which amongst other things includes the names of entities such as executables, data filenames/URLs, database names and pathnames.
- Explanations of particularly difficult-to-understand code sections.
- A careful division of the presentation of the code listing into appropriately labelled sections.

This section is marked according to two tables provided by AQA in the specification:

- 3.14.3.4.1 Table 1 Example technical skills: This provides three lists of programming algorithms and techniques. These are labelled A, B and C with A being the more complex through to C being the most simple.
- 3.14.3.4.2 Table 2 Coding styles: These show the level of proficiency that you might achieve when using the techniques described in the first table. You can be awarded Excellent, Good or Basic.

Your teacher will make a judgement on the complexity and proficiency of the code you have written using these two tables as a guide. You should refer to these tables before you implement your solution and try to use a selection of the technical skills or comparable techniques listed. You should also make your own judgement about whether to attempt more difficult techniques if there is a risk that you will not successfully implement them.

System testing

There must be evidence that testing has been carried out and that the tests were designed to ensure the system works as specified. You do not need to include evidence of every test, but you do need a representative sample that shows each type of test and you should ensure that the tests cover the main objectives of the project. This section should include:

- An overview of the test strategy including an explanation of the test data used. Test data should include normal (typical), boundary and erroneous data. As well as testing individual functions there should be 'whole system' tests that help to prove that the original objectives of the system have been met.

- Evidence that tests have been carried out including annotated hard copies.
- All possible outcomes should be tested with a table to show expected and actual outcome.
- Samples of screenshots or actual print-outs as evidence.

This section is marked according to two main criteria:

- Clear, well-presented evidence of testing.
- Evidence that the testing proves that the system is robust and works as intended.

Evaluation

It is important to be honest about whether the project has been a success or not:

- Copy the original objectives that you wrote in the analysis section. Go through each and explain whether you met the objective. If you met the objective, explain how effectively it was met and if you did not meet the objective, explain why not.
- Give your user a chance to use the system and ask them for general and specific comments. Don't invent the user/supervisor feedback – it will be obvious.
- Address the user/supervisor feedback explaining how you may incorporate any changes they have requested.
- Based on these comments and your own opinions, identify any ways in which the system could be improved or enhanced.

This section is marked according to three main criteria:

- Whether you have considered and addressed suggestions for improvements.
- Whether you have obtained real feedback from your user(s)/supervisor.
- Whether you have fully considered how well the solution meets its objectives.

KEY POINTS

- The A-level project involves working through the stages of the system development.
- You are expected to identify a realistic problem with a real end user or investigate a specific aspect of computing with reference to a project supervisor.
- You should also base your choice of problem on the tools and skills that you know are available in your centre, and on your level of expertise with different software.
- AQA need to see evidence of design for the 'key aspects' in a form that a third party could understand.
- 42 out of the total of 75 marks are available for the technical solution.
- Students must reference the AQA specification to see what skills are needed and at what level.
- The system should be fully tested and evaluated.

General advice on projects

The A-level project is a major undertaking if you want to get a good mark. Students who do well with coursework tend to get the best A-level grades overall. It's a comforting feeling when you go into the exam to know that you already have a good mark in your project. The reverse is also true – a poor project mark leaves you with a lot to make up in the exam room.

Students who do well in their project:

- Plan the project well.
- Stick to deadlines project.
- Ask their teachers lots of sensible questions.
- Refer to the specification and other resources provided by AQA, such as the Examiner's Report on last year's projects.
- Have a real user/supervisor and use real feedback.
- Consult with the user/supervisor throughout the project.
- Have an interest in the problem they have chosen to solve.
- Work on the project outside lesson time.

Glossary

Abstract data type: a conceptual model of how data can be stored and the operations that can be carried out on the data.
Abstraction by generalisation/categorisation: the concept of reducing problems by putting similar aspects of a problem into hierarchical categories.
Accepting state: the state that identifies whether an input string has been accepted. Also known as the goal state.
Address bus: used to specify a physical address in memory so that the data bus can access it.
Addressable memory: the concept that data and instructions are stored in memory using discrete addresses.
Addressing mode: the way in which the operand is interpreted.
Adjacency list: a data structure that stores a list of nodes with their adjacent nodes.
Adjacency matrix: a data structure set up as a two-dimensional array or grid that shows whether there is an edge between each pair of nodes.
Algorithm: a sequence of steps that can be followed to complete a task and that always terminates.
Alphabet: the acceptable symbols (characters, numbers) for a given Turing machine.
Analysis: the first stage of system development where the problem is identified, researched and alternative solutions proposed.
AND: Boolean operation that outputs true if both inputs are true.
AND gate: result is true if both inputs are true.
Application program interface (API): a set of subroutines that enable one program to interface with another program.
Application software: programs that perform specific tasks that would need doing even if computers didn't exist, e.g. editing text, carrying out calculations.
Arc: a join or relationship between two nodes – also known as an edge.
Argument: a value that is passed into a function or subroutine.
Arithmetic Logic Unit (ALU): part of the processor that processes and manipulates data.
Arithmetic operation: instructions that perform basic maths such as +, −, /, ×.
Array: a set of related data items stored under a single identifier and are accessed based on their position. Can work on one or more dimensions.
ASCII: a standard binary coding system for characters and numbers.
Assembler: a program that translates a program written in assembly language into machine code.
Assembly language: a way of programming using mnemonics.
Assignment: the process of giving a value to a variable or constant.
Association aggregation: creating an object that contains other objects, which can continue to exist even if the containing object is destroyed.
Associative array: a two-dimensional structure containing key/value pairs of data.
Asymmetric encryption: where a public and private key are used to encrypt and decrypt data.
Asynchronous data transmission: data is transmitted between two devices that do not share a common clock signal.
Attribute: a characteristic or piece of information about an entity, which would be stored as a field in a relational database.
Automation: creating a computer model of a real-life situation and putting it into action.
Backus–Naur Form (BNF): a form of notation for describing the syntax used by a programming language.
Bandwidth: a measure of the capacity of the channel down which the data is being sent. Measured in hertz (Hz).
Barcode reader: a device that uses lasers or LEDs to read the black and white lines of a barcode.
Baudot code: a five-digit character code that predates ASCII and Unicode.
Big data: a generic term for large or complex datasets that are difficult to store and analyse.
Binary file: stores data as sequences of 0s and 1s.
Binary search: a technique for searching data that works by splitting datasets in half repeatedly until the search data is found.
Binary tree search: a technique for searching a binary tree that traversed the tree until the search term is found.
Binary tree: a structure where each node can only have up to two child nodes attached to it.
Bit: a single binary digit from a binary number – either a zero or a one.
Bit rate: the rate at which data is actually being transmitted. Measured in bits per second.
Bit-mapped graphic: an image made up of individual pixels.
Black box testing: using test data to test for an expected outcome.
Block: in data storage it is the concept of storing data into set groups of bits and bytes of a fixed length.
Block interface: code that describes the data being passed from one subroutine to another.
BODMAS: a methodology for evaluating mathematical expressions in a particular sequence.
Boolean expression: an equation made up of Boolean operations.
Boolean operation: a single Boolean function that results in a TRUE or FALSE value.
Boundary test data: test data on or close to the boundary of the acceptable range.
Branch operations: operations within an instruction set that allow you to move from one part of the program to another.
Breadth first: a method for traversing a graph that explores nodes closest to the starting node first before progressively exploring nodes that are further away.
Bubble sort: a technique for putting data in order by repeatedly stepping through an array, comparing adjacent elements and swapping them if necessary until the array is in order.
Bus: microscopic parallel wires that transmit data between internal components.

Bus topology: a network layout that uses one main data cable as a backbone to transmit data.
Bus width: the number of bits that can be sent down a bus in one go.
Byte: a group of bits, typically 8, used to represent a single character.
Bytecode: an instruction set used for programming that can be executed on any computer using a virtual machine.
Cache: a high-speed temporary area of memory.
Caesar cipher: a substitution cipher where one character of plaintext is substituted for another, which becomes the ciphertext.
Call stack: a special type of stack used to store information about active subroutines and functions within a program.
Cardinal number: a number that identifies the size of something.
Cardinality: the number of elements in a set.
Carry bit: used to store a 0 or 1 depending on the result of binary addition.
Cartesian product: combining the elements of two or more sets to create a set of ordered pairs.
Certification Authority: a trusted organisation that provides digital certificates and signatures.
Chaining: a technique for generating a unique index when there is a collision by adding the key/value to a list stored at the same index.
Character code: a binary representation of a particular letter, number or special character.
Charge coupled device (CCD): in digital cameras it is a sensor that records the amount of light received and convert it into a digital value.
Check digit: a digit added to the end of binary data to check the data is accurate.
Checksum: a method of checking the integrity of data by calculating a sum based on the data being sent.
Child: a node in a tree that has nodes above it in the hierarchy.
Chip: an electronic component contained within a thin slice of silicon.
Cipher: an algorithm that encrypts and decrypts data, also known as code.
Ciphertext: data that has been encrypted.
Circular queue: a FIFO data structure implemented as a ring where the front and rear pointers can wrap around from the end to the start of the array.
Class: defines the properties and methods of a group of similar objects.
Class diagram: a way of representing the relationship between classes.
Client–server: a network methodology where one computer has the main processing power and storage and the other computers act as clients requesting services from the server.
Client–server database: a way of implementing a database where the database is put into a server and various users can access it from their workstations. The processing, for example, running a query, will take place on the server.
Client–server model: a way of implementing a connection between computers where one computer (the client) makes use of resources of another computer (the server).
Clock: a device that generates a signal used to synchronise the components of a computer.
Clustering: when a hashing algorithm produces indices that are not randomly distributed.
Code of conduct: a voluntary set of rules that define the way in which individuals and organisations will behave.
Codomain: all the values that may be output from a mathematical function.
Collision: when a hashing algorithm produces the same index for two or more different keys.
Colour depth: the number of bits or bytes allocated to represent the colour of a pixel in a bit-mapped graphic.
Commitment ordering: a technique to ensure concurrent transactions on a shared database are executed based on the timestamp of when the request is made and also the precedence the request takes over other simultaneous request.
Compiler: a program that translates a high-level language into machine code by translating all of the code.
Complementary metal oxide semiconductor (CMOS): is an alternative technology that performs the same functions as a CCD.
Components: the values within a vector.
Composition: building up a whole system from smaller units. The opposite of decomposition.
Composition aggregation: creating an object that contains other objects, and will cease to exist if the containing object is destroyed.
Compression: the process of reducing the size of a file.
Computational hardness: the degree of difficulty in cracking a cipher.
Computational security: a concept of how secure data encryption is.
Concurrence: the concept of two users trying to access the same data item at the same time.
Constant: an item of data whose value does not change.
Constant time: in Big O notation where the time taken to run an algorithm does not vary with the input size.
Context-free language: an unambiguous way of describing the syntax of a language useful where the language is complex.
Control bus: controls the flow of data between the processor and other parts of the computer.
Control unit: part of the processor that manages the execution of instructions.
Controller: in SSDs a controller is needed to organise data into blocks for storage purposes.
Convex combinations: a method of multiplying vectors that produces a resulting vector within the convex hull.
Convex hull: a spatial representation of the vector space between two vectors.
Copyright: the legal ownership that applies to software, music, films and other content.
Countable set: a finite set where the elements can be counted using natural numbers.
Countably infinite sets: sets where the elements can be put into a one-to-one correspondence with the set of natural numbers.
CRUD: an acronym that explains the main functions of a database: Create, Read, Update, Delete.
Cultural issues: factors that have an impact on the ways in which we function as a society.
Current Instruction Register (CIR): register that stores the instructions that the CPU is currently decoding/executing.

Data abstraction: hiding how data is represented so that it is easier to build a new kind of data object, e.g. building a stack from an array.
Data bus: transfers data between the processor and memory.
Data dictionary: a list of all the data being used in the system including name, length, data type and validation.
Data flow diagram (DFD): a visual method of showing how data passes around a system.
Data misuse: using data for purposes other than for which it was collected.
Data structure: a common format for storing large volumes of related data, which is an implementation of an abstract data type.
Data transfer operations: operations within an instruction set that move data around between the registers and memory.
Data type: determines what sort of data is being stored, e.g. integer, real, and how it will be handled by the program.
Database management system: software that enables the management of all aspects of a database including adding, updating and querying data.
De Morgan's Law: a process for simplifying Boolean expressions.
Debug: the process of finding and correcting errors in programs.
Declaration: the process of defining variables and constants in terms of their name and data type.
Declarative language: languages that declare of specify what properties a result should have, e.g. results will be based on functions.
Decomposition: breaking down a large task into a series of subtasks.
Decryption: the process of deciphering encrypted data or messages.
Definite iteration: a process that repeats a set number of times.
Depth first: a method for traversing a graph that starts at a chosen node and explores as far as possible along each branch away from the starting node before backtracking.
Design: the second stage of system development where the algorithms, data and interface are designed.
Development testing: testing that takes place during the development of the program.
Dictionary (data structure): a data structure that maps keys to data.
Dictionary-based encoding: a method of compressing text files.
Difference: describes which elements differ when two sets are joined together.
Digital camera: a device for creating digital images of photographs, which can be printed or transferred onto a computer to be manipulated and stored.
Digital certificate: a method of ensuring that an encrypted message is from a trusted source as they have a certificate from a Certification Authority.
Digital signature: a method of ensuring that an encrypted message is from a trusted source as they have a unique, encrypted signature verified by a Certification Authority.
Direct address: the operand is the datum.
Directed graph: a graph where the relationship between nodes is one-way.
Direction: one of the two components of a vector.
Distributed processing/computing: the principle of spreading large and complex tasks over a number of computers or servers.
Distributed program: a program specifically written to be used in a distributed processing environment.
Domain: all the values that may be input to a mathematical function.
Domain name: the recognisable name of a domain on the Internet.
Domain name server (DNS): a server that contains domain names and associated IP addresses.
Domain name server (DNS) system: a system of connected domain name servers that provides the IP address of every website on the Internet.
Dot product: multiplying two vectors together to produce another vector.
Dry run: the process of stepping through each line of code to see what will happen before the program is run.
Dynamic data structure: a method of storing data where the amount of data stored (and memory used to store it) will vary as the program is being run.
Dynamic Host Configuration Protocol (DHCP): a set of rules for allocating locally unique IP addresses to devices as they connect to a network.
Edge (programming): a connection between two nodes in a graph or tree structure – also known as an arc.
Edge: in a database graph schema, it refers to the link and relationship between two nodes.
Element: an single value within a set or list – also called a member.
Email server: a dedicated computer on a network for handling email.
Empty list: a list with no elements in it.
Empty set: the set that contains no values.
Encapsulation: the concept of putting properties, methods and data in one object.
Encryption: the process of turning plaintext into scrambled ciphertext, which can only be understood if it is decrypted.
Entity: an object about which data will be stored.
Entity identifier: an attribute which can uniquely identify each instance of an entity.
Entity relationship diagram: a visual method of describing relationships between entities.
Erroneous test data: test data that is clearly incorrect and should produce an error.
Ethical issues: factors that define the set of moral values by which society functions.
Evaluation: the final stage of system development where the system is judged according to certain criteria.
Event: something that happens when a program is being run.
Exception handling: the process of dealing with events that cause the current subroutine to stop.
Exponent: the 'power of' part of a number indicating how far a binary point should be shifted left or right.
Exponential time: in Big O notation where the time taken to run an algorithm increases as an exponential function of the number of inputs. For example, for each additional input the time taken might double.
Factorial: the product of all positive integers less than or equal to *n*, e.g. $3!$ is $3 \times 2 \times 1$.
Feasibility study: an analysis of whether it is possible or desirable to create a system.
Fetch–execute cycle: the continuous process carried out by the processor when running programs.
Field: an item of data.

FIFO: first in first out refers to a data structure such as a queue where the first item of data entered is the first item of data to leave.
File: a collection of related data.
File management: how an operating system stores and retrieves files.
Filter function: a method of creating a subset based on specified criteria.
Finite: countable.
Finite set: a set where the elements can be counted using natural numbers up to a particular number.
Finite state machine (FSM): any device that stores its current status and whose status can change as the result of an input. Mainly used as a conceptual model for designing and describing systems.
Firewall: hardware or software for protecting against unauthorised access to a network.
First-class object: any object that can be used as an argument or result of a function call.
Fixed point: where the decimal/binary point is fixed within a number.
Flip-flop: a memory unit that can store one bit.
Floating gate transistor: in SSDs it is a type of non-volatile transistor that stores data even without a power source.
Floating point: where the decimal/binary point can move within a number.
Flowchart: a diagram using standard symbols that describes a process or system.
Foreign key: an attribute in a table that is a primary key in another table and is used to link tables together.
Frequency analysis: in cryptography it is the study of how often different letters or phrases are used.
FTP: a protocol (set of rules) for handling file uploads and downloads.
Full adder: a circuit that performs addition using inputs from A and B plus a carry bit.
Function (maths): an expression that takes an input value from a set and returns an output value from another set.
Function (programming): a subroutine that returns a value.
Function application: the process of calculating the result of a function by passing it some data to produce a result.
Function composition: combining two or more functions together to create more complex functions.
Function type: refers to the way in which the expression is created, for example, integer of the domain and codomain, where f: A → B is the type of function.
Functional abstraction: breaking down a complex problem into a series of reusable functions.
Functional language: a programming paradigm that uses mathematical functions.
Functional programming: a programming paradigm that uses functions to create programs.
Functional programming paradigm: a language where each line of code is made up of calls to a function, which in turn may be made up of other functions, or result in a value.
Gateway: a node in a network that acts as a connection point to another network with different protocols.
Global variable: a variable that is available anywhere in the program.
Graph schema (database): a method of defining a database in terms of nodes, edges and properties.
Graph theory: the underlying mathematical principles behind the use of graphs.
Graph (maths): a structure that models the relationship between pairs of objects.
Graph (programming): a data structure made up of connected vertices and edges.
Half adder: a circuit that performs addition using inputs from A and B only.
Halting problem: an example of an unsolvable problem where it is impossible to write a program that can work out whether another problem will halt given a particular input.
Halting state: stops the Turing machine.
Hard disk (HDD): a secondary storage device made up of metallic disks that stores data magnetically.
Hardware: a generic term for the physical parts of the computer, both internal and external.
Harvard architecture: a technique for building a processor that uses separate buses and memory for data and instructions.
Hash table: a data structure that stores key/value pairs based on an index calculated from an algorithm.
Hashing algorithm: code that creates a unique index from given items of key data.
Head: the first element in a list.
Heap: a pool of unused memory allocated to a dynamic data structure.
Heuristic: with algorithms it is a method for producing a 'rule of thumb' to produce an acceptable solution to intractable problems.
Hierarchy chart: a diagram that shows the design of a system from the top down.
Higher order function: a function that takes a function as its inputs or creates a function as its output.
High-level language: a programing language that allows programs to be written using English keywords and that is platform independent.
HTTP (Hypertext transfer protocol): the protocol (set of rules) to define the identification, request and transfer of multimedia content over the Internet.
HTTPS: as above but with encrypted transmission.
Identifier: the name of a list.
Immediate address: the operand is the memory address or register number.
Imperative language: a language based on giving the computer commands or procedures to follow.
Implementation: creating code to produce a programmed solution.
Implementation: the third stage of system development where the actual code and data structures are created.
Indefinite iteration: a process that repeats until a certain condition is met.
Index: the location where values will be stored, calculated from the key.
Infinite set: a set that is not finite.
Infix: expressions that are written with the operators within the operands, e.g. 2 + 3.
Information hiding: the process of hiding all details of an object that do not contribute to its essential characteristics.
Inheritance: the concept that properties and methods in one class can be shared with a subclass.
In-order: a method of traversing a tree by traversing the left subtree, visiting the root and traversing the right subtree.

In-order traversal: a method of extracting data from a binary tree that will result in an infix expression.
Input size: in Big O notation the size of whatever you are asking an algorithm to work with, e.g. data, parameters.
Input/Output (I/O) controller: controls the flow of information between the processor and the input and output devices.
Instantiation: the process of creating an object from a class.
Instruction set: the patterns of 0s and 1s that a particular processor recognises as commands, along with their associated meanings.
Instruction table: a method of describing a Turing machine in tabular form.
Integer: any whole positive or negative number including zero.
Internet Protocol (IP) address: a unique number that identifies devices on a network.
Internet: a global network of networks.
Internet registries: organisations who allocate and administer domain names and IP addresses.
Interpreter: a program for translating a high-level language by reading each statement in the source code and immediately performing the action.
Interrupt: a signal sent by a device or program to the processor requesting its attention.
Interrupt register: stores details of incoming interrupts.
Interrupt Service Routine: calls the routine required to handle an interrupt.
Intersection: describes which elements are common to both sets when two sets are joined.
Intractable problem: a problem that cannot be solved within an acceptable time frame.
Irrational number: a number that cannot be represented as a fraction or ratio as the decimal form will contain infinite repeating values.
Iteration: repeating the same process several times in order to achieve a result.
JSON (JavaScript object notation): a standard format for transmitting data.
Key: in cryptography it is the data that is used to encrypt and decrypt the data.
Key/value pair: the key and its associated data.
Laser printer: a device that uses lasers and toner to create mono and colour prints.
Latency: the time delay that occurs when transmitting data between devices.
Leaf: a node that does not have any other nodes beneath it.
Legal issues: factors that have been made into laws by the government.
Library programs: code, data and resources that can be called by other programs.
LIFO: last in first out refers to a data structure such as a stack where the last item of data entered is the first item of data to leave.
Linear queue: a FIFO structure organised as a line of data, such as a list.
Linear search: a simple search technique that looks through data one item at a time until the search term is found.
Linear time: in Big O notation where the time taken to run an algorithm increases in direct proportion with the input size.
List: a collection of data items of the same type.
Load factor: the ratio of how many indices are available to how many there are in total.
Local Area Network (LAN): a network over a small geographical distance – usually on one site and typically used by one organisation.
Local variable: a variable that is available only in specified subroutines and functions.
Logarithmic time: in Big O notation where the time taken to run an algorithm increased or decreases in line with a logarithm.
Logic circuit: a combination of logic gates.
Logic gate: an electronic component used to perform Boolean algorithms.
Logical reasoning: the process of using a given set of facts to determine whether new facts are true or false.
Logical network topology: the conceptual way in data is transmitted around a network (see Physical network topology).
Logical operations: operations within an instruction set that move the bits around within the operand.
Loop: a repeated process.
Low-level language: machine code and assembly language.
Machine code: the lowest level of code made up of 0s and 1s.
Magnitude: one of the two components of a vector – refers to its size.
Main memory: stores data and instructions that will be used by the processor.
Majority voting: a method of checking for errors by producing the same data several times and checking it is the same each time.
Mantissa: the significant digits that make up a number.
Map function: a function that generates an output list from an input list by applying a function to each element in the input list.
Mealy machine: a type of finite state machine with outputs.
Media Access Control (MAC) address: a unique code that identifies a particular device on a network.
Member: describes a value or element that belongs to a set.
Memory: the location where instructions and data are stored on the computer.
Memory address: a specific location in memory where instructions or data is stored.
Memory Address Register (MAR): register that stores the location of the address that data is either written to or copied from by the processor.
Memory Buffer Register (MBR): register that holds data that is either written to or copied from the CPU.
Memory Data Register (MDR): another name for the MBR.
Memory management: how the operating system uses RAM to optimise the performance of the computer.
Merge sort: a technique for putting data in order by splitting lists into single elements and then merging them back together again.
Message: the name given to a packet of data being transmitted using the websocket protocol.
Method: the code or routines contained within a class or method.
Mnemonics: short codes that are used as instructions when programming, e.g. `LDR, ADD`.
Modelling: recreating a real-life situation on a computer.
Modular design: a method of system design that breaks a whole system down into smaller units, or modules.

Module: a number of subroutines that form part of a program.
Moral issues: factors that define how an individual acts and behaves.
Multi-core: a chip with more than one processor.
Mutable: changeable.
Naming conventions: the process of giving meaningful names to subroutines, functions, variables and other user-defined features in a program.
NAND: Boolean operation that outputs true if any of its inputs are false.
NAND gate: result is true if any of the inputs are false.
Natural number: a positive whole number including zero.
Nesting: placing one set of instructions within another set of instructions.
Network: devices that are connected together to share data and resources.
Network adapter/Network Interface Card (NIC): a card that enables devices to connect to a network.
Network topology: the layout of a network usually in terms of its conceptual layout rather than physical layout.
Node: an element of a graph or tree – also known as a vertex.
Node: in database modelling, it is an entity.
NOR: Boolean operation that outputs true if all of its inputs are false.
NOR gate: result is true if both inputs are false.
Normal test data: test data that is within the expected range for the system and should therefore produce the correct result.
Normalisation: a process for adjusting numbers onto a common scale.
Normalisation: the process of ensuring that a relational database is structured efficiently.
NOT: Boolean operation that inverts the result so true becomes false and false becomes true.
NOT gate: inverts the result so true becomes false and false becomes true.
Number base: the number of digits available within a particular number system, e.g. base 10 for decimal, base 2 for binary.
Object: a specific instance of a class.
Object code: compiled code that can be run as an executable on any computer.
Object-oriented language: a programming paradigm that encapsulates instructions and data together into objects.
One-time pad: a key that is only used once to encrypt and decrypt a message and is then discarded.
Opcode: an operation code or instructions used in assembly language.
Operand (maths): a value within an expression.
Operand (programming): a value or memory address that forms part of an assembly language instruction.
Operating system software: a suite of programs designed to control the operations of the computer.
Operator: the mathematical process within an expression.
OR: Boolean operation that outputs true if either of its inputs are true.
OR gate: result is true if either input is true.
Ordinal number: a number used to identify position relative to other numbers.
Overflow: when a number is too large to be represented with the number of bits allocated.
Overriding: where the methods described in the subclass take precedence over those described in the base class.
Packet: a block of data being transmitted.
Packet filtering: a technique for examining the contents of packets on a network and rejecting them if they do not conform to certain rules.
Packet switching: a method for transmitting packets of data via the quickest route on a network.
Parallel transmission: data is transmitted several bits at a time using multiple wires.
Parameter: data being passed into a subroutine.
Parent: a type of node in a tree, where there are further nodes below it.
Parity bit: a method of checking binary codes by counting the number of 0a and 1s in the code.
Partial application (of a function): the process of applying a function by creating an intermediate function by fixing some of the arguments to the function.
Peer-to-peer: a network methodology where all devices in a network share resources between them rather than having a server.
Physical network topology: the way in which devices in a network are physically connected.
Pixel: a picture element – the smallest unit that combined with other pixels makes up a picture, for example, a digital photograph.
Plaintext: data in human-readable form.
Pointer: a data item that identifies a particular element in a data structure – normally the front or rear.
Polish notation: another way of describing prefix notation.
Polyalphabetic: using more than one alphabet.
Polymorphism: the ability for different types of data to be manipulated with the same method.
Polynomial time: in Big O notation where the time taken to run the algorithm is a polynomial function of the input size, e.g. the square of the input size.
POP3: a protocol (set of rules) for receiving emails.
Port: an addressable location on a network that links to a process or application.
Port forwarding: a method of routing data through additional ports.
Postfix: expressions that are written with the operators after the operands, e.g. 2 3 +
Post-order: a method of traversing a tree by traversing the left subtree, traversing the right subtree and then visiting the root.
Post-order traversal: a method of extracting data from a binary tree that will result in postfix expressions.
Precision: how accurate a number is.
Prefix: expressions that are written with the operators before the operands, e.g. + 2 3
Pre-order: a method of traversing a tree by visiting the root, traversing the left subtree and traversing the right subtree.
Pre-order traversal: a method of extracting data from a binary tree that will result in prefix expressions.
Primary key: an attribute that can be used to uniquely identify every record within a table.
Priorities: a method for assigning importance to interrupts in order to process them in the right order.
Priority queue: a variation of a FIFO structure where some data may leave out of sequence where it has a higher priority than other data items.

Private key: a code used to encrypt/decrypt data that is only known by one user but is mathematically linked to a corresponding public key.
Problem abstraction: removing unnecessary details in a problem until the underlying problem is identified to see if this is the same as a problem that has already been solved.
Problem solving: the process of finding a solution to real-life problems.
Procedural programming languages: languages where the programmer specifies the steps that must be carried out in order to achieve a result.
Procedure: another term for a subroutine.
Processor: a device that carries out computation on data by following instructions, in order to produce an output.
Program counter (PC): register that stores the address of the next instruction to be taken from main memory into the processor.
Proper subset: where one set is wholly contained within another and the other set has additional elements.
Properties: the defining features of an object or class in terms of its data.
Properties: in database modelling, the items of information stored within each entity.
Protocols: sets of rules.
Prototype: a stripped down version of a whole system built at the design stage to test whether the concept works.
Pseudo-code: a method of writing code that does not require knowledge of a particular programming language.
Pseudo-random number generation: common in programming languages, a function that produces a random number that is not 100% random.
Public key: a code used to encrypt/decrypt data that can be made public and is linked to a corresponding private key.
Query: a search or sort carried out on data that retrieves the answer to a question.
Queue: a data structure where the first item added is the first item removed.
Radio frequency identification (RFID): a microscopic device that stores data and transmits it using radio waves – usually used in tags to track items.
Railfence cipher: a type of transposition cipher that encodes the message by splitting it over rows.
Random Access Memory (RAM): stores data and can be read to and written from.
Random number generation: a function that produces a completely random number.
Rational number: any number that can be expressed as a fraction or ratio of integers.
Read Only Memory (ROM): stores data and can be read from, but not written to (unless programmable ROM).
Read/write head: the theoretical device that writes or reads from the current cell of a tape in a Turing machine.
Real number: any positive or negative number with or without a fractional part.
Record: one line of a text file.
Record locks: a technique to temporarily prevent access to certain records held on a database.
Recursion: the process of a subroutine calling itself.
Reduce/fold function: a method of reducing a list to a single element by combining the elements using a function.
Regional Internet Registry (RIR): one of five large organisations that allocate and administer domain names and IP addresses in different parts of the world.
Registers: a small section of temporary storage that is part of the processor. Stores data or control instructions during the fetch–decode–execute cycle.
Regular expression: notation that contains strings of characters that can be matched to the contents of a set.
Regular language: any language that can be described using regular expressions.
Rehashing: the process of running the hashing algorithm again when a collision occurs.
Relational database: a method of creating a database using tables of related data, with relationships between the tables.
Relational operations: expressions that compare two values such as equal to or greater than.
Representational abstraction: the process of removing unnecessary details so that only information that is required to solve the problem remains.
Request to send/Clear to send (RTS/CTS): a protocol to ensure data does not collide when being transmitted on wireless networks.
Resolution: width × height of an image or number of pixels per inch.
Resource management: how an operating system manages hardware and software to optimise the performance of the computer.
REST (Representational State Transfer): a methodology for implementing a networked database.
Reverse Polish Notation (RPN): another term for postfix notation.
RGB filter: red, green and blue filters that light passes through in order to create all other colours.
Root: the starting node in a rooted tree structure from which all other nodes branch off.
Rounding: reducing the number of digits used to represent a number while maintaining a value that is approximately equivalent.
Route cipher: a type of transposition cipher that encodes the message by placing it into a grid.
Routine: another term for a subroutine.
Routing: the process of directing packets of data between networks.
Run-length encoding: a method of compressing data by eliminating repeated data.
Scalar: a real value used to multiply a vector to scale the vector.
Scheduling: a technique to ensure that different users or different programs are able to work on the same computer system at the same time.
Selection: the principle of choosing what action to take based on certain criteria.
Sequence: the principle of putting the correct instructions in the right order within a program.
Serial transmission: data is transmitted one bit at a time down a single wire.
Serialisation: a technique to ensure that only one transaction at a time is executed from multiple users on a database.
Service Set Identifier (SSID): a locally unique 32-character code that identifies a device on a wireless network.

Set: a collection of unordered, non-repeating numbers or symbols.
Set building: the process of creating sets by describing them using notation rather than listing the elements.
Set comprehension: see Set building.
Shift cipher: a simple substitution cipher where the letters are coded by moving a certain amount forwards or backwards in the alphabet.
Shift instructions: operations within an instruction set that move bits within a register.
Shortest path: the shortest distance between two vertices based on the weighting of the edges.
Side effects: in programming it refers to the fact that the value contained within a variable will change as the program is run, which has implications for other parts of the program.
Signed binary: binary with a positive or negative sign.
Single source: in Dijkstra's algorithm it means that the shortest path is calculated from a single starting point.
SMTP: a protocol (set of rules) for sending emails.
Socket: an endpoint of a communication flow across a computer network.
Software: a generic term for any program that can be run on a computer.
Source code: programming code that has not yet been compiled into an executable file.
Space complexity: the concept of how much space an algorithm requires.
Secure Shell (SSH) Protocol: a protocol (set of rules) for remote access to computers.
Stack: a data structure where the last item added is the first item removed.
Stack frame: a collection of data about a subroutine call.
Star topology: a way of connecting devices in a network where each workstation has a dedicated cable to a central computer or switch.
Start bit: a bit used to indicate the start of a unit of data in asynchronous data transmission.
Start state: the initial state of a Turing machine.
State transition diagram: a visual representation of an FSM using circles and arrows.
State transition table: a tabular representation of an FSM showing inputs, current state and next state.
State: in programming it refers to the state that the variables are in, i.e. the values that are currently stored.
Stateful inspection: a technique for examining the contents of packets on a network and rejecting them if they do not form part of a recognised communication.
Static data structure: a method of storing data where the amount of data stored (and memory used to store it) is fixed.
Status register: keeps track of the various functions of the computer such as if the result of the last calculation was positive or negative.
Stop bit: a bit used to indicate the end of a unit of data in asynchronous data transmission.
Stored program concept: the idea that instructions and data are stored together in memory.
String-handling functions: actions that can be carried out on sequences of characters.
Structure chart: similar to a hierarchy chart with the addition of showing how data is passed around the system.
Structured data: data that fits into a standard database structure of columns and rows (fields and records).
Structured Query Language (SQL): a specialised programming language for manipulating databases.
Subnet masking: a method of dividing a network into multiple smaller networks.
Subprogram: another term for a subroutine.
Subroutine: a named block of code designed to carry out a specific task.
Subset: a set where the elements of one are entirely contained within the other; can include two sets that are exactly the same.
Substitution cipher: a method of encryption where one character is substituted for another to create ciphertext.
Symmetric encryption: where the sender and receiver both use the same key to encrypt and decrypt data.
Synchronous data transmission: data is transmitted where the pulse of the clock of the sending and receiving device are in time with each other. The devices may share a common clock.
Syntax: the rules of how words are used within a given language.
Syntax diagram: a method of visualising rules written in BNF or any other context-free language.
System flowchart: a diagram that shows individual processes within a system.
System testing: a range of tests carried out on the system once it has been completed.
Table: a method for implementing an entity and attributes as a group of related data.
Tail: every element in a list apart from the head.
TCP/IP: a set of protocols (set of rules) for all TCP/IP network transmissions.
Terminal: a computer that has little or no processing power or storage capacity used as a client in a thin client network; on a syntax diagram it is the final element that requires no further rules.
Testing: the fourth stage of system development that includes a range of tests using a variety of data.
Text file: a file that contains human-readable characters.
Thick client: in a network where resources, processing power and storage capacity are distributed between the server and the client computers.
Thin client: in a network where one computer contains the majority of resources, processing power and storage capacity, which it distributes to other clients.
Time complexity: the concept of how much time an algorithm requires.
Timestamp ordering: a technique to ensure multiple users can execute commands on a shared database based on the timestamp of when the data was last written to or read from.
Top-down approach: when designing systems it means that you start at the top of the process and work your way down into smaller and smaller sub-processes.
Top-down design: related to the modular approach, this starts with the main system at the top and breaks it down into smaller and smaller units a bit like a family tree.
Trace table: a method of recording the result of each step that takes place when dry running code.
Tractable problem: a problem that can be solved in an acceptable amount of time.

Transition function/rule: a method of notating how a Turing machine moves from one state to another and how the data on the tape changes.
Translator: the general name for any program that translates code from one language to another, for example translating source code into machine code. There are three types – compilers, assemblers and interpreters.
Transposition cipher: a method of encryption where the characters are rearranged to form an anagram.
Traversal: the process of reading data from a tree or graph by visiting all of the nodes.
Tree: a data structure similar to a graph, with no loops.
Trojan: malware that is hidden within another file on your computer.
Truncating: the process of cutting off a number after a certain number of characters or decimal places.
Truth table: a method of representing/calculating the result of every possible combination of inputs in a Boolean expression.
Turing machine: a theoretical model of computation.
Two's complement: a method of working with signed binary values.
Unauthorised access: where computer systems or data are used by people who are not the intended users.
Underflow: when a number is too small to be represented with the number of bits allocated.
Undirected graph: a graph where the relationship between nodes is two-way.
Unicode: a standard binary coding system that has superseded ASCII.
Uniform resource locator (URL): a method for identifying the location of resources (e.g. websites) on the Internet.
Union: where two sets are joined and all of the elements of both sets are included in the joined set.
Unit: the grouping together of bits or bytes to form larger blocks of measurement, e.g. GB, MB.
Unit testing: testing carried out on just one module or component of the whole system.
Universal machine: a machine that can simulate a Turing machine by reading a description of the machine along with the input of its own tape.
Unsigned binary: binary that represents positive numbers only.
Unsolvable problem: a problem that it has been proved cannot be solved on a computer.
Unstructured data: data that does not fit into a standard database structure of columns and rows (fields and records).
Utility programs: programs that perform specific common task related to running the computer, e.g. zipping files.
Variable: a data item whose value will change during the execution of the program.
Variables table: a list of all the variables that a program will use, including names and data types.
Vector graphic: an image made up of objects and coordinates.
Vector space: a collection of elements that can be formed by adding or multiplying vectors together.
Vectored interrupt mechanism: a method of handling interrupts by pointing to the first memory address of the instructions needed.
Vernam cipher: a method of encryption that uses a one-time pad (key) to create ciphertext that is mathematically impossible to decrypt without the key.
Vertex (plural vertices): a point or node on a graph or network.
Virtual machine: the concept that all of the complexities of using a computer are hidden from the user by the operating system.
Virus: a generic term for malware where the program attaches itself to another file in order to infect a computer.
Von Neumann architecture: a technique for building a processor where data and instructions are stored in the same memory and accessed via buses.
Web browser: an application for viewing web pages.
Web server: a dedicated computer on a network for handling web content.
Websocket protocol: a set of rules that creates a persistent connection between two computers on a network to enable real-time collaboration.
Weighted graph: a graph with values attached to the edges.
Well-ordered set: a group of related numbers with a defined order.
White box testing: checks all pathways through the code, looking inside it and potentially adding extra commands to check what is happening.
Wide Area Network (WAN): a network spread over a large geographical distance.
WiFi: a standard method for connecting devices wirelessly to a network and to the Internet.
WiFi protected access (WPA/WPA2): a protocol for encrypting data and ensuring security on WiFi networks.
Wireless Local Area Network (WLAN): a LAN that does not use cables but connects using radio waves.
Wireless Wide Area Network (WWAN): a WAN that does not use cables, but sends data using radio waves.
Word length: the number of bits that can be addressed, transferred or manipulated as one unit.
Worm: malware or a type of virus that replicates itself and spreads around a computer system. It does not need to be attached to another file in order to infect a computer.
XML (Extensible markup language): a method of defining data formats for data that will be transmitted around a network.
XOR: Boolean operation that is true if either input is true but not if both inputs are true.
XOR gate: true if either input is true but not if both inputs are true.

Index

Q

R

S

T

U

V

W

X